THE CRUEL PEACE

THE

CRUEL

PEACE

Everyday Life in the
Cold War

FRED INGLIS

BasicBooks
A Division of HarperCollins*Publishers*

Poetry excerpt on page 310, © Peter Porter, 1983. Reprinted from Peter Porter's *Collected Poems* (Oxford: Oxford University Press, 1983) by permission of Oxford University Press.

Poetry excerpt on page 301 © J. V. Cunningham. Reprinted with the permission of Ohio University Press / Swallow Press, Athens Ohio.

Poetry excerpt on page 137 from Czeslaw Milosz "Spirit of History," *The Collected Poems* (New York: Ecco Press, 1988). Reprinted by permission of The Ecco Press.

Library of Congress Cataloging-in-Publication Data
Inglis, Fred.
    The cruel peace : everyday life in the cold war / Fred Inglis.
      p. cm.
    Includes bibliographical references and index.
    ISBN 0-465-01494-1
    1. Cold War.   2. World politics—1945–   3. Civilization,
    Modern—1950– I. Title.
    D842.I53   1991
    909.82—dc20   91-70059
                                             CIP

In memory of my mother and father

CONTENTS

PART II
THE BALANCE OF TERROR

PART III
HINGE OF AN EPOCH

PART IV
HISTORY AS FARCE

PART V
FINALE

CONTENTS

PREFACE

On 27 July 1945, the London *Daily Herald* bore the brave banner "Labour in Power." Further down the detailed front page another item announced world acclaim at this great victory and looked forward to "closer ties with Soviets." A column across, a smaller item reported an "Ultimatum to the Japs" with an accompanying warning about the enormous destruction that failure to surrender would bring.

The lineaments of cold war now stand out starkly on that page. A few days later, a cheerful little boy of eight in a happy household, I brought in the local evening paper from the doorstep. It announced the detonation of an atomic bomb over Hiroshima, and the big whitewashed "VJ" signs, proclaiming the Allied Victory in Japan, went up on the walls of bombed houses down the road to join "VE" (victory in Europe) and "Welcome Home Jack."

Forty-odd years later, in June 1987, I was fighting a Parliamentary election on behalf of my beloved Labour party in a seat safely held by the enemy. I was, for the fourth time, helplessly facing certain defeat, and could find no consolation but for the soft medieval loveliness of the city of Winchester in which I had conducted my genteelly doomed campaign.

I had been speaking at a packed meeting for all parties held in an ecumenical church in this most Christian of constituencies. Against the grain of the place—in Winchester, according to ancient European traditions, the Army goes hand in hand with the Church—I had made the case that the continued nuclear armament of Britain was too expensive, redundant, and out of control—*folies de grandeur*. As I left the crowded aisles a man my own age fairly gibbered at me. "How will you stop the Russians? Eh? How will you stop them? The home guard? Bows and arrows? Eh? Eh?" The cold war had by then been shaping the public forms of my life from mere boyhood until middle age. I was fed up with it.

I had grown up, as you might say, shaped first by the war virtues and then by the cold war virtues. An impressionable lad of the day could not but help being captured by the heroisms of the black and white war movies. My friends and I learned by heart the ardor for some desperate glory of which Wilfred Owen despairingly wrote in 1917. We had been, fortunately, too young to do anything about a world war that only touched our lives with a few downtown bombs and cups of hot chocolate to the tune of the air raid sirens. All we could do was feel and imagine with all the intensity of our little souls.

By the time I went away to one of those grand private emporia known in England as public schools, the Russians had moved to fill in the space left by the Nazis in popular consciousness for national enemies. Every Wednesday we did cadet drill, and without irony learned that the calling of the soldier was necessary, manly, and admirable. The Korean War filled the headlines, though God knew where Korea was. But if it came to it, when we left school at age 18, we would fight the yellow hordes in those barren hills.

As it turned out, when it *did* come to it, I became a very scared Parachute Regiment officer. I did it for bravado and seven-bob-a-day danger money. Taking with me a heavy dose of A. E. Housman's rue-in-poetry and a copy of *1984*, I went to war in the Suez campaign of 1956.

Apart from being terrified, homesick, and airsick, I assumed we were once more going to war for God and country. I returned a few months later, having spent some weeks in the desert north of Aden being pelted by hate-filled Arab crowds in the city, from a clearly pointless endeavor. A little later, after I'd been a freshman at Cambridge for the subsequent year, it had been bluntly made clear to me that there was a quite different account of what I had been up to in the Middle East. And so, by the time I left the university in 1960, I had settled for membership in a briefly unilateralist Labour party.

I hadn't forgotten the cold war; nobody could. One could only ignore it, or simply join the other side, as I had begun to do.

On the afternoon of 24 October 1962, at the very moment at which ships carrying missiles to Cuba touched the invisible line of Kennedy's quarantine five hundred miles out in the Atlantic, the whistle blew to start the rugby football game in which I was playing against my old university. There was a decent crowd and it was a lovely, soft October day. Even the football players, hardly noted for their political awareness, were a bit preoccupied. And yet, if it were the end of the world that impended, to be running on thick turf with a ball in my hands and my friends around me was what, at twenty-five, I'd have chosen.

Five years later, with a happy two-year-old daughter settled solidly on my shoulders, I was ambling alongside a couple of hundred token protestants past the chemical warfare research station near Salisbury, England, chatting amiably to a priest I'd been in the army with, called Bruce Kent.

The years turned, my daughter and her sister turned up at the marches on their own two feet, as the habits of dissent settled into the household, and I learned alongside many others the exhausting and exhilarating routines of campaign resistance run from your own sitting room and kitchen, of forlornly cheerful electoral sallies down the polite and private streets of English suburbs.

All the time the fortunes of cold war hardened and softened as the temperature was turned up or down in Moscow and Washington. All the time, in the homely, cluttered rooms of our private lives the television displayed the continuing spectacle of public life—all of it still going on in the windless spaces on the other side of the screen. Presidents and First Secretaries trod down the metal staircases from the bellies of great aircraft to shake hands on the tarmac and climb into big black cars. Tanks wheeled down blind streets and across muddy fields. Rolls of heavy wire netting were unwound and veered upwards against a dawn sky.

Occasionally, from somewhere across the world, an image sprang to the front of the screen and closed the gap between the events over there and our hearts within us. A man with his hands tied behind him is shot in the head on a Saigon boulevard, falling down clumsily, the black blood pumping from the hole in his cranium. A fourteen-year-old black, skinny beyond belief, is pulled roped and dead from the filthy swimming pool into which he was pointlessly tossed by passing guerrillas somewhere in Angola. An ancient Lada, loaded down on its axles by kitchen paraphernalia, sleeping bags, rattling with pots and pans like a tinker's cart, grinds through a checkpoint, two hefty Hungarian lads and a fair-featured waving young woman in the back, all three of them helpless with infectious laughter at their own daring, the frontier guards laughing back in spite of themselves.

Everyone has their memories of cold war; for a few on our side and millions on theirs, they were hard and harsh. For most of us, they were sights for sore eyes. We had *watched* the years turn, sometimes on the streets, mostly on television. Other people, representing us apparently, had done what there was to be done, and we had watched, in assent and in dissent. Sometimes we watched terrified, and on a few incredulous occasions, joyfully.

Then it was over. It had lasted forty-five years and had proved to be one hell of a story. Renewing itself endlessly, it had gripped our collective imaginations like a strangler.

When I left that election meeting in 1987, I wandered down the warm June evening behind the massive stone of Winchester Cathedral and recollected that Frank Thompson had been a boy walking these water meadows fifty years before, when I was born. I felt the historical commonplace all through my body, moving through me and past me like the river a few paces away—that war and then the cold war had held and made us, the men and women of three generations and most of a world, for all those years. And now, I knew, it was almost done.

One year later I was invited to spend three semesters in the unrivaled scholarly comforts of the Institute for Advanced Study in Princeton, itself a planning room for both assent and dissent among conscripts and volunteers in cold war. And there I began my history and looked for a form that would hold together epoch and event, private passions and the somber madnesses of public reason. While I did so, the cold war tumbled to its end.

My answer to the formal problem is the book itself. I have scattered the biographies of individuals variously significant in the history of the epoch through a chronicle of the age. I have chosen particular years and recounted the great events of those years because that is how the electronic wizardry of modern times has taught us to organize our memory. There is no sanctimony about this, still less any huffing and puffing over the ideological meaning of "events" and the discourse of narrative. Such things matter, no doubt, but not here, not now. I have written what I believe is a popular but not populist theory of history in which the ever-rolling stream of time bears its sons and daughters away, but not before they would swim with or against the current. Hence the ample space given here to the films, novels, plays, and poems of the day. They shadowed the actions and the events of the cold war and are as much texts in the contexts of political life as are the speeches and decisions of politicians and generals. I do not intend any fashionable collapsing of the category of fact into that of fiction, nor any betrayal of the historian's duty to tell the truth. I only intend to offer the banality that the fantasies of the age are no less telling a part of the chronicle than such more visibly formative actions as bullets and tanks. So this was the way it came.

Anybody who knows the Institute knows its generosity and magnificence. I hereby thank its doyens and honor its traditions as heartily as I may. In particular I thank Clifford Geertz, Michael Walzer, and Albert Hirschman for their hospitality with their ideas and their grace with learning. I also owe much to Lucille Allsen for her learning, her vigor, her great good humor, and her brisk way with academic vanities.

Naturally, I pay grateful tribute to all those whose biographies are told in these pages, and name them in order and in the hope that this prefatory acknowledgment adequately combines salutes for their help and apologies for however it may seem to them that I have not dealt justly with the fragments of their lives reported here: George Kennan, Willy Brandt, Freeman Dyson, Philip Agee, Neil Sheehan, Peter Carrington, Edward Thompson, Joan Didion, and Joan Ruddock. Each woman and man gave their time unreservedly, handsomely, and without any sense at all that it was for them to tell me what to say. They trusted the canons of history writing; I trust I have kept faith with them.

I also do proper reverence to the many friends who helped me. First, to Edward and Dorothy Thompson for the example of their own historical writing, which I cannot hope to have matched, and for the fighting clarion of their lives. They gave me access to their papers, their library, and their home, and I cannot repay them. If, a time or two, Edward Thompson catches an echo of his own prose in these pages, I hope it will sound as a note of homage by which his own principled, even crusty refusal of discipleship may be a little touched.

Phillip French and Phillip Whitehead gave valuable advice at the outset. There have been many other readers of parts of the thing who have kept me to the mark with their encouragement, notably Carol Marks, who typed it, and Jim Hunter, *il miglior fabbro*. The only person to have read it all before publication, with an eye both affectionate and caustic, was my most faithful, enthusiastic, sleepy, and determined reader, Eileen McAndrew.

Finally, as many times before, I invoke and honor two cherished friends: Quentin Skinner, without whose thirty years' example as a scholar and faithfulness as a comrade my life would have been so much more grimly darkened by the pall of cold war; and Michael Rae, most generous of hosts to this itinerant writer when looking for his sunny terrace on which to push a pen, a mile or two from the safe Naples harbor of the U.S. Sixth Fleet.

Research for and choice of the illustrations were almost all done by my friend Nick Cull whose reward, no doubt, will be in heaven but who is hereby cordially thanked. I am also grateful to Macmillan for permission to quote from Peter Porter's *Collected Poems*. Extracts from Frank Thompson's diaries and letters are quoted with the kind permission of Edward Thompson and Victor Gollancz Ltd.; from *Tinker, Tailor, Soldier, Spy* with the permission of John Le Carré Productions; from *Disturbing the Universe* with the permission of Freeman Dyson and HarperCollins; from *The Book of Daniel* with the permission of E. L. Doctorow and Alfred Knopf; from his *Collected Poems* with the permission of Czeslaw Milosz Royalties Inc.; from William Empson's *Collected Poems* with the permission of Harcourt Brace Jovanovich; from J. V. Cunningham's *Collected Poems* with the permission of Ohio University Press.

PROLOGUE

The cold war started a long time before the Second World War stopped. Amid the strong and general dedication of whole peoples to finishing off the bloody monster, Fascism, men who had been appointed to calculate the subtle balance of advantage to be pressed, riskily, here and there, tested opportunities for the future.

These men sat in high-ceilinged offices behind big desks in Washington, Moscow, and London. Their memoranda circulated among their colleagues along the marble corridors and down in the sandbagged cellars. In each capital the cream of a social class, each man in the prime of life and protected by comfortable middle age and seniority from the chillier aspects of war, adjusted, as nicely as possible, the claims of his nation's beliefs and values against its determination to get its own way, to win wealth and power and security out of the rubble and carnage of the day. After the war was won, there would then be peacetime, and that would have to be won as well.

The official view that emerged from a hundred thousand memoranda, minutes, diplomatic notes, and military planning documents was translated into invasion plans from west and east and into demands from the Allies for the unconditional surrender of the Axis forces. That view was reflected in the carpet bombing of the civilian populations in Hamburg, Frankfurt, Berlin, and Tokyo, in the Manhattan Project, and at the Los Alamos testing towers, where the first atomic bomb would flash brighter than a thousand suns.

But there was not only the laying waste of Germany and Japan to ensure, there was also the settlement of old Europe, with all its factious tribes and remnants, into a postwar order as amenable to one's own interests and values as possible. And so there followed in the wake of the vast and terrible crushing of the enemy on both frontiers of the Fatherland, a series of probes out toward the edges of the globe, where one's Allies might be off their guard, and where the horsemen of the Apocalypse might be turned to national gain.

Of course, the fighting there was hideous, down the peninsulas of Vietnam and Malaya, across Manchuria, along the Baltic straits, and in the ranges of the Balkans. Murderous enmities were being settled by local wars, and national and class allegiances of countless colors and intensities were gathered into the gigantic field of force that held the world tight to its twin poles: Fascism against anti-Fascism. "He who is not for us is against us": that necessary judgment ruled the lines between armies and, on the anti-Fascist side, gathered up the motley of democrats, liberals, nationalists, Communists, old patricians and new freethinkers, workers, peasants, and people into one brave cause.

The further these battles were from the centers of action in Europe, the more the one true cause of anti-Fascism was shot through with the threads of local loyalties. But the war in hand was total war. The industrial systems that begot it had produced for the first time in history the totalitarian governments capable of fighting such a war. Such total government, learning its early lessons from mass production and then, by 1918, from mass destruction, pervaded and invaded the lives of its subjects in all the great powers. It gave them, even in its liberal manifestations, thoughts to think, beliefs to hold, and work to do. Totalitarianism proved, as never before, totally convincing.

By the beginning of 1944 it was clear that the Fascist beast was beaten. There would still be appalling slaughter, especially in the tracks of unconditional surrender, but the job would be done. Meanwhile, the bureaucratic classes and power elites of the prospective victors began working on winning the peace and, in so doing, wresting from their Allies as near total control as possible of those great stretches of the world they counted as rightfully theirs.

Of course, dark suspicions fell between the Soviets, the Americans, and the British, the three mighty and puissant nations heading for Berlin and Tokyo. At this distance, half a century later, it is a bit hard to recollect that in 1944 the British empire was at its largest extent; that the Union of Soviet Socialist Republics was as one with Mother Russia, and that Communism was still a spiritual creed whose nobility and idealism held millions in the line of duty and death; that the citizens of the United States of America were loved and welcomed worldwide as just the nicest, richest, most generous, and easygoing conquerors ever—conquerors who came not to subjugate but to set free, to bring a new world out of the old.

But the suspicions threw long shadows behind the banners of the Allied crusade. Back in the high-ceilinged offices, the dark-suited officers of state power maneuvered for gain and tested their suspicions with the lives of men.

The readiest testing ground lay along the frontiers that were to mark the uneasy reaches of each Ally's power. Somewhere in the state departments and foreign offices of the great powers there is always an office of euphemism (the mark of whose wordsmiths is the creation of phrases like "the great powers"). This office came up with the genteel phrase "spheres of influence" to designate the swathes of the world that were to be yours or theirs or ours. In his memoirs, Winston Churchill, remembering the preliminaries to February 1945, uninhibitedly renders the cynicism of that final carve-up. He had been particularly robust in his proposals for the Balkans (clearly not an American pigeon):

> "Let us settle about our affairs in the Balkans," I urged Stalin. "Your armies are in Rumania and Bulgaria. We have interests, missions, and agents there. Don't let us get at cross-purposes in small ways. So far as Britain and Russia are concerned, how would it do for you to have ninety percent predominance in Rumania, for us to have ninety percent of the say in Greece, and go fifty-fifty in Jugoslavia?" Pulling out a sheet of paper, I jotted down a list of countries and the percentages of influence to be accorded Britain and Russia in each:
>
> | Rumania | |
> |---|---|
> | Russia | 90% |
> | The others | 10% |
> | Greece | |
> | Great Britain | 90% |
> | Russia | 10% |
> | Yugoslavia | 50–50% |
> | Hungary | 50–50% |
> | Bulgaria | |
> | Russia | 75% |
> | The others | 25% |

Stalin studied the sheet for a moment and then silently pencilled a large check by the figures. After a long pause, I had second thoughts.

"Might not it be thought rather cynical if it seemed we had disposed of these issues, so fateful to millions of people, in such an offhand manner? Let us burn the paper."

"No, you keep it," said Stalin.[1]

It might, to an exquisite sensibility, seem a little bit cynical, if we hadn't got Korea and Vietnam and the Middle East still to come. But after all, both men had seen in 1914 what the Balkans could do to Europe. Each knew the savagery with which the different peoples of so small an area—Serbs, Croats, Bulgars, Macedonians, Greeks—had treated each other for centuries. Churchill undoubtedly had in his makeup the ruthlessness and cruelty that all who wield enormous power must be able to call on. Stalin was a jealous killer and a tyrant. Churchill had arranged, only a month or two before this meeting, for the British airborne to put down the Greek liberation movement in Athens with maximum roughness because it had threatened to strike off on its own left-wing paths. Stalin—well, Stalin was by that date the author of one damned massacre after another; there was no one could teach him ruthlessness.

The three leaders sat at the seaside in 1945 and posed in a row for that famous photograph—all of them old, Roosevelt the idealist a few weeks from his sudden death, and the two old cynics, watchful, genial, patriotic to a fault, fiercely hanging on to their power but ready also to do deals. At a phone call's length behind them, their policymakers and subordinates waited, then took the communiqués and translated them into war plans and orders.

The huge political voltage of a summit meeting is gradually harnessed and directed down the channels of power until finally it is felt as the charges which energize biographies.

Roosevelt, Churchill, and Stalin had been gathering the voltage released at Yalta since hostilities began. Yalta marked the first of many moments when the older generation—holding power, as parents must, over their children's future—froze hope for the future in the iron frost of the cold war. Old men, anxious for their people's safety and greedy for their own gain, froze solid the passions for generosity and for revenge surging uncontrollably around Europe in 1944 and 1945. Innumerable, obscure, and ugly executions—as murder, by default, or in pointless street actions—slowly fixed the frontiers of the armistice, and formed the character of a fifty-year epoch.

PART I

Waging Cold War

I

The Short Happy Life of Frank Thompson

A biography is a matter of life and death. It is the readiest way we have of getting some purchase on our history and understanding who has been responsible for it. A good biography, however brief, should *signify;* it should not be just a tale told by an idiot.

And so, if we look up a few of those deaths in the last year of the Second World War, we shall be searching and researching for something left of the passionate charges flowing across Europe and buried with the bodies. To recover the best meaning of the sudden ends of those young men and women is to do them belated justice. More than that: perhaps we can rediscover the hopefulness and idealism that have lain dormant for so long and, it may be, release once more those energies for good and for happiness that were stopped dead by a bullet or a torturer or mere fatigue in the broken cities and unkempt countryside on the rim of Europe fifty years ago.

II

As Europe turns momentously round after its iron immobility, each man or woman who is alive to the rush of good hope goes back to look for his and her most dauntless past. There is homage to be paid. There are desserts that

3

are due. I am looking for a tall soldier with a heavy mop of hair standing on a hill above the Bulgarian frontier on 10 May 1944.

I know he was there at that date, looking at the roof of the countryside. The mountainside sloped gently downward where he stood; but below it was steep and he could see the stream running along the valley bottom, flashing white where it broke on the stones. Lifting his eyes, he could see the forests of the frontier and the lumpy, endless hills they covered. Beyond them, there opened the long, irregular ridges of the mountains, fold after fold, reaching as far east as he could see, deep into Bulgaria. The ridges were covered in heavy snow, and beneath his battered boots patches of snow and ice creaked and split as he turned in the sunshine, wriggling his shoulders as the heat hit them.

It was noon on 10 May 1944. Behind him, at a rickety trestle table, sat another soldier, and the sharp, silent mountain air was broken by the whistle and howl of radio noise as the second man turned the dials on two heavy, rectangular metal boxes, painted dark green, which stood upright on the tabletop.

Both men wore the thick khaki of regular British battle dress, torn and battered-looking but still clearly military in outline and in the faded traces of rank where badges had been on sleeve and shoulder.

"It's supposed to be a very big supply drop tonight," said the radio operator, putting down the headphones.

The officer turned, holding the bowl of his pipe, his other hand in his trouser pocket. "That's what they've said before. But there'll have to be a big drop tonight, if they are still to believe that we're any good to them. I don't know what's up. It took eleven days for yesterday's reply to come from Cairo, and then they offer us a Sapper when there's absolutely nothing in front of us except the whole of Bulgaria to sap."[1]

He looked down the reverse slope of the hillside into the thick copse of larch, hazel, and beech saplings which crowded up to the edge of the bare knoll. There in the patches of sunshine lay groups of men and not a few women, raggedly dressed but also still recognizably uniformed in the details of their caps or breeches, with rifles lying carelessly about and heavy bandoliers of bullets slung around their shoulders—the spitting image of guerrillas.

III

I am looking for Frank Thompson, gazetted major in the British army a little before this date though probably without knowing it, British liaison officer to one of the special missions assigned in late 1943 to link up with Tito's partisan brigades in Yugoslavia. His instructions were to aid the operations of the

Bulgarian partisans then waiting on the frontier to launch back into their fatherland and rouse its people against its Fascist government. When Frank Thompson was standing on that hillside on the border between Bulgaria and Yugoslavia, somewhere near the little village of Kalna, I was within a few days of my seventh birthday, in a wonderfully safe and loving household in the north of England.

But the connection was pretty direct between Frank Thompson's life and the way boys of my class and generation were brought up. Even then, the cheap comics and two-shilling hardback novels were full of dashing young men who fell out of the sky in parachutes to be welcomed by other young men in baggy caps and girls in berets crouched below the trees holding, with bright eyes, tommy guns, revolvers, and secret papers. Such tales stirred countless young English-speaking men and women to the bottom of their hearts and shaped their lives. It gave history itself direction and meaning. Victory over Hitler and the Fascists, whose unspeakable abominations came in waves of news for years, was a simple victory for the good for the best principles of both old Europe and the New World: kindness, decency, generosity, gallantry; above all, for liberty.

When we look again at those tales, and then look closer for the true lives they were based on, they speak indeed of high ideals: of courage and good humor, self-sacrifice and sheer nerve. Between then and now, however, lies the monstrous bulk of the iron curtain and the arctic fields of the cold war. The years between have been eaten by locusts and worse, so that it is now hard to pile the tales of 1944 back on the bookshelf without admitting that those ideals and those lives were wasted—or were too good for us, or something.

Looking for Frank Thompson is like looking for our own best past. Where do you find it? You can go and stand roughly—within a few kilometers, at any rate—where he stood and wonder whether he went this way or that; whether it felt roughly like this; whether you could have done the same yourself (of course not); blankly, what it meant then and means now. History is what really and truly happened, and it is at the same time the endless necessary telling of the tale of what really and truly happened. It must not be a lie.

IV

The air drops came that night. At last Frank Thompson could send a signal—the last of all his signals, as it turned out—to the office in Cairo: "This is the biggest encouragement Bulgarian partisans have had since Mostyn arrived. After last night and a few more like it stream may well become torrent."[2] The stream in question was of local people coming forward to volunteer for the desperate life and hard death of being a Bulgarian partisan.

Thompson had been sent to join Maj. Mostyn Davies at Tsrna Trava just inside the Yugoslav border in January. They were both officers of the Special Operations Executive (SOE), which sent many brave and clever young men on always dangerous, sometimes necessary, sometimes crazy missions in the dark.

In September 1943 Davies had parachuted into the Macedonian mountains with orders to link up with the tireless and vehement director of partisan operations, Gen. Svetozar Vukmanovic-Tempo, the most trusted lieutenant of Marshal Josip Broz Tito, who was by then the uncrowned king of partisan resistance in the Balkans.[3] General Tempo commanded various forces scattered across Moravia, moving by ancient car, horse, train, or on foot from Yugoslavia, Serbia, the length of Macedonia, down to Greece, across to the city of Skopje, and back again. At every stopping point he had to organize, educate, agitate to try to produce unity of purpose and clarity of command among groups of soldier-partisans who spoke any of half a dozen languages and innumerable patois, who were fervently nationalist but suspicious of all other nationalist groups, and whose only common purpose was fighting the Fascists.

Partisans. Fierce, hot-eyed, laughing, murderous men and women; courageous to a fault, daring, disorganized, ill-disciplined; crouched below a bank waiting for the chuffing enemy train to reach the dynamited bridge; running clumsily in heavy boots down bombed village streets with the wooden-handled cylinder grenades of 1914 in one fist and a revolver in the other; tortured, shot, mutilated, and disfigured, lying in the rain at the riding-booted feet of an SS officer, the knuckles of his gloved hand gleaming tightly.

There is truth enough in these compelling images. But like all tableaux, they are posed and immobile. In his travels General Tempo found such men and women, certainly. Some were the proper soldiers—drilled, armed, correct, and dedicated—he helped train for Tito. Some were brigands who were carefree and ruthless but filled with a hatred of Fascism; these he instructed rapidly in the stratagems of guerrilla warfare and sent on their way. And some, indeed many, were Communists and nationalists, the most serious of all the partisans. Long used to being on the run, they were roughly schooled in the lessons of Leninism, poor, hard, grim and cheerful, keenly anxious that yet another uproar in Europe should this time give them and their countries the chance to make a better future, decent lives, and real freedom for their children out of the incomprehensible chaos moving through the hills and valleys of their homes.

Tempo had to vary his urgent lessons to fit local knowledge and national

Figure 1.1   Frank Thompson in North Africa, 1943

politics. In Yugoslavia, after the fall of the collaborationist government, there was a straight fight between Tito's partisans and the royalist Chetniks. Bloody as the struggle was, there was never much doubt that the partisans would win. In Greece, occupied by the Germans, the two arms of the resistance were the partisan left (ELAS) and the nationalist right (EAM). After the German retreat at the end of 1944, these two partisan groups fought it out in civil war until Winston Churchill threw in the British airborne against ELAS, ensuring the outcome. Even so, the fight continued another four years in the hills above the Aegean. In Italy, after the Fascist regime collapsed and the country surrendered in 1943, odd groups of Communists and deserting soldiers sympathetic to the partisans linked up with Balkan groups—for the food, or the comradeship, or to stay alive until the war was over.

When Tempo's men and women came across drilled troops and enemy occupying forces, the guerrilla strategy was always to fight and melt away, to dart at lines of supply and disrupt them, and in Mao's famous metaphor, to swim in the waters of local peasant support. Head-on fighting with the mighty *Wehrmacht*'s tanks, flamethrowers, bombers, would simply mean obliteration. The partisan actions which were best for morale, which kept the villagers on their side in spite of the hideous reprisals, and never let the Germans sleep peacefully were the darting, lethal attacks that broke up trains and convoys and terrified the ordinary, indolent men who drove trucks, put up camps, and mended aircraft.

In Bulgaria it was hard for Tempo and his Bulgarian junior, Apostolski, to construct even a tactic, let alone a strategy. In the first place, Bulgaria was the only Balkan country in the war that was not under German occupation. King Boris had joined with the Axis powers against Britain and the United States, but judiciously, because of long, historical links with Russia, declared neutrality toward the Soviets. The Germans could use Bulgaria as a thoroughfare but did not need to garrison huge numbers of troops to hold it down. Consequently, there was much less hatred toward German soldiers than in the occupied territories. The instruments of formal repression remained the Royal Bulgarian Army and the police, both outfits utterly uninhibited in their repressive methods and driven by unbridled hatred of Communists.

If Bulgaria was to be at all a part of the partisan Balkan Federation—whose 1943 manifesto was written by Tempo and the partisan general staff—it had to be armed and formalized and given a small but coherent action from which to build up support. Winston Churchill had learned very fully from his experience in the First World War the significance of the Balkans. U.S. leaders, governing a country thick with Balkan immigrants, knew little about these nations and hardly included the area in their war plans. Churchill sent a patrician brigadier, Fitzroy Maclean, Scottish laird and idol of his tenantry, to deal directly with Tito, even though a separate mission had already established liaison with the royalist Chetniks under Draža Mihajlović.

Maclean's flair and his connection with Tito gave the Balkan missions their new thrust. Maj. Mostyn Davies was parachuted into Macedonia to join Gen. Bulgaranov, an inflexible Leninist, in September 1943. On 25 January 1944, just before midnight, Capt. Frank Thompson, aged twenty-three, parachuted onto the frozen dropping zone that the partisans had prepared along the river valley bottom, a short way from the village of Tsrna Trava.

It is Frank Thompson in Bulgaria for whom I am searching.

V

He is the man who most fits the moment. He was born in 1920, the son of a freethinking Methodist who was strongly interested in Buddhist mysticism, intermittently taught Indian history at the University of Oxford, and worked in India as a teacher, novelist, poet, tractarian and, above all, supporter of Indian independence, which he remained all his life.

Frank and his brother Edward, born four years later, were brought up in a home near Oxford that was never rich but comfortable enough in the shabby, untidy, decorously countrified way that still characterizes the culture of the English intelligentsia. The casement windows, jammed bookshelves, open fires, and deep armchairs of the sitting rooms, the brilliant untidy profusion of the herbaceous garden borders, speak eloquently through the pages of innumerable memoirs of the argumentative peacefulness, the conversability, and the democratic good manners of such homes.

Frank, the eldest son, was a natural hero. To begin with, he was a quite extraordinary linguist. By the age of twenty-three he read and spoke with varying fluency the two classical tongues as well as modern Greek, French, German, Italian, Russian, Polish, Serbian, and—to the partisans' admiration—Bulgarian. He cheerfully said, as he fell to learning the language on his way to the mission in December 1943, "Bulgarian, I find to my delight, as far as reading it goes, is another of those languages like Italian that is handed one on a plate. It is simply Russian as a Turk would talk it—a surly Turk who wasn't sure whether his false teeth were going to stay in place."[4]

VI

Memories are naturally prompted by photographs. One photograph of Frank Thompson has appeared in some of the books that mention him, and it also turned up recently in a Sunday magazine in an article about Iris Murdoch, Thompson's sweetheart at Oxford and nowadays Britain's best-known woman novelist. Her reminiscence echoes the concern in her novels about the nature and palpability of goodness.

Frank was the person I thought about. We weren't engaged, but we hoped to be married. . . . He was a remarkably good classical scholar who if left alone might well have stayed in Oxford pursuing classical studies. He was very clever and versatile and a good linguist. He was a gentle quiet man who loved animals and knew a great deal about flowers and birds. He had no desire for worldly power, had a horror of violence and would never have dreamt of himself as a soldier or war hero. . . . But when the war came and soldiering was his job he did it well and was just a hero. That is, he was good and unselfish and brave and ready to use his talents for soldiering. But he was *forced* to become a fighter. He was not seeking glory.[5]

By 1939 the fighting had to be done, that was clear enough. A good and gentle man had to go to war, and at such a time, had best go on behalf of the best the Left had thought and said.

The famous photograph forever canonizes Frank as a soldier: a tall Englishman with thick hair, a calm gaze, long sturdy neck and steady mouth, of striking handsomeness, an emblematic pipe in his mouth, and dressed in an army shirt with lieutenant pips. Frank Thompson looks out at us, and the fine picture ratifies him as all we could possibly want him to be.

He was very brave, no question about it. It takes nothing away from that plain fact, however, to say that such a man, fighting at the time in such an army, was—like all his friends and fellow students—much encouraged in courage. He was a member of what his brother has called

a resolute and ingenious civilian army, increasingly hostile to the conventional military virtues which became . . . an anti-fascist and consciously anti-imperialist army. Its members voted Labour in 1945. . . . But . . . the merits of that moment did not lie alone with the Left, and the demerits alone with the Right. Conservative soldiers were brave and effective, not only through an inherited patriotic code of class: many acquired a more democratic code as well, a settled anti-fascist commitment. . . . If any moral assets from that moment are left, let's not haggle over portions. Let us leave it at evens.[6]

Frank Thompson's war writings—his prodigious diaries and poems and letters—are as good as any in the genre. His picture of the moment, even as he lived it, squared with his brother's. Strongly patriotic as well as strongly socialist and internationalist, he frequently saw his soldiers as the "best ambassadors and gentlest conquerors the world has produced." To Iris Murdoch, he wrote: "As a matter of fact, whomever one talks to—Pole, Greek, Frenchman, German Jew, Italian—this theme, 'the courtesy of the English' (and

English means British in European parlance), always recurs—'*bardzo szlachetny narod . . . sorridenti, scherzosi, cortesi*'."

By the time Frank Thompson went to Sicily the thousands of men in the Eighth Army had been fighting together for three years; their grandest accomplishment was a *culture* as much as a victory. This huge, polyglot mass of men devised a way of life that, with no great show of militarism or solidarity, was informal, efficient, and very strong on resourcefulness, bravery, and improvisation. The desert army was brave, competent, and casual. At this range, it is no doubt easy to sentimentalize such a culture. But it is worth reaching back over the mad shrieks of laughter in *Catch-22* and *M.A.S.H.* to revalue and review a different achievement: with a view of life that was essentially comic, certainly, Thompson's generation shared a common purpose of overmastering importance—"We've got to beat the bastards." Their sense of communality and purpose lay precisely in the urgency of getting the job done so that men and women could go home and resume the irreplaceable loves, the happinesses, and the tedium of ordinary domestic life.

Courage, in such an admirable, non-Homeric culture, is expected of you. Frank Thompson, once an ungainly, uncoordinated giant, now fit and very strong, did not disappoint his army.

VII

Frank Thompson's energetically creative use of his war experience suggests for us what a different, freer, and simply *better* outcome the war might have had. He was the kind of man who could have helped create a united Europe at peace with the United States, if only there had been more like him at the forefront of the ruined societies of the war.

Thompson's settled socialist convictions were compounded by just as strong a feeling that the world is a queer and contradictory place.

> I feel very strongly that a Socialist should confine his hatred to systems, and judge every individual on his own merits. Every class, after all, is as much a victim of its environment as any other, and you [his brother], if your father had been a retired colonel and sent you to Harrow, would have been very much the same. Furthermore, Socialists who think too much along these lines ("every member of the upper class is a cad, every worker a Galahad") get a nasty shock when they meet a lot of exceptions.[7]

His socialism was *livable*, in a way rarely to be found in the insulated divisions of our present lives. Egalitarianism and altruism are hard to enact in the world of the rich nations today. These virtues were more feasible, even within the hierarchies of the armed forces, in the Western desert or the Italian campaigns of 1942–44.

Thompson read prodigiously in the languages he dedicated himself to learning, shaming most of us in our paltry struggles with a couple of cassettes and a primer. But it was a rare bird, all the same, which taught itself, as Thompson did, Polish and Russian during 1942. He had learned the rudiments of Russian from a few lessons at school; later he embraced the project by immersing himself in the language and its literature, getting lessons, while he was stationed in Persia and Iraq, from an Armenian, ex-Menshevik wine merchant. Thompson used his powerful 50-volt radio, as he said, to "knob-twiddle" his way into one or another of the Russian broadcasts; he was able to follow along because of the dictation speed used for the benefit of news editors from Estonia to Georgia, who were transcribing the news broadcasts word for word for the next day's newspapers.

He kept his faith and his principles vigorously balanced, more than some postwar, sedentary socialists did. "It was an eye-opener to me to learn how many people Stalin has personally poisoned. . . . Old Bolsheviks, it seems, never die. They only get bumped off by Stalin."[8] (This was written on 18 December 1942, an early date for such epiphany.) Thompson's robust, nice way with politics and the individual life came out in his continual celebration of the courage, endurance, and triumphs of both the Red Army and the Russian citizens. He foresaw the phantoms of hatred and the coming emptiness of the cold war. Heartened by the public display in Westminster Abbey of the Sword of Stalingrad, the huge jewel-encrusted symbol (made of the best Sheffield steel) given by the British to the Russian people in tribute to their courage in that battle, Thompson turns angrily on those many members of his own nation, to say nothing of the legions of the Red-fearing classes across the Atlantic—who even at that date still fell comfortably to cursing the Soviet peoples—

> Oh damnation. Isn't it time to admit in all humility that a regime which has inspired such sober courage, solidarity and self-sacrifice in 160 million people cannot be so terrible, after all? . . .
>
> . . . Of course I don't expect us to applaud or foster Bolshevism. But we might have the courage to realise that Europe's health can only be restored by the working-class movement, and do what we can to win its alliance and respect.[9]

Perhaps the human value most renewed and reaffirmed by the Second World War was solidarity, or cooperativeness. Solidarity was what Thompson found and was most moved by, at any rate—that and the politics of politeness.

I never fail to be amazed at how nice people are, especially those who have the most to put up with. This Brigade, for instance, came up to the desert shortly after we did, and has been in every tough show since. You might expect them to be a little browned off or a little contemptuous. In point of fact they are kind and courteous almost to a man. I think I never met a body of men so civil, men who had mastered so perfectly the key principle of our makeshift army—"mucking in."[10]

"Mucking in" is a powerful bit of military English. In a nice mix of the demotic and the scatological, the phrase enjoins everyone to get down to the job, however mucky, with everyone else and *do it*, regardless of rank. It means not standing on rank or status. It means *joining*, embracing membership one of another.

For the next two months, from mid-May to mid-July 1943, Frank Thompson was in training for the landings in Sicily at Salerno and Avola that would announce the return of the Allies under Patton and Montgomery to European soil. He continued in the brigade liaison group he had been with since he came to the North African desert, working with a mobile, heavy-duty, and long-range wireless set on long journeys between units of the brigade and between different headquarters of the division.

VIII

On 5 July 1943 Thompson stood in the wardroom of a troopship—its decks laden with landing craft—while the maps of Sicily were handed out. Hymns were sung before landing, very proper too. One of them was the grand, tolling chorale, "All People That on Earth Do Dwell," an honest message for the day. Frank wrote of the occasion in his diary:

Rich Wykehamist memories . . . Winchester a good thing to remember at this time, when a great many of the officers here are certain not to be alive in a few days' time. The teaching we had there gave us next to no idealism, but instead, and equally helpful, a strong intolerance of all folly, especially of folly which involves cruelty. Further, its long history gave us some sense of proportion, reminding one at every turn of one's own unimportance. . . . There's a lot one can say against the Wykehamist tradition, but one can never say that it squares up with fascism.[11]

A little later he was called up on deck to gaze at Mount Etna, smoking tranquilly against a blue sky empty except for the brilliant white exhaust coils of a few Spitfires. As he wrote that night, he began thinking of Persephone waiting indifferently for the arrival of the dead of both sides. Below the rail he could see the crumpled landing craft and the medieval scaling ladders. It was very hot, and the smell of engine oil stronger than the sea.

He turned in for an hour, fully dressed. The heavy bombing of Syracuse by the RAF could be felt as a series of thuds on the side of the ship. The noise mounted. The drone overhead, vibrating a little in the bulkheads, came from the passage of the heavy transport planes pulling the whalelike bulk of the gliders behind them; a few miles from the dropping zone, the planes would ditch the towrope and the gliders would swim massively on, silent through the singing air, to crash-land on their bellies on the ungiving ground.

Thompson had a solid midnight meal: a lot of fatty bacon and dollops of mashed spud. Too solid. He felt sick.

They waited endlessly on deck, in a pale darkness lit by flares and gunfire, tracer bullets, and the distant, burning town. The minute set for disembarkation onto the crude and unwieldy barges came and went. Finally, well after dawn, they piled themselves, gingerly and recklessly, into the barges. Landing had started. The placid sea speckled with the curiously peaceful detritus of war: the graceful waterspouts of distant shells; two gliders prematurely crash-landed but still intact, floating, quiet Leviathans; an untidy scattering of landing craft up and down the beaches.

"Here you are. Rum up." Captain Thompson began dishing out the rum ration to the short queue of soldiers that formed round him. The juicy perfume of the Mediterranean shore, only fifty yards away, pierced through the friendly pungency of his pipe, he afterward remembered. Mint and lemon—"Europe's greeting to her returning sons."[12]

Just as Thompson and the five men of his little signals unit began pushing their cumbrous barrowload of wireless equipment down the ridged, precarious gangway, the Italian mortars found their range. There was the whistle of the air splitting apart, and the muffled crump of the explosion right in the ship's rear gun turret, cupped like a baseball glove to catch it. The gun crew were killed at once, bits of them blown in red spatters over the gun mounting.

On the beach, the mortar shells fell regularly, accurately. The sand was soft and coarse, the specks of the stone and quartz it once was glinting in the softness. It brought images of seaside holidays keenly to mind, even as the six men struggled, their encumbering packs on their backs and their legs in hot, baggy, drill trousers, to shift the barrow bodily and axle-deep into the safety of a low gully.

As they hauled ferociously at the dead weight, Frank Thompson hurried, bent, from one side to the other of the damnable barrow. A tall lad, a lance corporal, strode past him sea-drenched, but then staggered slightly; as though dipped of a sudden in a bucket, his trousers turned scarlet in a flash as the blood poured from a rip in his side. Half a minute later the same thing happened to another soldier a little way up the beach. His trousers went red and sodden, and he limped, moaned quite dully (no one screamed when hit), and fell forward. He was dreadfully wounded; one arm and one leg torn half off at the joint by the exploding mortar shell, the tendons and bones ripped, snapped, and gaping.

Tolstoy, in *War and Peace,* was severe on those people who claim to have understood the great course of history: "Nowhere is the commandment not to taste of the fruit of the tree of knowledge so clearly written as in the course of history. Only unconscious activity bears fruit, and the individual who plays a part in historical events never understands their significance. If he attempts to understand them, he is struck with sterility."[13] And yet, Frank Thompson thought of himself and his comrades as not only caught up in but fighting for a great cause: an idea, he said, proved by the self-sacrificing heroes of the Spanish Civil War, "that freedom and Fascism can't live in the same world, and that the free man, once he realises this, will always win."[14]

Well, free men, as many Russians can now tell us, may lose for a very long season. But free men won in Sicily and Italy, as well as in France and Germany. A few hours after that tangle of human cross-purposes at the landing, the little town of Avola was taken, and in the bald, generalization of military reporting, "the beachhead was established." Such a historical event in turn meant that, so far, the Allies were indeed victorious and that, given the impressive efficiency of their systems of communication and lines of supply, it was, being early afternoon, time for tea.

Eighth Army desert tea was brewed up on a biscuit tin filled with gasoline-soaked sand. The 48-hour iron rations included a rich brown powder which turned thick and sweet and orange in boiling water. You boiled the water in the flat mess-tins with the metal prongs for handles too hot to hold. Then you leaned back with the smell of crushed mint under your rump and the sun hot on your shoulders, and luxuriated happily in a choc bar of fruit-and-nut and a strip of yellow processed cheese on a hard-tack biscuit. The war, as wars will, had wandered away to somewhere else, and you could wait for it to come back.

IX

Thompson saw the Sicilian campaign through to Patton's racing victory. It was wonderful to be back in Europe. In the quiet times always left by warfare for boredom, both the benignly restorative and the deadly kind, Thompson savored the scents and deep Mediterranean colors of bougainvillea, oleander, tomatoes, mulberries. By the end of the month, he had reached a new linguistic milestone: in the previous year, he had "learned to walk," as he put it, in Polish, Russian, Greek, and Italian.

It is his Europeanness that is so striking, one aspect of which, as he would have said himself, was his anti-Fascism. Fascism was—and in its latter-day, more vegetable forms, still is—a part of bloody old Europe, its cruel, rapacious history, its rivalries and nationalisms. To brave, intelligent, and idealistic Europeans like Frank Thompson, the unconditional surrender of Fascism was the ground of the future.

Frank Thompson was dispatched back to Cairo to join the Special Operations Executive and to lie on his bed in Shepheard's celebrated hotel, the Cheapside of the Middle East, where if you waited long enough you met everyone you knew in the services. He was there to train for liaison work with one of the British missions in the Balkans then linking up with the partisans and trying to book supply drops for them. While he was taught a bit about guerrilla warfare, more about the situation in Yugoslavia, Serbia, Macedonia, and Albania, a lot about undercover wireless procedure, and how to parachute, he set himself to learn Bulgarian.

Thompson spent November and December 1943 in Cairo waiting for his posting to a British mission. His spirits veered between boredom and exasperation, as well as exaltation. Cairo he found lowering; its smelly, raucous mixture of squalor, dilapidation, throngs of many nations, whores, beggars, soldiers, and street urchins left him adrift. Only when he knew his destination did his strong sense of direction renew itself.

He was utterly an Englishman whose traditions drew on the best of English romanticism and the socialism that came out of it. The idea of the freeborn poet, scholar, and political fighter could be traced back to Red Shelley, striding down the High in Oxford with a poetry book in one hand and a pistol in the other. His deepest feelings and the idiom in which he speaks of them come from the young Wordsworth, the Milton of *Areopagitica*, the William Morris of *News from Nowhere* (as he said himself). With those ancestral voices in his blood, and his posting to the partisans on the desk in front of him, he wrote to his family on Christmas Day 1943:

> There is a spirit abroad in Europe which is finer and braver than anything that tired continent has known for centuries, and which cannot be withstood. You can, if you like, think of it in terms of politics, but it is broader and more generous than any dogma. It is the confident will of whole peoples, who have known the utmost humiliation and suffering and have triumphed over it, to build their own life once and for all.[15]

For the socialism which is the strong child of the Romantic movement is, as Frank Thompson so often insisted in his short life as a man, above all, international. And so, twelve days before he took off for Tsrna Trava, he wrote to his brother:

> There has never been any question of history being merely a "necessary progress." You only have to read, say *News from Nowhere* or Gorky's *The Mother* or one of Ehrenburg's articles in the daily Press, to find the most passionate possible idealism and regard for "really human morality." The truth is quite simple. *Until* we are conscious shapers of our own destiny there *can* be no balanced coherent goodness or beauty.
>
> It's pleasant to think about, however. My eyes fill very quickly with tears when I think what a splendid Europe we shall build (I say Europe because that's the only continent I really know quite well) when all the vitality and talent of its indomitable peoples can be set free for co-operation and creation. Think only of the Balkans and of the beauty, gaiety and courage which their peoples have preserved through the last 600 years—years which have brought them little else but poverty, oppression and fratricide. When men like these have mastered their own fates there won't be time for discussing "what is beauty?" One will be overwhelmed by the abundance of it.[16]

Just before midnight on 25 January 1944, he dropped out of a plane into the dark, hit the frozen ground of the river valley hard, and greeted, in Bulgarian, the partisan who rushed up to check on him.

Thompson was just one of the parcels that rained down that night from more than twenty Wellington bombers. As the aircraft droned away, the sounds that remained above the fading roar came from the crackling, flapping parachutes and the meaty thuds as each hit the ground.

Maj. Mostyn Davies, who felt the cold all through, wiped his raw, continuously runny nose and stumped through the dark to greet the lad, ten years his junior, who was the commander of the new mission. Around them, the lurid flames of gasoline-soaked bales of straw by which the dropping zone at Dobro Polje was marked out for the RAF pilots began to sink. As the heat and light of the fires subsided, the dark of the valley, glimmering in snow and crackling with the frost, came into relief. There were muffled shouts from excited partisans as they bundled away the canisters of supplies: rifles, sten guns, bren guns, no. 44 grenades in good oiled paper; 48-hour iron ration packs, shirts, trousers, boots, water bottles; boxes and boxes of ammunition, rigidly packed in dense, neat rows of brass and steel tips, the brown-painted box tops clamped tight with metal hasps, the sides stenciled in white, "WD: 1000 rounds 303."

There was, for a day or two, a lull in the ferocity with which the Yugoslav royalists, the Chetniks under Mihajlović, and the Royal Bulgarian Army were pursuing the Serbian, Macedonian, and Bulgarian partisans. The story goes that in early February there was a great potlatch made from the huge drop of supplies that came down out of the night sky with Frank Thompson and his signalman. The word went round the hamlets scattered near Tsrna Trava and the dropping zone, and the local Serbians turned out to celebrate. Old people still in the area, smilingly remembering through the by now rosy mists of nearly half a century, speak of columns of peasants weaving their way down the hillside paths, playing accordions and clarinets, the girls in their red embroidered waistcoats and full skirts and a few men in their dancing breeches and pumps, the hoarded bottles of the lush plum brandy and searing *rakija* passing from hand to hand. They say that trumpets sounded for the Englishmen and for the new partisan brigade that was formed on the spot. They say, slightly embarrassed, that many of the younger men, drunk on spirits, high spirits, and political solidarity, punctuated the singing by rattling off gunshots that echoed off the icy mountains, the precious cartridges wasted wildly.

That night there was heavy snow. The next day the missions of both Davies and Thompson, six men in all, joined the partisans on the run because there had been fighting with the Bulgarian army in the snow the night before.

"On the run" was hardly the phrase. The motley brigade, weirdly dressed in bits of Bulgarian and British uniform over peasant clothing, had to make such progress as it could off the main roads, along little-known, snow-deep paths, up sheer hillsides, and through the pine and birch and snow-covered brambles. It was never better than slow, heavy going.

The two missions had the standing SOE orders to make contact with the partisans, to arrange for supplies to reach them, to encourage and coordinate their actions against the enemy, and to report faithfully and constantly to SOE headquarters all these doings.

The heart and mind of a mission was its wireless set. Without it, a mission was crippled. The 50-volt sets, with great metal casings and batteries, were extremely heavy and cumbersome, as Thompson and his five helpers had discovered on the beach at Avola. In traveling with the partisans, the wireless sets were loaded onto mules and the Englishmen stumbled as best they could after their laden beasts, delaying the quick-footed men of the brigade, lost in the dark on pathless terrain.

Supplies ran low. For five weeks no new supplies came through the murky weather. Bulgarian army units, together with Yugoslav collaborator auxiliaries, had been mustered up to about 15,000 and stationed in a miles-long ring around the whole mountain district of South Serbia. On 18 March they attacked. The Bulgarian army units had no tanks or aircraft, only light artillery. Time and again they encircled and ambushed the groups of partisans and the village refugees with them, who fought their way blindly out. The fighting was done with rifles and portable machine guns, and it was hand to hand. There was terrible, primitive killing, the personal hatreds welling up into passionate life-or-death confrontations with knives and stones and bare hands. The wounded lay in the snow and died.

The six Britishers were split up during the fighting. Mostyn Davies, Frank Thompson, and a signals sergeant had half-buried and covered with brushwood their irreplaceable wireless equipment. After four or five days of this desperate, blind fighting and retreating, driven onto the shoulders of the mile-high Mount Cemernik, the missions dropped off a ridge into a small valley where, beside the oiled and frozen road, an old mill stood, its wheel frozen stock-still in the ice of the stream.

It was pitch dark. The partisan who had been assigned to them, after a whispered conversation in Bulgarian with Frank Thompson, set off along the road to the tiny village of Novo Selo to beg for food from villagers terrified by the dreadful record of police reprisals. There were corpses in the fields. Thompson and the signals sergeant lit a fire and made things a bit cozier with sacks and straw and boards for benches. With Davies, they had a small swallow from the *rakija* ration. The cold began to release its grip on them, and fatigue to flood in through the warmed ways of their bodies.

Then heavy boots kicked at the big doors that opened onto a short stone platform above the sunken storeroom where they lay. The door hasps held for a moment, and Frank Thompson grabbed his coat, leapt onto the low window ledge, and jumped out into the black, snow-glimmering night. Mostyn Davies, slower and heavier, clambered onto the ledge and, glancing over his shoulder, just glimpsed the grenades curving into the light of the fire before the short, splitting crash of explosion. He felt the blow of the bits of metal as they thudded into his back and heard the unbelievable loudness of the submachine gun. He subsided slowly to the floor, mortally wounded. In front of him lay his dead signals sergeant.

Thompson ran up the hill across the road and waited in the trees. He could hear shouts and boots on the flagstones outside the mill. He dared not wait; carrying his revolver, pulling his quilted and hooded parachute smock around him, he climbed alone to the eastern edge of the trees. There, where the snow was not too thick, he fashioned a kind of shelter from a little hollow inside a bramble bush, his back pressed against a wide, comforting tree trunk, his arms clasped about his knees, while all about him the frost tightened like a nut on a bolt.

Finally light began very faintly to become present in the cold gray air. He had been crouched against the cold for perhaps ten hours. Many years later a shepherd recalled that he came along the track a little way below and Thompson called to him in Bulgarian. Nervously, furtively, looking the other way, the shepherd dropped a hunk of fresh grey bread on the snow. Thompson hissed a message. "Tell the partisans I am here. Say, it is the tall Englishman." The shepherd's eyes widened in amazement, and he put his hand to his Serbian cap in salute.

It was a long wait for darkness, rescue, or capture. Thompson had no idea whether the shepherd would report to the police or the partisans.

In the event, it was the partisans.

X

Thompson got back to the so-called free territory on the border of Serbia and Macedonia near the village of Tergoviste; the partisan writ ran safely there. He met for the first time the celebrated general and chief organizer of partisan activity throughout the Balkans, General Tempo.

Frank Thompson's mission, like most SOE missions, was left to its own devices under the general injunction to cause as much mayhem as possible alongside the local partisans. For the honest soldier, the task in hand was simply the defeat of Germany and all its creatures. But to some centers of command, the Germans were not quite so uncomplicatedly the first and last enemy. To them, the overthrow of the Bulgarian government by revolution, which was the goal of partisan activity, smacked, in early 1944, of Jacobinism and worse. Some such commanders, it was now apparent, were in Moscow and wanted little of local and democratic governments. Some, as it lethally and obscurely began to seem a month or so later, were in London and Washington, and they wanted socialism even less.

People used to think of Frank Thompson, as he said himself, as "normally a noisy man," but Tempo, and many partisans in the Bulgarian Second Sofia Brigade, remembered Frank as quiet, listening, percipient. Tempo, another honest soldier, was also clear about what to do and what was going on. In a letter to Edward Thompson he said, "I decided to tell [Frank] openly what I thought about the true situation." In Tempo's judgment, the Bulgarian partisans entirely lacked the support of the peasants. Tempo did not believe the optimism of the Bulgarian partisans about what they could do; they had no record of the kind of military success that had brought floods of recruits into the Yugoslav cadres. "The peasants had applauded them, but had not joined them."[17]

That April, Tempo's objective was to arm and train enough Bulgarians to form a brigade capable of penetrating into Bulgaria, engaging in effective but strictly limited military action, and then withdrawing, certainly out of reach of the Bulgarian army, perhaps into the free, partisan territory across the Serbian border. But in spite of repeated radio requests to the nearest SOE headquarters—which had been moved from Cairo to Bari in Southern Italy and was pitifully understaffed with young women decoders ("cipherines") conscripted for the job—nothing like enough supplies were dropped in March and April (only eleven sorties out of seventy planned) to provide for the incoming volunteers.

Thompson got no lead from Cairo, and no praise either. Tempo would have been glad to see him stay in the free territory—Tempo strongly advised him to do so—for his own safety and, it may be, because that way the British supply sorties would have been flown into the eager hands of the well-disciplined Yugoslavs. The temptation to run the mission from the relative safety of Tsrna Trava must have been very strong as April gave way to May and the sun brought out the plum blossoms and turned the grass green again for Frank's first European spring since 1940. He could easily and respectably have chosen such a course. Frightful dangers most certainly lay in wait across the border in Bulgaria. He was not obliged to say his piece at the partisans' open session at Kalna when both general command and local commanders discussed what they should do. But he hung on to his orders. The Bulgarian brigade officers remember him forcefully urging them to go back into Bulgaria where, with properly organized supply drops from the British, they could begin the process of quiet infiltration. With good guides and enough supplies and weapons—fifty supply drop sorties were planned for May, so even a skeptic would expect at least twenty successful ones—the mission could do more than its duty and activate the Bulgarian partisans in their own country.

Looking again for Frank Thompson as those decisions were made at Kalna some time before 12 May 1944, we can fix upon one of the earliest moments at which the antagonists of the cold war were starting up their vast, abstract, and material traffic in lies, propaganda, and murder that would collide with a smack in one man's death. Thompson's life was partly endangered by the bitter quarrels between the local, national, and ultimate centers of Communist command, and partly by the murky Allied efforts to obstruct the Communist advance and hold onto unofficial support amongst the willing stooges of Fascist domination, which could be turned to British advantage in the politics of whatever was to come next. Among these duplicitous, endless shifts of states to gain an edge, the best tale of 1944 is still of those who retained the virtues and values that defeated the Fascists.

Frank Thompson was one, Tempo another. When Tempo spoke with Thompson, he was trying to hold off one General Bulgaranov, sometime member of the Comintern and first secretary of the Bulgarian Communist party, who was attempting to organize from a safe distance his party's armed resistance in the towns of Sofia and Plovdiv.

XI

Tempo had sharply criticized the general's plans to his face, causing that worthy to fall back on the ritual party threats to haul Tempo up before the Comintern to answer for his insubordination. Bulgaranov was hot and heavy-handed. Tempo noted tolerantly that "he loyally expressed Moscow's thinking" but calmly observed that the older man simply identified the "higher interests of socialism" with the victory of the Red Army. By contrast, Tempo subscribed to the Tito heresy that the arrival of the Red Army might not perfectly conduce to the socialist society, and do little for the free development of each and all.

Bulgaranov was Moscow's man, and in Moscow his grim boss, Georgi Dimitrov, sat and issued his uncompromising orders. Dimitrov fit the simplest cold war and James Bond cartoons of a Stalinist. His stooge Bulgaranov thought as he was told to think. Between them, they set in concrete the strategy of town-based insurrection to be triggered by the arrival of the Red Army. In early May 1944 Dimitrov sent his ridiculous orders via the underground headquarters in Sofia. He began by angrily and rhetorically asking the ill-armed, largely untrained Bulgarian partisans why they were not back in their own country fomenting discord in the towns and continued by instructing them to march back into Bulgaria forthwith and head for Plovdiv.

These orders dispatched the newly formed Bulgarian Second Sofia Brigade, which was thinly armed with weapons handed over at Tempo's command from his own Yugoslav partisans, to an exhausting march, horrible privation, and eventual slaughter. Bulgaranov's orders sprang from what Tempo had noticed as his rigid failure to think in any but the terms of the 1917 October Revolution in Russia. The Bulgarian Communist party, as old as any in Europe, was authoritarian and centralist. Dimitrov was indifferent to any cost in lives and ignorant of such military matters as supply lines, communication systems, and the relative weight of military technology—rifles rarely defeat tanks. What he wanted was what, after suitable surgery and cosmetic work on the relevant history and its dead bodies, he got: national insurrection in support of the Red Army when it crossed the Bulgarian frontier in September 1944, and a tightly disciplined party in thrall to Moscow, blessed with a glowingly heroic past and no inconvenient survivors.

As Dimitrov made his own arrangements for coming to power in the new republic, the British made theirs. In typical British fashion, they made their plans anonymously and obliquely, playing for time, inconclusiveness, and a thin line of advantage. The Bulgarian government, pro-Nazi trimmed with neutrality toward the Russians, had been putting out, as the phrase goes, political feelers to both London and the United States by way of the American consular office in Istanbul. England's Foreign Office was happy enough to see the baby king's regent government stay put and was pussyfooting around the negotiations. It was, in its pained way, doctrinally opposed to anything as vulgar as revolution and did not want to deal with a new republican government that would have the embarrassing moral strength of having defeated Fascism and being of robustly egalitarian persuasion. Better by far to set up a muddled postwar coalition riddled with compromise and corrupted with closet Fascists, in the good old diplomatic way.

While the English pursued indirection, the Americans played out their historic role. The advice of the U.S. consul in Istanbul was to avoid the disaster of postwar communism by keeping "the moderate and governing classes" sweet and helping them "take immediate steps to get Bulgaria out of the war."[18] The poor mutt presaged the indestructible myth of U.S. foreign policy throughout the cold war: there are always good guys in other countries who, by virtue of their anticommunism, share our values and should be put in place by force to run things our way.

As American diplomacy gave its blessing to perfidious Albion, Frank Thompson wrote the last letter anyone received from him. It ended:

> There isn't really any news about myself. I've been working hard, I hope to some purpose, and keeping brave company—some of the best in the world. Next to this comradeship, my greatest pleasure has been rediscovering things like violets, cowslips and plum-blossom after three lost springs. I'd quite forgotten how marvellously lovely leaf-buds are just when they're breaking—especially beech-leaf buds. All this makes me more homesick than ever before, because England, when you've said all you like about Greece or Italy or the Lebanon, is the only place where they know how to organise spring. But I want to see dog's-tooth violets and red-winged blackbirds once again before I go over the hill.[19]

As he did so, the oblique directives moved from desk to desk, in and out of cipher, until they reached Allied Command in Algiers: the Balkan operation was left to go on by itself as best it could. The last big sortie flew out on 11 May. Naturally, no one told Frank Thompson about the change of policy, so we are left asking whether this one officer was deliberately left in ignorance, or whether the old cliché about communication failure must serve as his epitaph. A last, ineffable Foreign Office memorandum is a kind of answer.

It is not dated. After referring to the Bulgarian partisans as "being in high proportion traditional Bulgarian brigands," it goes on with a casual English disdain quite beyond parody: "The risking of the lives of spirited young flying officers, not to speak of arms deliveries to most undesirable elements, is not worth the candle."[20]

XII

On 14 May Frank Thompson again stood on a ridge overlooking the Bulgarian frontier. The big drops had come on the nights of 10 and 11 May—half the containers for his people, half for the Serbian Yugoslavs. On 12 May news came that the Bulgarian army was on its way to Kalna. The Second Sofia Brigade, a tattered, exhausted, cheerful, and unbelievably motley body of a mere 200 men and women (counting the British mission), moved away on foot, in haste, and with very little concealment.

Thompson's plan—the limit of what he could do—was to build a small base of partisan activity in Bulgaria and make minimal preparations for rescue by the Red Army. With good guides and decent supplies flown from Cairo, they could make a go of it.

Once inside Bulgaria, stumbling often but at a forced pace, they moved in the open and in the sunlight for many miles. Their destination was the only partisan brigade then known to be in operation in Bulgaria, the Chavdar odred, holed up in the mile-high mountains around Mount Mingash, a bare twenty miles northeast of Sofia. The Comintern had blindly appointed that region to become a liberated territory from which the rising of the people could be organized.

As they climbed, the wind swung round and abruptly became chillier. Snow fell once more, as snow will in those mountains in Maytime, softly, heavily, unstoppingly. Hungry, cold, and drenched, the partisans, as ever, had little strategy beyond staying alive, their military tactics being no more than desperate sallies to pinch a bit of bread or buy some goat's cheese.

The long march followed a wide arc around Sofia. The roads were at least passable, but the Bulgarian police were out in large patrols, cutting off their access to food and warmth and a decent night's sleep. The partisans were cautious, also, after one brief hand-to-hand struggle in the dark. They were left with two dead and a prisoner shaking with terror who after questioning could not even be mercifully shot, for fear of the noise, but had his skull stove in with rifle butts.

It was pouring rain now, and what remained of the snow had turned to blackened slush. The partisans were moving only at night and in the twilight, nearly 200 sodden figures trudging the goat tracks, wheezing and squelching as they dragged boots and beasts out of the heavy, clinging mud.

They had been journeying into the darkness for two weeks when the trampling of a horse in the woods ahead of them broke through the veil of half-consciousness in which they all dozed as they walked. A single rider loomed out of the gloaming and greeted them openly as comrades. A solid, decently fed middle-aged man, he announced himself as a revolutionary from the briefly victorious coup d'état in 1923 but brought desperate news. The Chavdar partisans operating in the mountain territory, whom the Second Sofia Brigade was to join, had been utterly wiped out three weeks before. Frank Thompson had nowhere to arrange supply drops and no one to bring them to. The brigade had lost all destination. Its only hope was the Sredna mountains, which, from Moscow, Dimitrov had declared free.

The Englishmen had hung onto the very last of their rations—the choc, the raisins, a bit of hard, chewy salami. They had, for very solidarity, shared out a little with the partisans, though the shares went nowhere. Now the food was all gone. The villages were shut to them, and every wayfarer was a likely spy. "The moderate and governing classes" with whom the Foreign Office officials behind their great desks and under their high ceilings hoped to deal had a thorough way with villages that helped partisans. They burned them to the ground, shot and bayoneted an arbitrary number of their own countrymen, butchered the bodies, and raped the women, generally to death.

Feed the heart on fantasies, Yeats wrote, and the heart grows brutal on the fare. The Bulgarian police and army nourished a fantastic view of partisans and established the habit of brutality early on.

The brigade ground grimly on. The forward patrols found two brothers on horseback riding to the hearing of a lawsuit, in which they opposed one another, in Svoge, a few miles north of where they were. Partly coerced, partly chaffed into it, the brothers led the exhausted brigade to a dense, sheltered wood high on the shoulder of the mountain.

The partisans slept where they stopped a little way from the hamlet of Batulia, men and women crouched with knees up and their backs against the pine trees, cushioned underneath by the dry needles. Silently, the brothers slipped away separately and were in different local police stations before morning. Misled about numbers, the Fascist authorities sent out hundreds of armed police and troops. They found the brigade sleeping the sleep of the dead, and opened heavy, uncoordinated fire.

The first crack and thump of bullets overhead, the piercing, oscillating whistle of ricocheting shrapnel, woke the partisans. Mortar shells went off with their thick *crump* in the soft earth and pine needles. The partisans scattered, running. Two of the British sergeants, Walker and Monroe, were cut down by machine guns enfiladed along the stream. The noisy, egregious Trichkov, the battalion commander, was shot in the thigh. The one machine gun jammed. The wireless pony was captured.

They were routed. Heavily outnumbered, in a state of hungry exhaustion that fearlessness could do nothing for, the handful of survivors, which included Frank Thompson and his remaining sergeant, Kenneth Scott, labored up the mountains, hauling along wounded comrades with rifle slings, webbing, and will power. They followed the sheep and goat paths until they reached the broken walls and perilous chapel of the Mremikovski monastery, which was utterly bare and stripped of every last mouthful of food.

It was dark. The rats rustled, and in the scoured granary the moonlight fell in narrow shafts from the battered plank doors. Distantly, the partisans could smell the warmth and sweetness of the gone grain. Suddenly there was a half-human shriek; a weird, tatterdemalion figure capered into the courtyard and stumbled and muttered his way over to the broken water butt. The partisans recognized him as a lost member of the brigade, but the wretch was dying of hunger and had long since lost his mind.

The rest of them were in little better shape. Frank Thompson and Kenneth Scott had kept up a bit more strength from their special 48-hour packs, but these were long gone. Some loaves of fresh bread hung out for them in Litakovo had come at a high price—the two bakers' lives—and had been eaten a week before. They foraged for anything—sweet grass or unripe wild cherries, hard as stone and bitter. The snails were the worst: "filthy and sticky, raw as they were, they kept stomachs moving but left a gluey, thick flavour in the mouth. Frank Thompson lamented he had not paid more attention to the culinary aspects of botany, for such grasses and herbs as they found hardly took away this devastating taste which remained in the head long after it left the mouth."[21]

Thompson and Scott conferred. Their only chance, like that of the surviving partisans, was to split into twos and threes and get through to the Sredna mountains or back to Yugoslavia. There had been, inevitably, desertions. It was 31 May 1944, and the sun was warm and golden.

That afternoon, after hopeless, plunging runs downhill, the little band was surrounded by Fascist police and soldiers. Scott and Thompson hid in a thicket, but the enemy rooted them out, combing and pounding the woods until they were found. Scott, who was shot in the hand, remembered Frank Thompson firing his revolver at their captors until he was taken, and his hands tied tightly behind him with a leather belt.

The prisoners were taken down to the village of Eleshina where they were threaded through a large, booing and jeering crowd, apparently convinced that the Englishmen were responsible for summoning Allied air raids on Sofia.

It was perhaps the worst moment of the mission. Here were the very people for whose liberation and future Thompson had staked his life shoving each other around for a chance to belabor their would-be rescuers.

The two Britishers were interrogated twice; each time they stood by the Geneva Convention rule that prisoners of war who are in uniform and carrying their official documents need only state their name, rank, and number. They were taken away past the decapitated bodies of their partisan comrades in the yard of the town hall and driven to the Gorni Bogrov School to be locked with other prisoners in the cellar.

Here they encountered that caricature of all Second World War chronicles, the maniacally cruel Fascist police chief. Captain Stojanov was waiting for them.

Jordanka, one of the troop commanders in the little brigade of brigands and Communist civilians, had greatly impressed the mission. She was quick on the uptake, vivid, determined, and understood Tempo's guerrilla tactics. Because she was a woman, Stojanov knew about her. Frank heard Jordanka's screams from the room along the corridor, the sound of dreadful beating, and heavy thumps that he couldn't explain. Jordanka would have been raped many times. She was never seen again, not even among the severed heads skewered on the wall of the school playground by the "moderate and governing classes" protected by the captain.

Frank Thompson was kept at Gorni Bogrov and then for a while at the jail at Litakova. After the first brutal interrogations he was adequately fed, and his young, big, lithe body began to recover.

Somewhere in Sofia, Geshev, head of the Bulgarian police and a very nasty, clever, duplicitous piece of work, knew about the British partisan officer with a Communist background; he also knew that deals with both the British and the Americans were in the air, and that the Red Army could roll through to Bulgaria with little trouble in a few weeks.

By 6 June 1944, the day the Allies returned to France, Thompson still waited, his strength flowing back. The cell door stood open, and he lay propped on his rough plank bed, one arm behind his head, his boots on the blanket rolled at the foot. There was a tiny window, little light, plenty of guards, a busy building, and no question of escape.

He knew how matters stood. I look for him again at this moment, peering from behind Raina Sharova, who was in the building to collect some official papers. A cousin of hers was on guard and whispered to her that the famous English prisoner was in the room down the hall. She speaks herself.

> I was curious to see him because I had already heard about him. I opened the door. . . . He looked at me but did not rise and only leaned on his arm.

He looked clean and smart. His face was serious and there was not a sign of despair on it. I did not know that he spoke Bulgarian so I spoke to him in French, "The victory will be soon ours . . . Don't despair." He answered me in Bulgarian, "I don't despair but time flies very fast."

I was worried because the gendarmery could find me with him. It is hard to remember details after so many years have passed, but one thing I remember vividly—Thompson was clean, smart and was not dispirited.

In February 1945 I left for London to take part in the preparations for the World Trade Union Conference. I was happy to meet Frank's mother. His father was ill and I couldn't see him. I told Mrs. Thompson all I knew about her son. She seemed a firm woman. She didn't weep. Only when I told her Frank's words "Time flies very fast," I saw tears in her eyes. At our parting she gave me a bunch of dry flowers to put on Thompson's grave. I readily fulfilled her request as soon as I came back to Bulgaria.[22]

I can just pick him out in the gloom of the cell, a big handsome man, less than half my age but still immeasurably older, graver, more of a *man* than I shall ever be. I remember myself on the way to the Suez expedition in 1956, another young officer but with my friends all around me and fearful, homesick, soon lost and inept in the desert.

He knew how matters stood, but as he waited and his body recovered, he must have found the unkillable strain of hope herself the hardest thing to bear.

A few years before, and a few hundred miles west along the Mediterranean coast, Ernest Hemingway had placed another young soldier among Communist partisans. In great danger and only two days away from his own death, Robert Jordan weighs up his short fighting life.

> . . . You felt that you were taking part in a crusade. That was the only word for it although it was a word that had been so worn and abused that it no longer gave its true meaning. You felt, in spite of all bureaucracy and inefficiency and party strife something that was like the feeling you expected to have and did not have when you made your first communion. It was a feeling of consecration to a duty toward all of the oppressed of the world which would be as difficult and embarrassing to speak about as religious experience and yet it was authentic as the feeling you had when you heard Bach, or stood in Chartres Cathedral or the Cathedral at Leon and saw the light coming through the great windows. . . .
>
> . . . It gave you a part in something that you could believe in wholly and completely and in which you felt an absolute brotherhood with the others who were engaged in it. It was something that you had never known before but that you had experienced now and you gave such importance to it and the reasons for it that your own death seemed of complete unimportance; only a thing to be avoided because it would interfere with the performance of your duty.[23]

It seems that the Fascists hoped that the local people would join in a noisy expression of hatred for the partisans and their tall Englishman. A few months later Raina Sharova told a British journalist of a phony trial held in the village hall at Litakovo, probably with the unspeakable Captain Stojanov presiding.

The hall was packed with spectators. Raina Sharova saw Frank Thompson sitting against a pillar with his pipe in his mouth, as it always was, empty or not.

When he was called for questioning, to everyone's astonishment he needed no interpreter, but spoke in correct and idiomatic Bulgarian. He was asked his name, rank, race and political opinions.

"By what right do you, an Englishman, enter our country and wage war against us?" he was asked.

Major Thompson answered, "I came because this war is something very much deeper than a struggle of nation against nation. The greatest thing in the world now is the struggle of anti-Fascism against Fascism."

"Do you not know that we shoot men who hold your opinions?"

"I am ready to die for freedom. And I am proud to die with Bulgarian patriots as companions."

The crowd was deeply stirred, and an old woman broke from it. . . .

"I am an old woman, and it does not matter what happens to me. But you are all wrong. We are not on your side; we are on the side of these brave men."

The captain struck her to the ground. He saw that the crowd were against him, and the trial was hustled to a finish; it was all over in less than half an hour.[24]

Thompson was shot a day or two later.

The Bulgarians named him a national hero. A bas-relief of him in half profile is fixed to the wall of the little white railway station at Frank Thompson Halt, north of Sofia, near the site of his capture.

XIII

Is that it? The wasted death of a noble man? The end and aftermath of the Second World War was filled to the brim with such deaths.

Frank Thompson was left without orders from his own people. Caught in the tightening grip of the Fascists, he had no choice but to follow the badly equipped, planless, and misdirected brigade deeper into enemy-occupied country. Dimitrov and the Soviets didn't give a toss about what happened to a couple of hundred wild ones, so long as they could tell the tale of the gallant comrades now safely dead when they came to power behind the Red Army's tanks. Stalin and his grim henchmen did not want an active, nationalist, and democratic party leading the new Balkans. They wanted dull obedience.

Geshev, sitting in Sofia waiting for the end and restlessly turning this way and that, trying to get paid on both sides, had seen all this. The long fingers of the first moments of cold war reached out from Moscow and chilled Frank Thompson where he lay.

Geshev's government, the baby king's regency, was in touch with Britain and the United States and still hoped to ward off the Red hordes and hang onto its "governing and moderate classes." The cold war was already spreading from London and Washington and freezing whatever it touched. The European Allies wondered among themselves just how much of Europe should be allowed to go Communist. Not Greece, cradle of the civilization. Italy? The Americans floated the idea of splitting it across the middle: Rome and Southern Catholicism to us, Florence, Fiat, and Tuscan Communism to them. Meanwhile, the Bulgarian regency and its police chief had left Frank Thompson for ten days or so in the cellar while they translated his notebooks and hesitated over what to do with him.

So Thompson had waited, caught between the anguish of hope and that strong human impulse to put hope down in order to be ready to face the worst. He had also been caught exactly on the twistpoint of the inhuman interests of states: Communists this way, capitalists that. Forget about SOE in the Middle East, it was only a bloody nuisance anyway; Normandy was where the action was. And so, in a combination of willfulness and negligence, Geshev had been left to infer that it didn't matter what happened to his solitary British prisoner. The hateful Stojanov had been allowed to take Frank Thompson out of his cell in Litakovo on 10 June 1944 to a little grass-topped hill above the village, and to shoot him. Ever since that date all reference to the Bulgarian mission has disappeared from the official British histories, and the Records Office has been stripped of the relevant files.

In a letter home from his training headquarters in Cairo written on Christmas Day 1943, Frank Thompson allowed himself that rare moment of exaltation, a Christmas message as he said, "of greater hope than I have ever had in my life before."[25]

Nearly fifty years later, reading that Christmas letter dry-eyed, we should not be too quick to call it utopian or idealistic or even, in a burst of middle-aged realism, sentimental. The spirit that had been abroad in Europe was speedily put down. All the nations of the hemisphere took to speaking the demoralized idiom of defense-and-crisis management. Numb to the suffering and humiliation Thompson spoke of, that iron language and its systems held down a continent, and then half a world, from 1945 until 1989.

There are still politicians who, speaking as robots for the blank mechanics of the state, will say that enormous armies and heaps of weaponry, cold diplomacy and murderous espionage, and above all, nuclear weapons, the absolute emblem of the cold war, kept the peace for those forty-five years. This book is one small rebuttal.

We can still find a few monuments to the spirit abroad in Europe in 1944. In Great Britain, for many men and women of my generation, they would be such excellent institutions as the health service, the comprehensive school, and municipal housing. In Poland or Germany testimony to that spirit can be found in the wonderfully accurate reconstructions of Warsaw or Dresden. In Italy one such monument is the exemplary, convivial administration of the noble city of Bologna. In the United States there was the incomparable generosity and readiness of the Marshall Plan, ample evidence, in Thompson's words in that Christmas letter, of how the West could take "the marvellous opportunity before us—and all that is required from Britain, America and the U.S.S.R. is imagination, help and sympathy. This may look like an over-simplification, but it isn't."[26]

Once upon a time such feelings and motives were in free play; but we did not know what to do with them, and they died. Their remains lie in our history. But the beliefs and actions of our past do not lie there like corpses. They are more like the dormant energies of the globe itself—coal, oil, electricity, waiting for a channel through which the power of the past can surge again.

Such is the power of a good story. It summons our ghosts. The past comes alive, so that we feel those believers and actors move again beside us like a palpable presence. Frank Thompson and his comrades return as one such presence. The tales I will unfold, all vividly part of our English-speaking folklore, are others. For now, half a century later, the epoch is ending and the world is once more on the move. Which way it will go depends on which ghosts we choose to lead us.

2

EVENTS I

The Casting of the Iron Curtain, 1945–1947

*T*HEY made it, of course; they drew the "iron curtain" across the middle of Europe, partly to stop us looking in, partly to stop their own poor wretches looking out enviously at the boundless goods and comforts on our side. Behind the iron curtain were the hapless peoples held captive by the grim-faced Russians and their stooges in office in the satellites; in front of it were ourselves, expressing sympathy for the captives but apologetically remaining very thoroughly armed, in however subdued a way.

The phrase had been used once or twice before, but it was Winston Churchill who commandeered it. After his ejection from office in August 1945 in a British general election that installed a left-wing government with a huge majority, Churchill prowled unhappily about looking for the aura of lost power.

He had been Western Europe's spokesman when the Big Three, Churchill, Roosevelt, and Stalin, met at the seaside in Yalta by the Black Sea in 1945. There each man's proposals had disposed of the future of millions who only learned about it later when the tanks turned up. The three elderly men, accustomed to working around the clock and vigorous enough it seemed, but all in their sixties, allocated their postwar spheres of influence—"Greece to us, Bulgaria to you," and so forth. They were not sure what to do with Germany;

Figure 2.1   Stalin, Roosevelt, and Churchill, 1943: Casting the Iron Curtain. (AP/Wide World Photos)

for the time being they would look after it themselves, one portion each and a fourth to France.

By their next meeting, in the grimmer environs of a Berlin suburb, Roosevelt was dead and Truman came in his place, and Churchill had been sacked by an electorate that judged his great gifts to be less applicable after the armistice, so Clement Attlee came instead.

Truman invited Churchill to speak in the United States, at a little college in Fulton, Missouri, whose president had access to *the* president. On 5 March 1946 Churchill drew on the full armory of his rhetoric and on his intuitions about power and politics in the only continent he understood.

> From Stettin in the Baltic to Trieste in the Adriatic an iron curtain has descended across the Continent. Behind that line lie all the capitals of the ancient states of central and eastern Europe. . . . But . . . I do not believe that Soviet Russia desires war. What they desire is the fruits of war and the indefinite expansion of their power and doctrines. . . . From what I have seen of our Russian friends and allies during the war, I am convinced that there is nothing they admire so much as strength, and there is

nothing for which they have less respect than for weakness, especially
military weakness.

He appealed for "the fraternal association of the English-speaking peoples"
based upon their new and frightful weapon and upon their soldierliness
against all "communists and fifth columnists," who "constitute a growing
challenge and peril to Christian civilisation."[1]

In time that association was formed and would include many who spoke
innumerable other languages. In its earliest stages, however, the iron of the
curtain was an alloy compounded of four elements: Stalin, the atomic bomb,
the Marshall Plan, and the Truman Doctrine.

Stalin stood for power—power without human value, power as a huge fist
that came down and crushed the life out of anyone who objected to it. The
bomb, however, was to stand as the bodyguard of the munificence and good
intentions of the world's new plenipotentiary, the United States of America.
Its remote and silent presence would prevent the darkness of Stalinism from
ever threatening the rest of a world ready to endorse and replicate America's
good and open society. Such a society would live by the principles of sound
business, its trade doors always standing open to the flow of goods, its checks
written on the honest dollars of Marshall aid. After Stalin's refusal of the
aid—which was formally regretted but had been confidently expected—the
Truman Doctrine fused the dollar and the fission bomb into an ideological
icon that commanded the servility of ruling classes and the ready allegiance
of all those within the sphere of influence the United States had taken upon
itself. The British government feebly attempted a motion or two of independence—a nuclear device here, a Byzantine adventure there—before taking its
obsequious place as chief satrap to the Pax Americana.

Faced with an enormous, barren geography and a dearth of industrial plant
either at home or in his new territories, Stalin turned all Soviet production
toward weaponry. Beyond the iron curtain, its heavy pall ratified by the
Truman Doctrine, the Western nations, fueled by American dollars, began
producing and spending as no one had ever produced or spent before. And
so it happened that the capitalist countries of the world began to produce the
weapons of mass destruction in order to protect the way of life of mass
consumption and the Communist countries of the world produced the same
weapons in order to keep out the consumers' way of life. Meanwhile, the
nondenominationally poor countries of the world either went on dying of
starvation, as they always had, or cadged some food off one or the other of
the rivals in return for looking after their weapons.

II

The most masterful hand at creating this insane world system was Stalin. He was born in Georgia in 1879 as Iosif Dzhugashvili. His mother was a peasant, and his father a drunken cobbler. He proved to be a bright pupil at the clerical school and, as hardly more than a boy, went to train as a priest in Tbilisi, where priestly training meant being introduced to a ferment of revolutionary ideas with a strong Georgian-nationalist tinge. The boys at the back of the class had Bibles on their desks and Marx open on their knees.

In 1908 his beautiful young wife Yekaterina died, and the hard young revolutionary said of her (as he never said of anyone else), "This creature softened my heart of stone. She died and with her died my last warm feelings for people."[2] Indeed, he was already well known in the Bolshevik party for his indurate commitment to the revolutionary cause and his tough, reticent, unyielding nature. He had proved himself both as a theoretician and as an undercover agent: his pamphlet on method and leadership had caught Lenin's attention, and he had been prominent as a strike agitator and demonstrator as well as a bank robber, stealing funds to extend the party's reach.

So much might be said for many of the future leaders of the revolution. But in taking the self-publicizing nom de guerre Stalin (man of steel), he not only followed party practice (Molotov means "the hammer"; Lenin is also a sobriquet) but named something inhuman in himself. He was as cruel as steel, and as unyielding. He was also secretive, suspicious as a Borgia, ruthless, rigid of will, and morally indifferent. His character was made hideous by his despotic office, but the traits were there in his peasant origins, in Russian history, in party dogma, as well as in him.

By 1917 he was in charge of *Pravda*, the party newspaper, and in spite of accurate criticisms of his cold imperious way with other comrades, he had become a full member of the Politburo-to-be. He remained fairly obscure during the October Revolution of 1917, though nobody was allowed to say so later on, on pain of death. In 1918, after the Bolsheviks seized power, he was put in charge of food distribution.

It was his first senior post, and he instantly displayed his notorious murderousness. He killed on mere suspicion of counterrevolutionary tendencies; he took no prisoners in the civil war. During the war's dreadful famine, made worse by the first collectivizations of agriculture, Stalin shifted closer to the center of power. In 1922 he became general secretary to the party, initiated the early purges, and developed the *Nomenklatura*, a new class of faithful party servants whose existence and reason for being would be party-identified.

Lenin noticed the gradual, unobtrusive concentration of Stalin's power, as

well as his ruthless, solitary, and punitive disposition. Lenin warned the party
in his "Testament," but after a stroke paralyzed him in 1923, Stalin worked
by the secretive methods and with the hard will he was master of to isolate
Leon Trotsky and ensure alliances that, when the time came, would set him
immovably at the summit of state power for the rest of his life.

By the time he was settled on the Bolshevik throne after Lenin's death in
1924 the New Economic Policy had returned Russian agriculture almost to
the desperation of the famine in which five million died in 1921. Stalin had
only crude, mechanical economic models with which to work. He determined
to solve the food shortage by the forced collectivization of all farms and by the
"elimination as a class" (Stalin's own words) of the smallholding peasants
known as *kulaks*.

Six and a half million people were dispossessed. With no provision made
for them, a good half died of cold, hunger, misery, and thirst.[3] The OGPU,
the state police, removed all food in the villages, accusing the peasants of
hoarding. The grain seizures, the massive impossibility of farming on the new
party terms brought famine. Its existence was denied. To mention famine was
to ensure arrest, deportation, and death. So, too, if anyone stole a pan of grain
to feed their children. Like all famines, the problem was as much a matter of
entitlement and distribution as of shortage. People died within reach of well-
stocked granaries. The army and the OGPU ate well; another seven million
died.

Stalin ignored the suffering he had caused. In the midst of the famine he
launched the first Five-Year Plan, which would bring forced industrialization
paid for by the forced farming that had already collapsed. He also began the
program of terror and denunciation, lasting until he died in 1953, that sentenced his enormous country to a hateful polity and a culture in which lies
were the only public statements, treachery was its routine public relation, and
abject fear the daily frame of feeling. Events in the Soviet Union in the
1930s—the pointless executions, the millions incarcerated in the camps of the
gulag, the utter distortions of truth and justice that left so many dead, so many
families homeless, so many children starving—composed the image of communism and of Russia whose frightful shapes filled the imagination of its allies
and enemies alike and impelled the nightmares of the cold war.

As the events of those dreadful years were slowly revealed, they turned the
wits of many a decent communist in the West and gave a generation that had
supped full of Nazi horrors the new, general enemy of totalitarianism. As the
tales came out, the Russian ally slipped easily into the space vacated by the
Nazi enemy, and fitted it comfortably.

The roll call of the monstrous deeds reduces, these many decades later, to
the statistics of death: Stalin was the direct cause of how many dead? Robert

Conquest guesses twenty million.[4] Later guesses have been even higher; nobody will ever really know, neither here nor in the Soviet Union. Stalin is long gone, and the generation that will be mismanaging the world in the next millenium was being born as he died. What should we remember of the man whose malignancy, suspiciousness, and nonchalant cruelty did more than anything else to shape the political feelings of the epoch?

The eye lights upon the figures: say that the police arrested nineteen million people between 1935 and 1940. Say from 1930 to 1940, seventeen million or so either died from famine or the effects of collectivization or were shot. Say that by 1941, when Hitler reneged on his treaty with Molotov and invaded Russia, nine million people were in the labor camps; forced labor was an integral part of the economy, and calculation of its costs included the certainty of its killing thousands. When the 1937 census revealed a sudden drop in the anticipated levels of population, the explanation was all too plain. Stalin suppressed the figures and ordered that the statisticians and data collectors all be shot.

As the Soviet Union revolved so momentously after Mikhail Gorbachev began his policy of glasnost in 1985, new details about executions in the closing years of the thirties came out. In Koropaty forest, 250,000 people were shot and dumped in mass graves. The neighbors remembered in 1989, "The earth would breathe. Some people weren't actually dead when they were buried, and the earth breathed and heaved and the blood came through." Such sites are all over the Soviet Union.

> Once we came here to play and we saw that there was a van. And from the van they were unloading absolutely naked bodies. We wanted to get closer but they wouldn't let us. We stood there waiting until they unloaded the very last body. There were three people, wearing rubber gloves, dark overalls and rubber aprons. They were taking them from the car with hooks. They were removing them with hooks. We saw this twice on that day. I also saw this the next day.[5]

On 12 December 1938 Stalin and Molotov personally signed 3,182 death warrants before leaving for the movies.

The killing drive for industrialization produced a heavily industrial economy, there is no doubt. Feeding an army while the peasants starved produced the men who would defeat the Nazis with weapons manufactured out of forced labor. The sums of cruelty may appear as credits in the audit of war.

As war approached, Stalin gripped power more tightly in his pudgy hands. In a sweeping vote at the party congress of 1934, 270 delegates voted for Sergei Kirov to replace Stalin as general secretary; Stalin polled just three.

The stooges destroyed the inconvenient ballots as a "misunderstanding," and Kirov was mysteriously shot a little later.

From that moment Stalin never relented in his pursuit and execution of anyone who might challenge him in any way at all, no matter how trivially. Something like 40,000 people were arrested when Stalin launched a purge of "the enemies of the people"; he supported it by a vast propaganda campaign: statues, films, posters, and broadcasts all circulated endlessly the picture of the warm "Little Father" of his people, omnipotent, awesome, loving, and severe.

The show trials of 1936–39 clinch the image of Communism in the twentieth century. The abject confessions tortured out of "the people's enemies," the reversal of history, the disappearance of honored names as though they had never been, the arrest in the night and the knock on the door, the betrayals, the stooges, the spies: these are the world-hated, world-feared signs of Stalinism, and therefore of communism.

Molotov's pact with Hitler was typical of Stalin's foreign policy. Its secret protocols handed over eastern Poland and the Baltic states of Latvia, Estonia, and Lithuania to the Soviet Union. In 1941 Stalin signed another peace treaty with Fascists; he said to the Japanese, "We are both Asiatics." On 22 June Nazi Germany invaded the Soviet Union. For four uncomfortable years, the Communists became the allies of Western capitalism, and at Stalingrad they won the most crucial victory of the war.

Stalin wavered and then sat things out with his people in Moscow. His sojourn confirmed him in the eyes of his people as "Little Father." The vast tide of the war turned, and Stalin's armies swept westward, to occupy Rumania, Bulgaria, Hungary, and Poland; finally, they met up with the American army in Berlin. In the spring of 1945, when Churchill and Roosevelt came to the warm little seaside spot of Yalta, Stalin was emperor of half a continent, the despot who ruled from the Baltic to the Mediterranean, from Berlin eastward to Manchuria.

And still he kept up his suspicions, his casual killing, his elimination of anyone who might oppose him at any level of society. His country was in ruins, his people starving again—picking over the rubble for scraps of food and dead vermin—while he concentrated his merciless mistrust upon the job of executing the returning Cossacks and "uncooperative elements" in Poland and elsewhere, crushing any alternative government in his new subordinate territories, and installing his own android facsimiles in charge.

Stalin tightened surveillance and censorship, which had never relaxed very much. The atomic bomb had intensified, if possible, his paranoia, and his scientists were set full tilt at making a Soviet version. He gave the word to his obedient leaders in the satellite states, and they turned down Marshall aid. He tried and failed to corral Yugoslavia; Tito went his independent way and instantly became a target for vilification by Moscow propaganda. Stalin gingerly tried to blockade Berlin and test the nerve of his old allies and new enemies. The terrible Lavrenti Beria, head of the police, sat at his shoulder, reporting on all his colleagues and cronies, claiming to see a U.S. agent in any servant or old friend. The execution rate began to rise again.

He drew in upon himself more and more in old age. His paranoia deepened and darkened. He had his sisters-in-law arrested; he shot all the leading members of the Leningrad Communist party, the second largest in the country; he eliminated—the Stalinist verb for *murder*—the last prominent liberals in Czechoslovakia, Hungary, Bulgaria. It was 1950 and, as Phillip Whitehead says, "the old man in the Kremlin, surrounded only by yes-men, had come to believe his own myth in the bloated terms in which it was purveyed."[6]

In his reclusive state, he sent unrefusable invitations to his henchmen, Beria, Malenkov, Bulganin, Khrushchev, Voroshilov, the dreaded names of his always fearful subordinates. They came to share his phony, watchful geniality: drinking vodka like water, the whole stag party would be ordered by Stalin to dance cheek to cheek, to sing, or to declaim snatches of poetry, to be cornered for unwelcome confidences while the rest of the party went through its tense frolics.[7] Malenkov and Khrushchev jostled each other warily, waiting for the end.

On 1 March 1953, after a heavy night's drinking, Stalin stayed all day in bed. His bedroom was locked. The Politburo phoned one another very cautiously, unable to say much because of the listening stooges plugged into every line. His housekeeper finally plucked up the courage to go in and found him speechless on the floor. No one dared fetch a doctor, for Stalin had in the last excesses of his mistrust begun to arrest senior medical men on farfetched charges. Anyone who fetched a doctor would, if Stalin recovered, be under suspicion. At the point of death, as one of his generals later said, "he was the hostage of his own system."

For four days they watched the dying man, Beria at one point groveling at his feet. Once or twice his eyes opened, and the watchers shrank fearfully. But to their relief, at last he died on 5 March, and the country fell into heartfelt mourning. The fiend was gone, leaving a legacy that would weigh like a nightmare on the brain of the living for forty more years.

III

On 16 July 1945 the extraordinary team of scientists assembled at Los Alamos, New Mexico, by Gen. Leslie Groves and Robert Oppenheimer detonated its prototype atomic bomb, Trinity, and for the first time men saw the sequence of atomic explosion whose images would dominate public life and private fear for the duration.

Truman had been president barely three months. On that day he was doing his best to puzzle out the demands of the job in the most exigent circumstances; he was at Potsdam, a more or less undamaged suburb of the devastated city of Berlin, carrying forward the discussions launched by his all-wise predecessor, and the nation's rightful hero, Franklin D. Roosevelt. Truman bore up well enough but confided to his diary that Stalin was a genial sort of fellow (which he could be) and would surely keep his word (which he often would not). Groves had been asked by Secretary of State James Byrnes and the venerable secretary of war, Henry Stimson, to make sure the President knew, while he parleyed with the Red peril, that his secret weapon would work.

Groves set the detonation for 2:00 A.M.[8] By the summer of 1945 Los Alamos had become quite a company town, and although amazing security was maintained for the two years of the settlement's life, the scientists' first display of their gadget could not be kept secret in their community. That gripping feeling of excitement that precedes a really impressive bonfire or pyrotechnic show was abroad. But most of them knew that this explosion would exceed any other explosion in history; it would be a man-made miniature of the first seconds of the Creation, thrilling, certainly, but also unimaginable. The atmosphere was strongly apprehensive as well; a few of the scientists were seriously concerned that if the bomb went out of control, the globe itself might unravel.

Long before two o'clock, the camp meteorologist forecast a thunderstorm, and Groves, the rain streaming off his coat and pouring off the peak of his cap, was patrolling the muddy dirt road, looking for a break in the weather. He found Oppenheimer, and they put back the time to 5:30 A.M.

The spectators began to pile onto the viewing platforms twenty miles away; huddled under the short wood-and-felt shelter, cold, drenched, tense with anticipation, and trampling on the duckboards to keep warm. Ten miles nearer, crouching or lying in shallow trenches behind a low revetment, were the designers themselves holding medieval welding visors in front of their sunglasses.

Verey flares curved into the black night at five minutes to go. The rain had stopped. A horn gave the one-minute warning, and dead on time the monster woke. It broke into dazzling light.

Shortly before, a lad called Donald Hornig had thrown a switch that powered the automatic starter he had invented. At 5:30 the starter discharged the impulse that fired the first detonators of ordinary high explosive in the tubby mine-shaped thing suspended in its cradle ten miles away. Thirty-two detonators exploded and lit the packing around the second stage, a special cake made of wax baked in with a fierce explosive, all cooked up by George Kistiakowsky, the house explosives expert. The cake exploded and hit the Baratol walls made, in turn, of a counterexplosive, which drove the detonation waves inward. Implosion followed, the waves assuming the shape of a sphere. The inward-moving waves then hit a second cylindrical cake of explosive that drove them inward faster. The waves accelerated.

They now reached the penultimate component, the parcel of uranium that surrounded the plutonium core. The uranium melted to become part of the shock waves, always moving inward upon themselves until they struck the core and carried on squeezing. The plutonium sat in its den in two spherical halves, each face gold-plated. The plutonium, named after the god of the underworld, had been manufactured in atomic piles. It was held within their massive bulk and forced, under delicate and dangerous control, to absorb as much radioactivity as possible without becoming so hot the process got out of hand. The gathering of radioactivity began to happen spontaneously, and what scientists so accurately call meltdown followed.

Meltdown is the process of chain reaction gone out of human control. It is what an atom bomb seeks to initiate—the sequence of molecular unwindings that occur when heavy-laden radioactive lumps, plutonium, become so hot they release their components and the thongs that bind the atom spring open, flinging their neutrons outward.

What has happened is that the force that holds matter in its place has been loosened, at first in a single atom. The neutrons rush out (only a few are needed) and unfasten the energy bonds around the next atom, and the next, and the next. Chain reaction has begun.

Chain reaction in the Trinity bomb took place as the tamping parcel of uranium squeezed the plutonium bullet at its heart. Two nonferrous, poisonous metals, beryllium and polonium, brewed together, mixed in violent conjunction, the polonium bullying the neutrons out of the beryllium until a handful of the deadly dots assaulted the plutonium and the splitting—fission—started. The energy surged outward as X rays, as flash, as heat, as blast. The vast heat consumed everything in its reach, soared on its own unbelievable temperature, and swept away. Below, the vacuum created by the explosion rose, filling with blast, taking with it huge clouds of dust each speck of which gathered a lethal little charge of radioactivity. Then the whole monstrous confection coiled upward and grandly spread and drifted and killed.

The watchers in New Mexico were the first to see the now familiar beast in its metamorphoses: the flat white ellipsis flashing blindingly sideways; the tumbling orange fireball endlessly rolling in upon itself and yet expanding upward; the tall column of purple smoke with a bulging, purple, mushroom head; the slow darkening of the dust and smoke as they spread along the horizon; the stupendous gale flattening trees and blowing into the experimental houses, which swelled a moment, split and rushed in pieces in the wake of the wind; the wind's howling, its storm of dust. All this they saw, their souls shaken; Oppenheimer quoted Vishnu, god of the Bhagavad Gita: "Now I am become Death, the destroyer of worlds."

Three weeks later Trinity, in a marked diminuendo, became Little Boy. The B-29, one of the newer bombers the United States had designed, was named *Enola Gay* after its temporary pilot's mother (making the regular pilot very cross). It took off carrying Trinity's dreadful offspring during the night of 6 August 1945. At six o'clock in the morning the aircraft picked up its escorts over the Philippine island of Iwo Jima, where the previous February thousands of doughboys had died overcoming the Japanese army's suicidal last stand.

The *Enola Gay* flew over Hiroshima and dropped four tons' worth of Little Boy just before 8:45 A.M. It was a beautiful day, sunny, mild, moist, about 80°.[9]

Little Boy behaved as little boys should, exemplarily. He flattened the city. He killed, roughly, 100,000 people at once or very quickly, and perhaps as many again lingeringly. By the "standardised casualty rate," a British measure, it was an unprecedentedly efficient bomb, 6,500 times more efficient than everyday high explosive.

The blinding light of the three explosions, at Los Alamos, Hiroshima, and Nagasaki, illumines the horizon of the epoch and casts a fifty-year shadow. The awful stories in the hosts of books, beginning with John Hersey's first-rate report, *Hiroshima* (1946), are as fresh and piercing as ever. Not the least of the grim realities of the Chernobyl accident of 1986 was that we knew so much about what to expect for its victims after their irradiation. Whatever the moral judgment on the decision made by Truman, Byrnes, and Stimson, the facts of atomic explosion moved very quickly into everyday political consciousness. And even now, few people really question the rightness of the decision to send Little Boy to Hiroshima. (His stout brother Fat Man, dropped on Nagasaki a week later, is almost forgotten.)[10]

Yet the short, secret, and intense conflict at the center of American power in the summer of 1945 fixed the position of nuclear weaponry in the cold war. It confirmed these weapons, on each side of the iron curtain, as the responsibility of politicians and the military. It exonerated them from public judgment. And the power conflict denied science any role in deciding on the meaning of the weapons in human life.

At the time, scientists called for a more careful reckoning. They knew that science could thrive only in an international culture. The notion of a scientific secret is, except in the very short run, as absurd as the notion of a private language. To be used as a language, speech must be social and public; science is a special but public language. Niels Bohr, Einstein's great collaborator and the doyen of nuclear physics, foresaw the certainty that the Soviet Union would not only develop nuclear power but would do so in retaliatory forms if it was kept out of the developments at Los Alamos. He knew the magnitude of the discovery; he had the necessary magnanimity to think about the nuclear capability without the politicians' insistence that it be turned to advantage, a type of thinking he dreaded because he realized how it would lead so certainly to the arms race, to an avoidable intensification of suspicion and danger.

In July 1944, having failed to persuade Churchill that the new weapon would completely change the world, Bohr wrote a long memorandum to Roosevelt. He understood that "we are in a completely new situation that cannot be resolved by war," that the old notion of pursuing military victory as the means of resolving intolerable conflict would simply disappear. As Rhodes puts it, "Bohr saw that far ahead—all the way to the present, when menacing standoff has been achieved and maintained for decades without formal agreement but at the price of smaller client wars and holocaustal nightmare and a good share of the wealth of nations—and stepped back."[11]

So Bohr put to Roosevelt his vision, not an artless or ingenuous vision but a merely practical one, of a postwar world in which the nations certain to deploy this new power so soon—and he accurately anticipated its proliferation as an inevitable consequence of science's undammable international culture— would consider together how to avoid letting their rivalry over it bankrupt them. He wrote five years later, "It appeared to me that the very necessity of a concerted effort to forestall such ominous threats to civilisation would offer quite unique opportunities to bridge international divergencies."[12] And he recommended a system of mutual inspection and exchange (of a sort finally, grudgingly prepared in the 1990s) as the condition of a safe world.

Roosevelt was cordial and appreciative. When he next saw Churchill, however, he capitulated entirely to the politician's defense of his secrets and advantages. Knowledge sharing was canceled and Bohr was lucky not to be visited with a knock on the door in the night.

Bohr stood as spokesman for the international college of science; he was probably its noblest as well as its most politically intelligent representative in the United States or Britain at the time. Leo Szilard, the discoverer of the origins of nuclear chain reaction who, from his eminence in the subject at the University of Chicago (and in the world), ranked only just below Bohr, wangled an appointment with James Byrnes.

Like Bohr, the much less diplomatically gifted Szilard tried to persuade the secretary of state that once the bomb had gone off the Russians would join the race, would see the challenge as a race. In the event, Szilard could have saved himself the effort, and yet, at this distance, it is proper to understand once more how things might have gone differently, and better, if only one set of decisions, plainly available at the time, had been taken rather than the other.

Szilard irritated Byrnes. Just as Churchill, bald as a coot, had taken against Bohr "with his hair all over his head," so Byrnes said stiffly of the second great scientists, "His general demeanour and his desire to participate in policy making made an unfavourable impression on me."[13] Groves had told Byrnes that the Soviets did not have the uranium ore to make the bomb. Szilard explained why this assumption was wrong, but Groves said the same to the senators, who believed him, preferring an honest soldier to all these mad boffins.

Of the politicians, only Stimson had imaginatively grasped the momentousness of the weapon. At the crucial meeting of politicians, scientists, and military officials on 31 May 1945, he spoke with due gravity of the responsibility now settled on their shoulders to decide for the sake of the future—for the human race—what to do next. Oppenheimer recommended, like Bohr and Szilard, that the scientific books be opened to international inspection. Even Marshall suggested inviting two Soviet scientists to watch Trinity go off.

Byrnes, an astute committee man, quietly heard these recommendations out and then killed them. He was the man who made the key decision first, that the new bomb gave his country an incalculable superiority over both its present and putative enemies. It was his duty to build foreign policy that would maintain that superiority. In this view, he and the president were of one mind. The arms race was on. Competition for world destruction had begun.

As one would expect, there were plenty of people who thought that the weapon could be held, as Truman's secretary of defense, James Forrestal, put it, as a trusteeship for the United Nations. Forrestal, like all of his sort ever since, made much of the pointlessness of appeasement. Acheson prophesied "armed truce." Groves blustered and said that the Soviets wouldn't have the bomb for twenty years as long as what he called "certain scientists of doubtful discretion and uncertain loyalty" could be corralled. Only Stimson and the leftish commerce secretary, Henry Wallace, ever favored any exchange with the Soviets. They never had a chance with Truman.

On 3 September 1949 an American B-29 on high-altitude patrol flew into a sudden squall of radioactivity. The Soviets had tested a highly satisfactory nuclear-fissile device, four years and three weeks after Hiroshima—just about the time Bohr had given them.

IV

The reconstruction of Europe after 1945 is one of the most amazing feats in history.

The basics of survival simply weren't there: water, food, energy, capital—the greatest deficiency was capital. Every European country ran a colossal deficit on its balance of payments. Even by the end of 1946 industrial production was 20 percent below prewar levels, in Germany 36 percent. Moreover, in the weird economic aftermath of war, the beneficiary of this unprecedented European need for imports was the United States. The European harvests failed in 1946; there was a Siberian winter across the Continent in 1946 and early 1947, and the coal ran out; there were universal shortages of clothes, paper, soap, gas, furniture, internal combustion engines, let alone iron ore and minerals. The Europeans dug their gold out from under their beds, turned it into dollars, and spent their last beans on what the Americans would sell them.[14]

European manufacturing equipment was either worn out or wrecked. The labor forces were hungry and bloody-minded; the Communists, after their exemplary fight against Fascism, had the moral credit to lead the unions and were backed by the confidence of the victorious Soviet Union. The party had been eradicated in Germany by Hitler, but it was resurgent in Italy, strong in France, and at civil war in Greece, quite apart from its impending assumption of power, legally or not, in the several states of the Balkans and in Poland, Hungary, Czechoslovakia, and the Soviet zone of quadripartite Germany. Even in Britain, the staunch old ally, there was an odd sort of crew in power calling themselves socialists who were, to a partial eye from Georgia or Idaho, indistinguishable from Communists anyway.

The problem was capital, but it was also liquidity. Starvation, subversion, and underproduction turned less on unalterable shortage than on effective planning and the restoration of markets. The always exemplary Belgians, after a (fairly) quiet war and a prompt devaluation, were making plenty of steel; there was no point in the United States selling Pennsylvania steel to France if it could be had more easily from just up the road. So, too, with French grain and grapes for Germany; the huge farms of the north, the vineyards of the Herault and the Rhone, were in fine shape. Even Britain, which had long since cooked up vigorous preferential trading deals with its imperial territories, suffered acute shortages. She was still stuck with wartime rationing and the deplorable egalitarianism that rationing enforced. In fact, the British could not afford to run their zone in occupied Germany. The whole continent faced the certainty of out-of-control inflation unless it chose either Communist or American-capitalist economic solutions.[15]

The early loans of American dollars on a country-by-country basis had run out in no time. Even on a minimal level European recovery depended on buying American goods with American dollars. They had none left. The dollars had been spent the moment they arrived. When the United States made credit dependent on the British making sterling convertible to dollars on the money markets, everybody sold their pounds for dollars and the Bank of England was scraping the bottom of its reserves in a few weeks. Bankruptcy was barely avoided.

Normal business practices could not possibly rescue Europe, any more than in 1919. Reparations then had led directly to the Second World War, which was only, as Keynes had predicted, the economic consequence of peace. With no cash and no regard for future stability, the Soviet Union presented the ruined Reich with a reparation invoice of $10 billion and pinched as much of its defeated enemies' industrial equipment as it could get away with. (Thousands of miles of German railroad track had been removed by the end of 1946, hardly the action of a far-sighted expansionist power.)

Step by step the Americans moved toward the conclusion that their ideology—that is to say, their deepest assumptions about the political world—predisposed them to take. Gen. George Marshall, most honest of Americans, had taken over from Byrnes at the State Department, had concluded that Stalin was waiting to dominate Europe and, moreover, that the tyrant believed all he had to do was wait. Fifty years later we might more readily think that all Stalin wanted was a Soviet Union within whose much-invaded frontiers even a paranoid would feel safe, together with a check to rebuild his shattered land. But in 1947, under the strong compulsion of having a new totalitarian enemy to replace the old, Marshall and Truman, Attlee and Ernie Bevin, the Labour foreign minister and socialist commie-hater, were in agreement. They began to plan for a merger of their two occupied zones in one new country with the Army-exercise-and-musical-comedy name of Bizonia.

Thereafter, on 5 June 1947, George Marshall gave the Harvard University commencement speech that launched the plan commemorated by his name for the rest of history.

He invited all European nations to take part in a continentwide program of reconstruction funded by the vast American budget surplus. The program had not yet cleared Congress, which was brimming after Republican electoral victories with anticommunist hatreds. Nonetheless, there were no exclusions from the invitation, although plenty of conditions.[16]

Stalin and Molotov were both hesitant and rattled. They desperately needed the capital, but they could not possibly join a scheme that would so abruptly open up the hoarded darknesses of thirty years of postrevolutionary Stalinism. Molotov came to the preliminaries in Paris backed by eighty-nine counselors, then took himself off in a theatrical huff almost at once and thwarted Polish and Czechoslovak efforts to attend as well, the latter at the last minute. Molotov knew that the Marshall Plan meant no reparations from Germany. Acceptance of American money would mean the loss of the Soviet grip upon their new satraps. In any case, the Soviet Union would have to send raw materials to Western Europe.[17]

There seems to have been a relishable schadenfreude among American dignitaries over the Russian vapors: "They asked for what they got," "I told you so, they want us to be their enemies," that sort of thing. The Soviets went home to their blasted and impoverished land, and the remaining European countries put in for $20 billion over four years, some of it to Bizonia. Thereby was Europe divided by the dollar.

The Marshall Plan prevailed in Congress only after some hustling. The Republican bigots, ignorant of both economics and European politics, thought it was all some kind of dole and complained that "there seems to be no grasp of any business principles in connection with this situation."[18] Marshall appealed to national security and fears of the Communist alternative. The drums of anticommunism began to throb as the competition of world systems started up along the line of the iron curtain and capitalism set itself to defeat communism.

V

Bevin observed to one of his Foreign Office staff at some point during 1946, "Our trouble is that the Russians are frightened and the Yanks bomb-minded,"[19] which is to say, the Russians were paranoid and the Americans were gunslingers. It was a lethal opposition. When the Russians moved to claim their defensive rights over the straits into the Black Sea, as had been sketched out before Potsdam, all the alarm bells rang in Washington. The great bear wanted his warm-water ports; what would he want next? He *said*—but, of course, the bear would say anything—that the defenses and gun emplacements would be nominally shared with Turkey. Everybody knew, however, that in no time at all Turkey would be bear-hugged to death.

It was a small enough incident. The Soviets had sufficient, if rather old-fashioned reasons for wanting to match the guns of Gibraltar with their own at the other end of the Mediterranean. They also had the Potsdam agreement behind them. But in their present state of mind, the Americans chose to see the move as a menace. Truman sent the Soviet Union a nasty letter, and Forrestal set a fleet on guard at the eastern end of the inland ocean—where they made themselves very comfortably at home in Naples and stayed for half a century.

Every little turn of events served further to set in concrete America's anticommunism, both in the power elite and in popular consciousness. And as the concrete hardened, so the iron of the curtain entered deeper into the soul of the Kremlin. Jimmy Byrnes, Yergin tells us, canceled a check to the Czechs because he saw them sycophantically clapping a routinely anti-American speech by Vyshinsky at the 1946 Paris peace conference.[20] The United States delicately timed an atomic test on the Bikini coral reef for just the moment at which the United Nations was to debate the establishment of an international atomic energy authority. The U.S. proposal, constructed by the businessman Bernard Baruch, was that such an authority would store all the data that were presently known or developed by any nation about the weapons. The United States would stick to its arsenal while the Soviets told the world how much they'd accomplished. Naturally, Andrey Gromyko turned down the idea flat, and his refusal became another instance of Russian obduracy.

The document that best encapsulates the many dimensions of the Americans' anticommunism, their suspicion of the Soviets, and their unreflective certainty of their own admirableness, is George Kennan's famous Long Telegram of 1946. There could hardly have been a single more cited, more influential, more completely expressive statement in the epoch.

Kennan had long been an American diplomat on the Soviet front, beginning as an observer of the aftermath of the civil war between the Whites and the Reds; posted to Riga in Latvia, he built for himself the most assured and architectonic version of anti-Bolshevism in the Foreign Service. He saw collectivization and the terror from fairly close up while in Moscow with the American mission, and dispatched his telegram after another two years' service in the capital from 1944 to 1946 as chief of mission and Ambassador Averell Harriman's consultant. He was, at the age of forty-four, the American magus of Russian affairs, fluent in the tongue, soaked in the nineteenth-century version of its culture, and unremittingly anticommunist.

Subsequently, the paper was published in *Foreign Affairs* in 1947 as "The Sources of Soviet Conduct" and circulated simply everywhere.[21] The most mannered touch in its mannerist presentation in this publication was that it appeared under the cypher "X," although the authorship was unmistakable to anyone in the trade. Under his alias, Kennan announced in magniloquent prose that the advent of the cold war was America's historic and unrefusable opportunity to assume leadership of what would shortly be described as the free world.

It was a simple, stirring tale—the essay recalls the form of the short story—whose occlusions, caricatures, and comic-strip plot all nicely appealed to political executives on each side of the Atlantic. The plot was boldly unsupported by facts in any form, most notably lacking the deadly statistic of the Soviet death toll between 1941 and 1945. Indeed, the Second World War is mentioned only as the occasion for the dislodgement of the British from the throne of power and the assumption of its heritage by the United States.

In a rude trope, Kennan ascribed to the old Bolsheviks an "insecurity," a "brand of fanaticism," a "Russian-Asiatic skepticism" that led to their psychological preference for Marxism and its relish for "bloody revolution," a preference that was no more than a "highly convenient rationalization for their own instinctive desires." Settled in power after the revolution, these dark and shadowy characters taught by "ideology" that "the outside world was hostile and that it was their duty eventually to overthrow the political forces beyond their borders."[22] Nothing is said by Kennan of the Russians' rather direct experience of outside intervention (including intervention by the British and Americans in 1921) or of the slaughter and starvation which left for dead twenty or more million Russians during the war. A man of his experience must have known how *unlikely* a tale he was telling. But he was telling it in its bold, stark outline to a callow president and his incorrigibly ingenuous, stiff, and upright team of secretaries. They were gentlemen and Americans; they were going to run the world, but they didn't yet know how. The Kennan story rendered the momentous makings of policy intelligible.

Kennan began with the "fanatic" and subversive will to power of Lenin's Bolsheviks in the past, detailed the present obduracy of the Soviet Union's policy robots ("like the white dog before the phonograph, they hear only the 'master's voice' "), and for the future raised the satisfying possibility, "in the opinion of this writer it is a strong one, that Soviet power, like the capitalist world of its conception, bears within it the seeds of its own decay and that the sprouting of these seeds is well advanced."[23]

Set aside the appalling metaphors and note the characteristics of Kennan's dramatis personae: the Soviets' "iron discipline," deafness "to the appeal [of] common mental approaches" (why can't they be like us?), and the "caution, circumspection, flexibility and deception" whose "value finds natural appreciation in the Russian or the oriental mind."[24]

These are the cartoons to which Molotov, Vyshinsky, Gromyko, and company lent themselves for the epoch's crazy legend. Opposite them, powerfully implied in Kennan's narrative, are Americans, who, with their "love of peace and stability," need only this timely admonition to maintain a policy of "firm containment." The United States has created "generally the impression of a country which knows what it wants, which is coping successfully . . . with the responsibilities of a World Power, and which has a spiritual vitality capable of holding its own among the major ideological currents of the time."[25] (They have ideology, we have "spiritual vitality.")

Kennan's snappy story matched the feeling of the day and gave it a plot: *containment.* The Soviet Union must be "contained" in its relentlessly expansionist drives; it must never be appeased. Clichés became rife. Dean Acheson warned against one "rotten apple" of a country infecting "the whole barrel."

In 1947 it looked as though Greece might be going rotten. In 1944 Stalin had stood by an agreement with Churchill to allow the British to put down the partisans' liberation front. The leftists who were not killed took to the mountains, and the British returned a pro-Nazi police force and a venal royal family to power. Civil war followed inevitably, the Yugoslavs lending a hand with supplies to the leftists in the hills. The British shoveled in what money they could spare, and their nominated government in Greece skimmed off the gold at the rate of a million real sovereigns a week.[26] The British were almost bankrupt as it was and coal ran out and food became more and more scarce. In February 1947 they confessed to the United States that they could no longer pay up in Greece.

Truman took over. With the support of the new anticommunist tough eggs of his administration and in Congress, where they were led by Sen. Arthur Vandenberg, he also took the opportunity to pronounce the new policy, the Truman Doctrine, for the times. Every country of the world had to choose between two ways of life:

> One way of life is based upon the will of the majority, and is distinguished by free institutions, representative government, free elections, guarantees of personal liberty, freedom of speech and religion and freedom from political repression.
> The second way of life is based upon the will of a minority forcibly imposed upon the majority. It relies upon terror and oppression, a controlled press and radio, fixed elections, and the suppression of personal freedoms.[27]

Put in these terms, it would be an easy choice to make; even Kennan was horrified at this kindergarten version of containment. Truman was choosing to make the belly and loins of Congress tremble with righteous rage. The President described the wretched Greeks, inventors of democracy, being held to ransom by a few thousand ragged Communists and declared, "it must be the policy of the United States to support free peoples who are resisting attempted subjugation by armed minorities or by outside pressures."[28] Congress sent the Greeks, or rather, a few of them, 300 million bucks with which the left was finally wiped out in 1949 and the cradle of civilization was, as usual, made safe for old corruption and American military bases.

Once the Truman Doctrine was enunciated, the process whereby the monoliths of cold war institutions and industries were cemented, first across Europe, then across the globe, can be traced through the headlines alone. Late in 1947 Bevin mooted the idea of an anti-Soviet defense pact. Marshall backed it on the condition that the first outline plainly come from Europe. Dead on cue was poor Jan Masaryk, the last liberal in the Czech government, murdered as Czechoslovakia went Communist in the 1948 coup. The papers were crammed with rumors of war. Truman began to wind his defense budget ever higher and tighter toward the 1948 presidential election, in which he crushed Wallace on the left and defeated Thomas Dewey against the odds. The troops and weaponry were all in place in Europe; all they needed was the North Atlantic Treaty and its bureaucracy.

The treaty was signed on 4 April 1949. Signatories included Italy, Norway (hesitantly), Denmark, Portugal, and Iceland (but no nuclear weapons), as well as Britain and France. The same year the Soviets fired off their first atomic gadget, and in April 1950 the huge acceleration in U.S. defense spending was rationalized in the second most important document of the cold war after Kennan's telegram, National Security Council memorandum number 68 (NSC 68), mostly written by Paul Nitze at the start of his forty years on the job. This document committed the country to continuous increases in military strength for the unforeseeable future.

It had taken no more than a couple of years to set the pattern upon which the texture of the epoch would be woven. As never before, one political doctrine was counterposed to another. In the old struggle of powers and principalities, the balance of interests was replaced by the clash of ideologies. Each great event became a victory or a defeat in this struggle. Each celebrity became a hero of its outcome. Each cold war story, true or false, was fitted to the pattern of the times and shot through with the colors of ideological allegiance.

By 1949 the war, vibrating with hatred and never colder, was ready to flash to a heat capable of destroying whole countries. China was lost, Korea was about to burst into flames, and the opening engagement of the cold war had been won, at its very center, in Berlin.

3

EVENTS II

The Berlin Blockade, 1948–1949

BOTH the simple and the complicated versions of the Berlin Blockade story start with the Soviet government's blank uncooperativeness; its lies, cant, and evasions; its murderous ways with those on its own side who objected to its activities; its equation of its Marxism with its foreign policy, so that there seemed to be no sure diplomatic distinction between what Stalin and his gangsters would regard as their hard advantage, and what they would count as ideological pressure, to be kept up routinely on the Western nations as an expression of Soviet power.

If the same was and is true of the liberal and democratic powers—as seems to have been the moral of the Greek episode—it remains true that for sheer bloody-mindedness and barefaced hypocrisy Soviet negotiators took the collective breath of our side clean away. So much so that when in March 1948 the Communists captured power in Czechoslovakia after very shady goings-on, the political elites of the Western nations took it to heart.

The Czechoslovak commotion looked very like an augury. Since the end of the war the country had believed itself to be a multiparty parliamentary government, with the Communist Klement Gottwald in charge but with every prospect of not being under Moscow's thumb, of aiming for a happier version of the Yugoslav kind of independence. But as its economic condition deteriorated through 1947, and the Soviet government made its allies reject Marshall aid, the signs became clear that in the general elections of 1948 the Communists would be correctly blamed by the electorate for all that was going wrong, and they would lose badly.

The Czech Communists reacted by stacking the police with their people, infiltrating the opposition, and mounting scare propaganda. There was the usual rush of arbitrary arrests, quickened by the discovery of an evidently genuine, much-needed plot against the Communist executive. The party shrewdly and properly carried through popular reforms—equalization of incomes, reform of farm ownership, nationalization, and the like. The opposition ministers resigned, and Gottwald moved swiftly.

At public meetings with authentic mass support, he announced a general strike. Soviet troops and tanks were highly visible at the border. The independent-minded and honorable Jan Masaryk, the foreign secretary, was murdered by being shoved out of a window at the ministry. Gottwald, in a perfectly constitutional action, replaced the resigned ministers with Communists. After the election, the strictly edited list of candidates confirmed a one-party state. Another Communist monolith was planted at the eastern perimeter and would not budge, not even in 1968. Its people could only brace themselves to sit things out.

II

Czechoslovakia was a charged word in 1947, only ten years after Neville Chamberlain dealt it away to Hitler without a word to Prague. It was the only context needed for the simple story of Berlin during the same year. Gottwald's coup confirmed all that the Western nations suspected of communism, and most of what they had experienced at the hands of the Kremlin and its implacable leader since Berlin fell. When trouble followed in Berlin, it was only what they had expected.

The Berlin settlement divided the city at the Brandenburg Gate and east-west, the Americans, British, and French each taking roughly one-third of the western half, the Russians the rest, the whole city, of course, remaining within the Russian zone.

Its anomalous position in the Russian zone, the only access to the Western powers being the autobahn and the air corridor, put the capital under arc-lights for the whole cold war period. The puzzle about what to do with the defeated Third Reich is at the heart of the era. That question was the intersection between them and us, between nuclear weapons and the free passage of the dollar, between Soviet fear of the rest of Europe, Germany in particular, and America's vision of endless open markets stretching across the European continent, irrigated by infinite capital.

The iron curtain split the very country. Berlin therefore became the stage upon which Apollyon wrestled with Christian. Neither could permit so public a fall as to surrender the city. The blockade was the first bout.

The Western story of the blockade is brisk. After Czechoslovakia, on 17 March, the West European allies quickly signed the defense pact that had been under discussion for a while. The unnamed enemy was the Soviet Union. Three days later the Soviet group walked out of the four-power Control Council in Berlin. (Walking out of key meetings was to become a frequent and infantile gesture in all East-West negotiations from then on; the diplomatic vaults of the cold war echo to the crash of doors slammed in a huff by assorted uniformed grandees of the international elite.) By the end of the month, *une petite guerre à coups des épingles*, as the French delegation called it, had begun in Berlin: petty obstructions at crossing points, endless delays before trains could move past signals, fussing over papers, roadblocks.

While the first, small airlift was bringing food to the Allied troops, the Allies (the term now excluding the Soviet Union, the former ally) met regularly to devise protective measures and to set out the European Recovery Program, which together would hold the Soviets at bay, bring the western half of the continent together, and lead to the new federated state of West Germany, with "the minimum requirements of occupation and control."[1] The first, symbolic step toward this fresh consolidation of the West would be the reform of the currency and issuance of a deutsche mark valid in all three western zones.

This agreement was clinched in London on 7 June 1948. Over the next two weeks the Soviets intermittently stopped all traffic between the eastern and western sectors, swept out of the kommandatura—the committee that ran the day-to-day business of the city—announced unilaterally a new East German currency system that would also be used in Berlin, and on 24 June closed all roads and railways from the Soviet zone to the western sectors of the city. The full blockade had begun.

The stock account is at this point probably the correct one. The Soviets worked urgently to block the formation of a federal state of West Germany, and they also worked to keep the Soviet zone under their thumb. In any case, they wanted to force the Allies out of Berlin, which was at once a powerful symbol and a rich trophy. Their method was crude, and apparently insuperable: they kept food and fuel from reaching one and a half million people barely recovered from semistarvation. Their means were simple, and to hand: they just closed all the ground routes and waited for the Allies to cave in and either abandon Berlin or indefinitely postpone all plans for a federal West Germany.

The two generals commanding the British and American zones were, respectively, Brian Robertson and Lucius Clay. Clay was an ex-engineer of formidable gifts, including diplomatic daring and a capacity for both imaginative sallies and the odd show of a stinging military irascibility. He was Eisenhower's deputy, a crisp, sharp, imperious man with a powerful sense of duty that kept him at a post he deeply and understandably disliked.

Robertson complemented Clay. Son of another general but himself a soldier of the two world wars, he spent the years between as a senior business executive in Dunlop's South African offices. He was a man of exemplary calm and quiet good judgment, reserved, astute, studious, and not inclined to encroach on political matters that really were not his affair. John and Ann Tusa touchingly report that throughout the blockade, Ernie Bevin, the British foreign minister, always asked first, "What does the General think?"[2]

Together they made determined and united cause against Soviet bullying and obduracy. The three Western powers resolved that, after Czechoslovakia, there was no knowing where the Soviets would stop. Chief executives Harry Truman and Clement Attlee were clear that Berlin should remain a quartered city if its retention was at all possible, as a sign that they would stand up to the Soviets for the agreements that had been reciprocally signed (as the Berlin settlement was agreed to at Potsdam) and were equitable in their general outline. All parties to Western deliberation—in Berlin, in the zones of Ger-

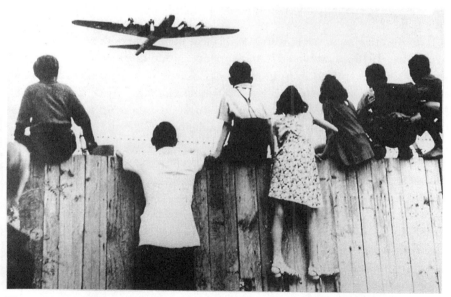

Figure 3.1   Children waving airlifters into Berlin during the blockade, 1948. (AP/Wide World Photos)

many, and in the capitals of London and Washington (and, a little obliquely, Paris)—committed themselves to withstanding the siege. The airlift began.

It *is* a heroic story. From the start it was a triumph of improvisation. In June any permanent relief looked impossible: there were not enough aircraft, personnel, and landing space to fly in the food and fuel needed to provide for one and a half million people. There was perhaps five weeks' supply of food and fuel in the city, through which rumors swept that the British and Americans were leaving.

The aircrews flew around the clock, and the figures began to tell the story. The city needed 2,000 tons of food and fuel a day in summer, 5,000, including coal, in winter. By 15 July the aircrews were bringing in almost 1,500 tons. Bringing in coal would have to wait until winter, and it was the winter upon which, as in so many of their historic campaigns, the Soviets were convinced they could count. The British and American newspapers openly discussed the danger of war over Berlin. However deadly that prospect certainly was only three years after the cessation of total war, the nerve and determination of the Western powers were unfaltering.

During July Truman sent to Berlin, without fuss but in full view of the Soviets, the B-29 bombers that were then the only planes in the world capable of carrying the kind of atom bomb that had wiped Hiroshima and Nagasaki off the map. During assiduous but firm approaches to Molotov and Stalin in early autumn, the stock of freighter aircraft was built up, the flights found a more sensible, less overworked, underslept routine, and the thing began to look feasible.

The Berliners themselves were utterly clear about where they stood. On 9 September 1948 an estimated 250,000 people—one in six of all souls in the western sectors—rallied in front of the old Reichstag to hear the admirably rousing speech of their staunch Social Democratic Party (SPD) leader Franz Neumann as he pledged not to capitulate to the Russians. In a fine gesture, the Red flag was torn down from the Brandenburg Gate; the offenders were whisked to trial by a Soviet military tribunal and with that utterly hateful and unbelievable Soviet callousness that was to become synonymous in the cold war with the grim name of the gulag, they were promptly given *twenty-five years* of hard labor.

Berliner solidarity held. Talks in Moscow collapsed. Small, motley civil airlines were called in from Britain to help, flown by ex-bomber sergeants and squadron leaders who had left the RAF in 1945 with their hard-won flying and drinking skills and had set up on their own with a couple of ancient, recommissioned Wellingtons. The day the talks folded, 7,000 tons of freight reached Berlin from the combined operation.

By mid-October six new airfields had been built, rapidly but safely, with runways largely made out of wartime experience and the heavy-duty steel combat grids with which Allied air forces made a landing strip out of mud or sand or whatever was there.

Winter came, but with life-saving mildness. The new Soviet foreign minister, Andrey Vyshinsky, arrived with it and turned out to be more stubborn and arbitrary than Molotov the Hammer. In December the Soviets sealed off their sector of the city with roadblocks; residents had to apply for visas to visit their relatives two blocks away. The shadow of the wall that would be built in 1961 fell already across the checkpoints.

As the warm winter held, the airlift consolidated. More of the big freight aircraft, the C-54s, came from the United States to help the smaller C-47s and the British motley. An aircraft landed every three minutes; the air above the city was never silent or empty. The steady, thrumming roar of at least four propellers was the background music of Berlin.

Western unity in the face of the Red threat solidified alongside the aircraft. The draft for the North Atlantic Treaty was published on Christmas Eve; the constitution for the new state of West Germany followed. The airlift battle honors—a battle fought to save, not to kill, to deliver flour, meat, vegetables, chocolate, coal, gasoline, blankets, not bombs and bullets—are measurable in tonnage: one million tons loaded, flown, and unloaded, all by hand, by mid-February 1949; 13,000 tons delivered on Easter weekend; 2.4 million tons delivered by September 1949.

By April 1949 the Allies knew they could hang on indefinitely, and the Soviets knew they were beaten. True to form, they conducted protracted, often mischievous negotiations at arm's length through two fairly junior officials. On 11 May electric power was restored from the Soviet sector, and in one brilliant minute all the city's lights came on. At midnight the Helmstedt Gate, the checkpoint from the British zone that led onto the Berlin autobahn, was informally opened by a Provo corporal and the blockade was over.

The next day 200,000 Berliners gathered in thanksgiving outside the city's assembly hall, the Rathaus, and heard their honored *oberbürgermeister* Ernst Reuter salute their courage and call for the restoration of their sovereignty. As an alliance the Western Allies now included the West Berliners; they had won the first, bloodless engagement of the cold war.

III

The Berlin Blockade is a rattling good story, and largely true. The Soviets were indeed the bad hats of the plot, and the West Berliners showed their unequivocal commitment to remaining in the Western camp by bearing the unevenly distributed pains of the blockade with little complaint and some humor. The generals were exemplary and earned all the great public affection showered on them.

The aircrews were marvelous. For them, perhaps, the airlift was largely a great lark. Killing hard work, of course: all the loads were hefted into the belly of the aircraft by hand, and in the first weeks of the airlift they flew twenty-four hours a day, kipped down anywhere, head down on a heap of sacks in the hangars, then staggered out, yawning cavernously, to start off again as dawn edged up over the city and the huge combustion engines sputtered, caught, and crashed one by one into whirling life. The powers of improvisation in the early days spoke eloquently of those civilian-military, informal, laconic, but wholehearted virtues nourished in the best, most democratic quarters by the Second World War itself, and best exemplified in the unhierarchical, technology-loving, slangy way of the American and British air forces.

For such men, and not a few women, the Berlin Blockade soon became the happiest days of their lives. The crisis—and it was a crisis—required that everyone muck in, and the going was often indescribably mucky.

The airlift started in freak weather: torrential and bitter cold hail and thunder storms in July. Above a few hundred feet, the airplanes iced up and had to land far more heavily than was safe, with great slabs of ice drifting back in the slipstream and crashing on the runway as they came in. Everyone was drenched. Sodden flying boots froze to rudder bars. On the ground, the runways at Gatow, Tempelhof, and Wunstorf shone wetly, surrounded by shin-deep, muddy swamps. People trudged through the slush to clattered-together Nissen huts to steam in the heavy smell of drying battledress by the stove.

It was bloody dangerous all through. Given the number of flights and aircraft, the tiredness, the primitive landing controls, and the layers of aircraft stacked upstairs waiting for a turn to come in, it's a wonder that more were not killed. But as it was, the year's work was punctuated by twenty-odd crashes and several dozen deaths, which lent bitterness and gravity to the jumbled pleasures and truculent merriment of men and women devotedly working together to rescue ordinary domestic life.

And it was for life they were plainly working. The citizens they so recently had dropped firebombs on were those they now risked so much to protect. It was a restorative change. It is no cynicism to say that acts of generosity make us feel benignant toward those we help. Berliners became more than allies: they became dependent friends, helplessly grateful for our self-sacrificing labors. And, of course, the enemy was at hand, acting up in thoroughly inimical, even villainous ways, so that it was easy to transpose the wartime feelings from 1945 to 1948 and to substitute the Soviets for the Nazis. That is why the blockade was so much more than an ideological victory, a phrase that betokens a limited piece of propaganda; it was a long, effortful triumph of life-saving, and way-of-life saving.

The Berlin Blockade lasted from mid-June 1948 to mid-May 1949; even then, Russian go-slow tactics compelled the airlift to continue until September 1949.

Over those months something important happened to people's feelings about the Soviets. The great leaders and their officials signed treaties, enacted laws, and were photographed in suits and nondescript, grand rooms accomplishing those matters. Elsewhere, across Europe, over the Channel and all the way up Britain, beyond the Atlantic where the currents from Berlin stirred old hatreds among the Armenians, Latvians, and Poles in Pennsylvania or Wisconsin, the Soviets stopped being our comrades—the brave defenders of Stalingrad and the first troops into Nazi Germany—and became Ivans—either booted, fur-hatted soldiers or expressionless, heavy-coated crowds. They became our enemy.

IV

Which wasn't, as we can see now and some could see then, quite fair. The Russians, their townscape still in ruins, their economy beggared by war, urgently and at all costs wanted reparations from Germany. They feared the restoration of that mighty nation on their own doorstep. Its version of capitalism was what had devastated their land. As the two most powerful foreign policy officials of the United States and Britain, George Marshall and Ernie Bevin, agreed in London in the early part of 1948, occupation of Germany had to end; its salvation could only be won by its own economic good works. They were sure at the same time that the Soviets would have none of it.

This was hardly surprising. The Soviets, as George Kennan said in memoranda at the time and in his memoirs later, "would hold out . . . for arrangements that would place them in a favorable position either to block German recovery or, if it could not be blocked, to assure that it operated to their benefit and not to the benefit of European recovery generally."[3] The Soviets not only felt beleaguered, they still *were* beleaguered. They would certainly refuse to ratify a noncommunist German government of the kind unconvincingly submitted by Marshall and Bevin while they still occupied the Soviet zone of Germany. When the Allies proposed through the London Program, promulgated without the Soviet Union's participation, a new deutsche mark for the western zones together with a schedule for a West German constitution, the Soviets had their customary dark suspicions that everyone was against them fully confirmed. They attributed to the rest of the world a conscious, rationally ordered, and fully intentional policy of unmitigated hostility whose final purpose was to cause the collapse of Soviet socialism, in spite of the historical inevitability of capitalism's own demise.

It is fair enough to say that the Western Allies, though their policies were and are far less horribly poisoned by collective paranoia, held a reciprocal view of Soviet foreign policy. George Kennan saw that view developing in the moves toward the making of the West German state in 1948:

> As the process went forward, it gained steadily in momentum and in the aura of legitimacy. People's *amour propre* as well as their enthusiasm became engaged. There was growing personal commitment to what was being accomplished. Increasingly, as the months and weeks went by, the undertaking assumed in many minds an irrevocable character; and the idea of suspending or jeopardising it for the sake of wider international agreement became for these people less and less acceptable. Once again, as is so often the case in American diplomacy, what was conceived as an instrument became, little by little, an end in itself. What was supposed to have been the servant of policy became its determinant instead.[4]

What the Soviets *were* right about was the purpose of U.S. and British foreign policy: to make American-capitalist settlement of Europe the character of the epoch. Think how the West must have looked to the Soviets peering across the Elbe River or through the Helmstedt Gate. The richest portions of Germany were in Western hands, especially in the Ruhr; France was comparatively undamaged by war and would soon reassert its economic strength; Britain had been uninvaded; the Atlantic became narrower as flying time decreased and every tide brought new waves of dollars. Berlin, the anomaly, was the Western powers' only weak spot: the only purchase the Soviets had on the new state—which represented the certainty of European military alliance being turned against them, the likelihood of weaponry being back in German hands within a few years—was to put the squeeze on Berlin. That way they had just a chance of holding off the founding of the new state and its shining new money.

The only exit from the city (and entrance) for the Western Allies was the air corridor. General Clay, with his tendency to gunslinging imperiousness, was prepared to shoot his way up the autobahn, but Marshall wouldn't let him. It seems likely, however, that if Soviet Yak fighters had shot down any of the air freighters, the Western response would have been quick and bloody. The time from such an engagement to all-out warfare with the Soviet Union, utterly unable as it would have been at that time to withstand it, would have been short. But as Wilfried Loth puts it, "The myth of Soviet strength . . . led most Western politicians to overlook how risky and how close to the limits of the feasible this Soviet manoeuvre really was."[5] It was a myth of extraordinary pulling power and grim, unkillable vitality, as we shall see.

Nobody was going to go to war over Berlin. To be sure, Winston Churchill, out of office and power and therefore, as such people are apt to be, slightly hysterical for attention, had suggested in a much-reported speech that the Soviets leave Berlin and East Germany or else the atomic bomb would be used to raze their cities, but the State Department took no notice of such irresponsibility. In steadier council rooms, it looked as though the Soviets would win their demands: either the Allies would have to quit Berlin, which would be a deep humiliation, or they would have to reconvene the Council of Foreign Ministers and all plans for the establishment of the new German state and new currency would have to be either postponed or dropped. Kennan's policy planning group was asked to advise the president. He came up with Plan A: the withdrawal of all occupying forces to garrisons on the edge of Germany, national elections for a provisional government, and the demilitarization of Germany. Berlin would be restored to the Berliners.

Kennan was horrified at the steady confirmation that Germany would be split. Having two Germanys became so taken for granted in London and Washington that we can too easily forget how desperate the circumstances were there for the past fifty years. Ordinary life in the two Germanys included masses of troops facing each other across an arbitrary border made of barbed wire; the far deadlier nuclear weapons on trucks and in silos and at 30,000 feet threatening unprecedented destruction on people of the same language, nation, and, until 1948, history; and every year, the brutal reminders of these powers as the soldiery of NATO and the Warsaw Pact displayed their tanks and rockets on the harmless farmlands that had become their military playgrounds.

Kennan's deathly vision, as we all know, was amply fulfilled. George Marshall and Brian Robertson thought Berlin could not be held; Bevin and Clay hung on, Clay in particular remaining strongly committed to founding the new West Germany to ensure that the Allies could not then revoke the anticommunist settlement. Public opinion seems to have been that the blockade was an act of simple piracy by the Soviets, who wanted full control of the German capital, to say nothing of as much of the rest of Germanys as they could add to it. Ordinary conversation in the West took little account of Soviet fears of German revival. The war was over: the Soviets had already assumed the enemy's familiar place in political sentiment. The yearlong blockade crisis settled that.

V

At its end, the monoliths of Western anticommunism and cold war defense were entrenched like Stonehenge in the soil of Europe. The North Atlantic Treaty, the Western Defense Union, the Marshall Plan, and the Federal Republic of Germany stood as thick and stout as the iron curtain facing them. Berlin remained an utter anomaly. The defeat stuck in the gorge of the Soviets, particularly in the throat of Nikita Khrushchev. Thirteen years later he once again steered Soviet weapons as close to the edge of crisis as he dared, and in so doing, dared John Kennedy to stop him. The humiliation of the blockade smarted hotly, and the unattractive note of lofty triumph that Dean Acheson, Marshall's successor, sounded stung where it was meant to.

> So although I was a mere observer in 1948, the choice of the airlift seemed to me the right one. It showed firm intention to insist upon a right, plain beyond question, and gave the Russians the choice of either not interfering or of initiating an air attack, which might have brought upon them a devastating response. To say, as secretary Murphy has done, that the decision to use the "airlift was a surrender of our hard-won rights in Berlin" seems to me silly. One can as well say that to put one's hands up at the command of an armed bandit is to surrender one's hard-won right to keep them down. One regains it, as we have regained and are now enjoying our hard-won rights to Berlin.[6]

Berlin became the emblematic city of the cold war. The Wall, the checkpoint furniture, the wire and searchlights, the white helmets and jeeps, the bridge, and the dark river below loom out of two dozen thriller movies. The city contained all the contradictions of our history, waiting for an end.

All the same, the Berlin Blockade remains a terrific tale. Its heroes are Clay and Robertson, certainly, but also the honest, earnest, brave trade unionist, Franz Neumann, who led the SPD and set his face like flint against the lies and gerrymandering of the Communists. Ernst Reuter, who finally became *oberbürgermeister* when the western sector held unilateral elections, did more than anyone—before the accession to the chancellorship of West Germany of his great deputy Willy Brandt—to restore Germans to popular affection and respect in the hearts of their former enemies.

There were more transient, utterly engaging heroes along the way, like the American lieutenant Halvorsen who arranged for parachute drops of chocolates for the children of Berlin at Christmas. There were satisfyingly nasty villains, like Vyshinsky and the good-looking Marshal Sokolovsky, whose metal-featured military police operated such a consistent policy of bullying uncooperativeness, paper-scrutinizing, threats, indifference, cruelty, and laughably trivial intrusion, creeping by night into other sectors and having to be teased out like kids. There were also gorgeously comic satisfactions, like the sangfroid with which the French commandant, General Graneval, blew up the two Radio Berlin transmitters that pumped out nonstop Soviet propaganda and stood in the French sector. They were in the way of incoming aircraft. His removal of them cheered up everyone enormously and was denounced in a later, trembly broadcast from much further east as "an act of brutality against a cultural wireless station."

It was all a jolly good show, and it held the line. But defeat and victory in Berlin left legacies, each inscribed in an ideological codicil. The West quite consciously turned the city into a glittering shop window for the display of its most voluptuous goods. The ostentations of very expensive sex, leisure, and possession were nowhere more lavish or hypnotic.

For their part, the Soviets turned their little realm into a memorial to 1945. Time stopped in Karl Marx Platz. They used the apparatus of repression, of spies and *Vopos* (from *Volkspolizei*, "people's police," a bitter misnomer) and mean streets to insist, absolutely insist, on their *will*: on keeping things their own grim way, on *stopping* everything—traffic, progress, human excitability, time itself.

They lost in the end; but it was a long tunnel to the light, and many people lived and died in the dark.

The Berlin problem wasn't solved, just silenced for a season. The danger it represented after the blockade, focused by the commonplace apparatus of the checkpoints, stood as the single most dominant emblem, at the heart of a divided Europe, of the cold war and the permanent occupation of a continent by the two superpowers. The ideological fracture of the century ran through Berlin. When the cold fever of the great fear of nuclear war rose in 1953, 1956, 1961, and 1980, it was at Berlin that each of us looked most apprehensively. In 1961, as the deadly ritual was played out in the Caribbean over Russia's Cuban missiles, the dark shadow of the new Berlin Wall fell all the way across Europe and the Atlantic. All it would take was an injudiciously fired gun at the flashpoint and, as movies and spy thrillers and nightmares reminded us, the world might be in ashes.

4

EVENTS III

China and the Korean War, 1949–1953

Europe is, no doubt, Eurocentric. Exactly because its new frontiers were so touchy, so they were walked circumspectly by their custodians in the two great new empires. Behind the *grands seigneurs* of the superpowers trotted their lesser European clients—Bevin behind Acheson, Ulbricht behind Vyshinsky. However, at the rim of world geography things were decidedly less settled, and, indeed, the natives less clear that they were obliged to defer to either power. For the United States and the Soviet Union, the new epoch was defined by its new world order; yet here were portions of the globe planning their own future without recourse to either Moscow or Washington.

On the edges of the globe, however, the typical forms of the new order were not easily adopted. Frontier disputes in Korea, infiltration and guerrilla subversion in Vietnam, trade embargoes in China, economic aid to Formosa, border skirmishes, and sabotage—these tropes and actions, gradually but inevitably, found a theater in every corner of the world. China opened the second front of the cold war. The great fear of communism—which lasted, from 1949 until (at the earliest) 1960—was fueled by the furious heat released by, as it was proprietorially put, "the loss of China."[1]

China had been bloodily contesting the advent of modernity since the insurrection that blazed up in Shanghai in 1925, burned brightly here and there—in Shanghai, Canton, Hunan—and was murderously repressed after the Communists seized power in Shanghai. As they waited for their ally, the Nationalist Army, to join forces with them there, Chiang Kai-shek wiped them out. Financed by an urban bourgeoisie terrified of expropriation, he turned on the poorly armed insurgents and slaughtered them.

It was the start of the White Terror, during which hundreds of thousands were killed and the Communist party was left broken and scattered. Where once Chiang's Nationalist party (the Kuomintang) of would-be democrats had joined forces with the proletarian left to finish off the warlords and the landowners, to throw off all foreign exploitation and unify the Chinese people as a single nation, the Kuomintang now formed an alliance of the grandee warlords, the bankers and industrialists, the officer corps, and the many cohorts of counterrevolution, and ripped into the opposition. The trade unions, peasants, and working class were uncoordinated and had few weapons; they were cut to pieces. Chiang, incarnating old reaction, assumed the presidency in Nanking, and Mao Ze-dong[2] led his remnant to the mountains of Kiangsi Province where, in 1931, the Communist party proclaimed the Chinese Soviet Republic, with a tidy population of three million.

Chiang came to get him. Mao and his most important recruit, the Red Army commander Zhu Teh, began to devise the rural guerrilla tactics that would ultimately bring him victory and, forty years later, would defeat the Americans in Vietnam.

But Chiang's forces hugely outnumbered Mao's. Chiang detested the Communists; utterly outgunning them, he was indifferent to casualties among his own men so long as the enemy was killed in even larger numbers. In 1935 Mao and the Red Army were driven out of Kiangsi, lost local support in their new fiefdom, and began the Long March to Yenan, which has since become the Chinese Communists' great and self-justificatory epic.

The party had been at violent odds with itself for years, the Bolshevik stooges waiting for the word from Moscow, Mao's men devising heresies, both sides bumping off each other. Chiang moved in to finish them off, but Mao dodged, fought, and feinted his way to refuge in Yenan, leaving nine-tenths of his army dead or captured on the way. Only the youngest and fittest survived, and under their indomitable, puritanical, ascetic, and ruthless leader, they created the military force and political symbolism that brought Mao heroic status and Chair of the People's Republic in 1949. Mao's small army had learned the fearful steeliness of the guerrilla, his discipline, his entire lack of mercy, his abstract idealism, his accurate justice.

These severe weapons became the cutting edge of patriotic resistance when the Japanese, after three decades of probes and imperial penetration, finally launched a full-scale invasion of China in 1937. After capturing Chiang

Kai-shek in his pajamas and without his teeth, Mao made common cause with his rival, shelved the Red Army's mammoth claim of vengefulness and hatred against the Kuomintang, and fought heartily to beat back the Fascist Japanese and build a mass army of peasant Maoists at the same time.

Maoism was formulated during the retreat to Yenan and the Japanese campaign. The leader was indeed an extraordinary man: salty, talkative, a prodigious and prodigal writer—of poems, essays, meditations, aphorisms, axioms, *pensées,* theories—yet a vehement anti-intellectual, a womanizing puritan, a flexible dogmatist, a killer and a hero.[3] Against all the canons of Marxist-Leninism, he carried off a revolution in which the peasants led the proletariat, and the rural population overwhelmed the cities.

Tempered by years of war against the Japanese, inspired by the amazing bravery and dedication of the survivors of the Long March, thrilled by patriotism, and rightly convinced of the justice of its cause, the Red Army handsomely defeated Chiang Kai-shek's Nationalists when civil war inevitably followed the anti-Fascist war.[4]

The Americans could see it all coming; they were horrified at the prospect of Mao's ascendancy, but in the three years of civil war before Chiang Kai-shek relinquished the presidency, they could do little to prevent it. The Europeans were less moved; to them the Far East looked simply like a repetition of Central Europe, but with much less interference from Stalin and a good deal more local justice. The Chinese peasants were shockingly poor: millions had starved in the famines in 1927 and 1936. Most Europeans believed that the Chinese landlords deserved the treatment they were getting from the Red Army, that Confucianism would mitigate Marxism, and wished good luck to the People's Republic.

But by 1949 the front lines of ideological and literal entrenchment in the cold war were deeply dug and wired. The Truman Doctrine had been uttered, the bomb was the custodian of what the United States had baptized the "free world," and Stalin had proved himself the most satisfactory monster any virtuous nation could want for an enemy. The Marshall Plan was working, and the omnipotent dollar was lubricating the machinery for a new realm of free trade all across Europe. Freedom itself had been happily reconceptualized as freedom to trade.

The loss of China to the dark forces of communism was therefore a shocking affront not so much to the political consciousness of the president, his cabinet, and his international allies as to the political imagination of the American people. China's going Communist stimulated new frenzies of commie-hating and wild visions of hooded hordes running ashore in Honolulu. When it was needed, that nightmare was quite consciously mobilized to push through defense budget increases.

I have no doubt that between 1946 and 1949, the years of Chinese civil war—in Central Europe the coldest season of all the forty-five years—some of the best men in the world were at the centers of decision in Washington and London. The public servants in George Kennan's policy unit and Bevin's Foreign Office were as serious, upright, resolute, and civilized as their particular traditions of moral thought would be likely to make them. They were determinedly internationalist, both in their own nations' best interests and in the best interests of the world as seen from their high windows. But nationalism, a much hotter potion, distilled itself into rotgut anticommunism in the United States. There was no holding back a nation not only addicted to it but with the money to sell it to poorer nations patrolled by its millions of open-handed troops.

In short, the loss of China became the node of hate and fervor from which anticommunism pulsed into popular U.S. politics. Sen. Joseph McCarthy, in his stentorian, calloused accents, presented the magi of international liberalism—Acheson, Marshall, Kennan—as the tools of international communism and its plot to take over the world. China was *the* case in point. Never mind that there were no American troops there, and precious little American capital. Never mind that there were no threats to the Philippine colonies, and that Japan was completely under the thumb of Douglas MacArthur. The barbarian was abroad. He would devour us all if he wasn't stopped.

Thus did American nationalism arm itself with a brave ideology. The armed services won for themselves a new Seventh Fleet to patrol the China seas and to keep Chiang safe in his asylum in Taipei on Taiwan Island, where the duplicitous old has-been was held up as the emblem of Chinese resistance to the Communist barbarians. In the meantime, the wide coalition of anti-communism in the United States worked itself into a frenzy of preparedness to do battle if the time came. Its hard-eyed prophet John Foster Dulles went about bearing witness against the "evil doctrine of containment," urging upon his country the kind of ruinous overspending on weaponry that may yet bankrupt it and in fact has already bankrupted the Soviet Union.[5]

The detonation of the first Soviet atomic bomb in the fall of 1949 sent a thrill of corroboration through every anticommunist American heart. The millionaire anticommunist spokesmen, safe from any danger of having to do the fighting themselves, called for a "rollback" of the Asian encroachments. U.S. "advisers" flew in and out of the Honolulu and Guam military stations in their hundreds, initiating a hidden war upon the Orient that lasted, alongside the official conflagrations in Korea and Vietnam, until the day before yesterday.

As 1949 wore on, Mao Ze-dong, until then a far from obedient comrade, decided that he needed a foreign policy; he went to Moscow to conclude a

painful and ambiguously worded alliance with Stalin on 16 December. From
then on, the European internationalists in Washington—Acheson and company—were crippled. Communism had by the simplicity of treaty stretched
the iron curtain from Stettin to Shanghai, with only a little gap in Asia Minor.
Across some of the decidedly shaky frontiers in India, Malaysia, and Indochina, Communists were nevertheless hard at work agitating, educating,
propagandizing. For a State Department whose first concept of foreign policy,
inherited from the nineteenth-century European diplomats, was protecting
the frontier, the domino theory was a ready-made description of the anticommunist nightmare.

So whoever started the war in Korea—and I. F. Stone makes it quite clear
that its beginnings are not clear—South Korea became the first domino and
had to be propped up at all costs.[6]

The costs were indeed enormous. They are complaisantly estimated in the
Truman administration's National Security Council memorandum 68 (NSC
68) and included fueling the superpower arms race as well as drafting an
American army disguised in U.N. berets to fight a war in a country no one
in Europe and few in the United States could mark accurately on a map.

Figure 4.1   John Foster Dulles on the 38th Parallel in Korea, 1951. (KMAG PHOTO)

Korea erupted like a naked baby onto the stage of global ideological war. It was the pastless, unpopulated terrain upon which the ideological giants of modern progress—Christian on our side, Apollyon on theirs—would settle the first proper fight for freedom in the new epoch.

II

For our purposes, it all began in 1910, when the Japanese coolly annexed Korea into its enormous northeast Asian empire, which was also to take in the wide arc of all of Manchuria. The ambitions of imperial Japan long predated Germany's and rivaled Britain's enough so that the two countries made a little treaty in 1919, allocating the welfare of other countries to each other: Hong Kong for Britain, Korea for the Emperor. Japan had ended a millennium of Korean self-government and gave the Koreans in return the sublime fruits of modernization.

It was not until 1930, however, that Japan launched an intense program of industrial revolution that ended only with its military defeat in 1945. Over those fifteen years, it installed at every level of Korean society a disproportionately enormous bureaucracy: 250,000 clerical officers for a population of less than 9 million people by the outbreak of war. In the 1920s and 1930s, Korean development was in the hands of a class of landlord-entrepreneurs encouraged by the Japanese, an alliance which inevitably produced the bourgeois class that collaborated with the Fascists during the war and was the elderly recruiting ground for the forces of the right after 1945.

The traditional Korean peasantry was transformed by the brutality of sudden industrialization and forced labor. One and a half million Korean workers, including women and children as young as nine, were transported to Japanese labor camps for the war effort; 250,000 were abruptly conscripted into the coal mines, a radicalizing form of occupation. Communism, already widespread in part because of its appeal to desperately poor Asian peasants, had a good soil in which to grow in Korea.

Suffering under the blows of wholesale anticommunist repression by the Fascists throughout the Pacific islands, peasants and the forced labor class admired the discipline, courage, silence, and sternness of party members—the grim virtues, hardened in grim resistance, that were the real foundations of the grim postwar Communist state. Korea's Kim Il-Sung was yet another of those enduring Communist leaders who took over in 1945 and did not let go of power until death pried open their fingers. Like Hoxha, Tito, Ho Chi Minh, Mao Ze-dong, and Ceausescu, he was still in power at well over seventy.

He had commanded a group of partisans in guerrilla warfare against the

Japanese since 1931. There was a price on his head, but he was hardy and lucky; he was never captured. He had the wits to keep faith with noncommunist nationalists, and in 1945 he was in the right place at the right time. He had come back from Manchuria with enormous numbers of returning villagers. They returned on foot from fighting with the Chinese army against the Japanese, from forced labor camps in Japan, and from refuges all over Korea. They were a strong political formation: many of them had been powerfully radicalized by their wartime experiences under the Japanese and had learned a vigorous political language. But they were still tied to the traditional networks of village and district. These characteristics nurtured an active democratic and nationalist culture.

This newly politicized working class quickly ran up against its future opposition. When hostilities ended in 1945, the pressure for land reform was irresistible and instantaneous, and two groups were immediately formed to promulgate it: the Korean People's Republic and the modestly titled Committee for the Preparation of Korean Independence. They were nationalist-radical in their colors, apprehensive, like all Koreans, of Russian domination, and eager to reestablish a free Korea after thirty-five years of colonial rule, and a thousand years as their own masters.

They were immediately opposed by the Korean Democratic party, an unappetizing group of well-off worthies who had survived the war comfortably, were based in Seoul, and aimed to become the ruling party from the capital simply by organizing a national magistrature to keep the peace. Their leader, from the moment he returned from exile in the United States in October 1945 and announced his importance to the new Korea at a meeting put on for him by the Americans, was the notorious and legendary Syngman Rhee. These groups provided the social bases and enmities of the two countries that are now North and South Korea.

There had been reason for more optimism only a month or so before Syngman Rhee and Kim Il-Sung came home. In early August, just before Little Boy went off, the Soviets, in agreement with the United States, entered the Pacific, declared war on Japan, and promptly moved down into Korea from Manchuria. The Americans were preparing for landings on Japan and had just finished the terrible slaughter on both sides that was the cost of retaking the Philippines. Taking Japan was certain to be worse. Korea was of no strategic importance; the Soviets were welcome to it.

In the event, the Soviets met little Japanese resistance and advanced south easily and steadily until the atomic bomb went off, when the world stood still before changing course forever. The Japanese surrendered suddenly, and just as suddenly the Americans rediscovered their active interest in Korea. They radioed the Soviets, told them to hang on until they got there themselves, and bundled two young staff officers into a room with thirty minutes to draw up a two-power division of Korea until the postwar occupations in the East were settled. Half an hour later, Maj. Dean Rusk, later secretary of state, came out with his colleague to say that the division should be sort of in the middle of the country along the 38th parallel, thus giving the Americans the advantage of Seoul. Rusk and his colleague handed over their paper, and U.S. troops, perfect strangers to the doubtful Koreans, were sent off to occupy half of the country.

The Soviets silently and, as we would now think, surprisingly acquiesced. Although there were few Japanese in their way, they dutifully stopped at the 38th parallel, which was only a line on the map, with no significance in the bleak little hills of mid-Korea. The Americans sent Gen. John Hodge, a redoubtable, plain-spoken, craggy, crew-cut, touchingly honest, and heartily uncomprehending sort of chap, to be U.S. commandant and, in effect, governor-general. He held the post for three years; from 1945 to 1948 the Americans operated a full-blown military government in the South, a little detail of the Pax Americana not often remembered.

Before the Americans got there, the defeated Japanese, of course, had sacked and looted and printed a riot of madly inflationary yen. The Koreans themselves, as little consulted as Poles or Czechs or Greeks about their future, set up a national government in 1945 that was divided in its formal coalition between left and right. The new political leaders on the left were all under fifty; the representatives of the older guard on the right averaged sixty-six. Inevitably, each half of the coalition was ideologically defined strictly according to its conduct under the Japanese. The leading figure of each group accurately fitted the ideological mold: Kim Il-Sung the tough partisan and party stalwart of thirty-three; and Syngman Rhee, a Princeton Ph.D., exile, and Washington lobbyist for thirty-odd years who was dictatorial, stubborn, fiercely anticommunist, immune to snubs, and ruthless in his retention of power.

Ruthless is a word we tend to reserve for the enemy. In Korea it applied even-handedly. But ruthless is as ruthless does. Franklin Roosevelt had his ruthlessness too, debonair and cocksure as he was. In his vision of the international order, dominated by the liberal conscience and the open-door free market, Korea was to be a four-power trusteeship. Anticipating the division between American and Soviet world interests somewhere along a Korean frontier, Roosevelt had blithely judged that the Korean people, self-governing for a millennium, were not yet ready for self-government. Thereupon he had prepared the plan that lapsed with his sudden death and Churchill's departure from power in Britain.

In the plan that did take effect, the two-power division of Korea, all

American decisions were colored by their darkly suspicious view of any party on the left. In the short space of time before the Americans arrived, during which the defeated Japanese did so much damage, the Koreans had energetically set up a network of "people's committees," loosely federated (but not organized) from Seoul as the system of national government. In the South the committees did not last long. General Hodge landed at Inchon amid a parade of Japanese police; the Americans imposed a curfew, rearmed and put the Japanese back in police cars as well as behind the desks of the bureaucracy, and began to disband the offices of the Korean People's Republic without their connivance.

In the North, on the other hand, the land reform so long awaited went ahead vigorously but equitably, driving dispossessed landlords south. In the South, the nonideological Americans tried to transform the rice economy overnight into a free market, and the subsistence-and-barter rice farmers into go-getting, surplus-producing entrepreneurs. The result was roaring inflation and a rice shortage.

In the as yet unwritten history of international resentment, American policy in Korea during 1945–46 would have a central place. The Americans preferred to work with the hated Japanese; the only leading Korean they would deal with was the morally very dubious figure of Rhee. When the Americans set up bodies intended to pave the way for a national parliament—an advisory council and a Republican-Democratic group—the ratios of membership were so hopelessly in favor of the right and pro-Japanese collaborators (ten to one in each case) that the groups were boycotted by the left. After all, as an army secretary wrote to the State Department in September 1946, "the Koreans are not psychologically or technically prepared to undertake self-government." General Hodge preceded this note with a less temperate one of his own: "We are dealing with poorly educated Orientals . . . who stubbornly and fanatically hold to what they like and dislike, who are definitely influenced by direct propaganda, and with whom it is almost impossible to reason. We are opposed by a strongly organized, ruthless political machinery designed to appeal to millions of this type."[7]

Western powers had their own stubborn fanaticisms. For instance, the United States demanded a joint declaration from the Soviets on the desirability of free speech in Korea, while at the same time it suppressed the people's committees, closed newspapers, forbade trade unions, and instigated widespread arrests of members of the left during 1946. Korea set a model in which American power, embodied in its new military mandarins, entrenched the right in local power but set the left impossible, even ludicrous tests of reliability, responsibility, and loyalty, according to strictly American criteria. Bruce Cumings comments dryly that, in 1945 and after, "it seems that many Americans had difficulty conceiving of *any* Russian interests worthy of American respect."[8]

The same seemed to go for local Korean interests. The Americans made no distinction between the Communists and the traditional elders who were in charge of different people's committees. John Hodge was a pure Washingtonian centralist like Clay, LeMay, Westmoreland; he was, Averell Harriman declared, when he went to Korea to check, "our best type of commandant, and man."[9]

And so he was, for the foreign and military policy machine being bolted into place from 1945 onward. It does no good, all this time afterward, to beat our breasts and blame the Americans for the massive rigidities of what they believed was the liberal conscience in action. Nothing in history is inevitable, doubtless; but it is not hard from here to see why the exasperated spirits on either side of the gaping ideological abyss of the era proceeded from wrong to wrong.

A case in point is Cheju Island, just off the South Korean mainland. When the forced migrants came home in 1945 to their island, they naturally established and lived by a people's committee. After all, their dialect was so particular it was almost another language, and they had an indigenous, self-sufficient industry as well as a long history of independence. For three years the islanders maintained that solid independence—from both mainland Korea and Moscow. Then the North-West Youth, a weird mainland group of virulent anticommunists, joined the local police and, in uniform as well as under American command, began an islandwide campaign to put down the Communists, who were concentrated in a loose-knit guerrilla army of about 4,000 men, as completely as possible.

By April 1949 the island was in ruins, one-third of the population had been shifted to policeable camps by forced draft, and an unknown total of what the U.S. reports formally called rebels were dead, although guesses at the number fluctuated between 20,000 and 30,000.

Such an action was not just an augury of Vietnam. It reflected in every detail standard cold war practice all over the East. In Korea, Quemoy, Indonesia, Malaya, Borneo, and Vietnam, the guardians of the liberal conscience—the great moral citadel built out of nineteenth-century history—supervised mass murders, cruel extirpation, and evictions from home in the name of freedom and resistance to the Communist threat. In return, small groups of hard, pitiless, and disciplined men and women strove never to rest or relinquish arms until they were either dead or victorious.

III

During the first two years of American occupation of a country that was never a belligerent let alone an enemy, the U.S. forces never failed to put down not only the left but also any other kind of popular movement. In the autumn risings of 1946 and the general strike that followed, the Americans endorsed the brutal police action, which was carried out by men who spent the war as Japanese police and were so utterly hated for it that some hospitals refused to treat police wounded. Hatred fuels hatred. The police arbitrarily battered to death or shot 1,200 people.

The north-south division of Korea deepened inexorably into the hard east-west division. After the first and dreadful depredations of the Red Army in the North, the indigenous northerners carried out their calm, bleak reforms: good schools, wholesale land reform, the abolition of harlots and Christianity, and the full Marxist-Leninist treatment from Kim Il-Sung, whose government taught that in the South lay the class enemy. A resentful bourgeoisie trickled steadily down into Seoul, passing along the way the aspirant working class that was bringing its new political consciousness northward through the hills.

By the end of 1946 the glowing opportunity that had lit the first moments of the peace had gone. The United States had forced its interests and its ideology on Korea with the aid of the Japanese-trained bureaucracy, the Japanese-trained police, its own military, and dollars. It had set in place a strong police state apparatus and a dollar-driven economy and had eliminated guerrilla hideouts (the hills were stripped of trees in the larger provinces). By the beginning of 1947 the forces of reaction were fully installed in the South. Kim Il-Sung jeered that the South "feared the North wind and when it might blow." A country that was at once deeply divided and still of a piece nationally, geographically, and linguistically began to germinate civil war spontaneously.

Given that the Americans had made of Korea what they had, their principles required them to hand it over to its rulers. During half-baked elections that were widely boycotted by all the nationalists—who understood that such elections confirmed Korea's status as two states—the hated police shepherded voters to a poll that inaugurated the new Republic of Korea in August 1948. Syngman Rhee was elected president, and a financial-industrial notable called Kim Song-Su became his unofficial deputy. The Americans hung about for another year or so, agitated by the victory of communism in China, and left their usual underground of advisers in military, political, and economic matters. They also left behind officials of the new, not-quite-soldierly trade of defeating guerrillas without going to war known as counterinsurgency. The North was sending agitators to work in secret in the South Korean Labor party. The southern left was still full of energy and not at all quiescent. Rhee himself, though over seventy, was recklessly antileft and determined to remain in power. The whole country was busily boiling over, what with guerrillas on the regular rampage for supplies and the massive police redoubts and pill-boxes under siege.

The story of the war usually begins with the invasion by the North of the South on 23 June 1950. This version of the story overlooks the virtual liquidation by the Rhee regime and its U.S. advisers of the southern guerrillas during 1949, an act that apparently roused the Communists in the North to stand up for comradeship by fighting running battles along the 38th parallel. The U.S. advisers known as the Korean Military Advisory Group had difficulty holding down their 100,000 trainees, full of beans over the fine new weapons the Americans had given them and not so discreetly being urged to look for trouble by President Rhee, who consistently tried to pull the Americans into a full-blown alliance with the South against the North. In the end, of course, he succeeded.

He needed to. In May 1950 Rhee lost heavily in a rather more representative election for the National Assembly. In the months preceding the election, the Truman administration had been drafting what is perhaps the second most significant document of the cold war, canonized as NSC 68 and largely written under the gimlet eye of Paul Nitze, a banker who became Washington's Gromyko—long-lived, small, tough as old boots, opportunist, and an inveterate cold warrior.

Behind NSC 68 loomed the Soviet atomic bomb, first tested in the autumn of 1949, years ahead of American prediction. Truman needed a justification to authorize the beginning of work on the hydrogen bomb, as sketched out by Edward Teller—to set in motion the arms race itself. There were fierce rows about it; Louis Johnson, the defense secretary, left one senior official in tears after a quarrel. Only Kennan was by now convinced that much U.S. foreign policy was even more stereotyped and less flexible than that of the Soviets; he simply did not believe in the NSC 68 premise that the Soviet Union's grand design was world domination, Stalin's ambitions notwithstanding. But Truman and the new militants were dogmatic, and the political economists backed them up. The document stated:

> The Kremlin is inescapably militant. It is inescapably militant because it possesses and is possessed by a world-wide revolutionary movement, because it is the inheritor of Russian imperialism, and because it is a totalitarian dictatorship. . . . It is quite clear from Soviet theory and practice that

the Kremlin seeks to bring the free world under its dominion by the methods of the cold war.

The Soviet Union's "fundamental design" necessitated the destruction of the United States, and thus, the Soviet Union "mortally challenged" the United States. The National Security Council approved NSC 68 at once, although, as one of Truman's senior advisers recalled, "what I read scared me so much that the next day I didn't go to the office at all."[10]

The document proposed that enormous increases in defense expenditure, somewhere between three and four times the budget already allocated for 1950, were instrumental to "a more rapid build-up of political, economic and military strength and thereby the creation of confidence in the free world." The Republicans combined a noisy commitment to reductions in state expenditures with accusations that Truman was soft on communism and not fighting the cold war belligerently enough. There was no saying what an ambivalent Congress was going to do with the proposal. John Foster Dulles, waiting just offstage to succeed Acheson as secretary of state and become the architect of the grand facade and ornate ceilings of the cold war, departed for a visit to Japan and stopped off to peer at North Korean soldiers from a safe distance on the parallel, clad in his homburg hat and a pinstripe suit. Rhee sounded off to Dulles about all that he could and would do to the North, and Dulles went home.

IV

The war began in the early hours of 24 June. From that moment, NSC 68 became the foundation of American foreign policy, and there were few limits to what budget the president could ask from Congress for the next thirty-odd years. The tale of the Korean War fits the myth of global Communist expansionism and American determination to contain it on behalf of democratic freedom, but only to go to war when provoked. For a confident journalist and genial ideologue like Max Hastings, it's a strong but easy narrative. The North Koreans surprised the South by invading it. Acheson immediately ordered American military support and took the matter straight to an emergency session of the United Nations. Truman backed him fully. The United Nations, left alone for whatever reasons by the Soviets, voted to support U.S. "police action" in Korea. It was taken for granted that Stalin was testing the nerve of the Western allies through his stooge Kim Il-Sung.[11]

The North Korean People's Army walked easily into Seoul. The only decent opposition was put up by the Marines; but U.S. forces were outnumbered, and they fell steadily back before the advance until they were just hanging onto the southeast tip of the peninsula around the little port of Pusan. What Hastings so distinctively calls "the bleak, mustard-drab battalions" of Kim Il-Sung's army had all but won a lightning victory by 1 August 1950.[12] American battalions of the 24th and 25th divisions were overrun, and some were disgraced and routed. The North Koreans encouraged others by mercilessly killing several batches of American prisoners with a bullet through the neck.

For six weeks the Americans, helped by a handful of British—who had been rushed out, accompanied by lousy equipment—hung on, besieged, around Pusan. Supplies were short, discipline desperate, battle experience nil. When the wounded were removed and flown to Japan, the hospitals there could barely cope. There have been many terrible tales of chronic incompetence, terror, lack of training, lack of bullets, and lack of any knowledge by the soldiers about what was going on and what made this battle worth fighting.

The North Koreans, however, overstretched. Their lines of communication were too long, and they could not bring up reserves in time to break the last stand being made by the Americans.

There was a lull. Gen. Douglas MacArthur, supreme commander in the Pacific and uncrowned successor to Emperor Hirohito in the new Japan, gathered his enormous pride, his great tactical gifts, his vividly American strengths—guts, dedication, energy, dash, and kitsch—and planned a waterborne invasion well up the west coast, at Inchon a few miles from Seoul. It would be a surprise landing, all right, for he could hardly have chosen a topographically less promising beachhead. Surrounded by a high, thick seawall, it had a tidal lift of over thirty feet and lay below steep hills.

Against all advice and all odds, MacArthur's astonishing sally won the day. With the full weight of the American military force available in the Pacific, he restored the lost American honor. After huge naval bombardment, the First Marine Division went ashore. Kim Il-Sung's defense force was utterly taken by surprise. There was little fighting: North Koreans either surrendered or took to the hills to prepare a different kind of war. Twenty thousand of them remained to defend Seoul itself, but the Marines, in their own reckless way, recaptured the capital, in utter ruins, by 25 September.

The recapture of the South had left over 150,000 South Koreans and nearly 10,000 Americans killed or missing; the North never announced its dead. The next move was obvious to Syngman Rhee, overpoweringly obvious to MacArthur, and somewhat less so to the more cautious Truman administration: the North Koreans, who had withdrawn across the 38th parallel, had to be pursued and their regime and their country destroyed in the way hateful enemies ought to be destroyed. George Marshall, a sufficiently cautious man,

Figure 4.2   A Korean refugee near Inchon, 1950. (The Hulton Picture Company)

was back in office in Truman's Defense Department; but the Americans were determined to dodge Soviet objections at the United Nations and to finish the business in Korea as triumphantly as in Berlin. The war was a communist adventure, and must be seen off conclusively. It was, Dulles told the National Assembly in Seoul in June 1950, part of "the constant struggle between the forces of good and evil."

South Korean soldiers advanced briskly northward along the east coast, captured Wonsan, 150 miles north of the 38th parallel, and encountered few enemy soldiers. They had melted away once more into the hills. The Americans in the west had a harder time but by mid-October had captured the northern capital, Pyongyang. British and American newspaper and radio reports, accustomed since early in the Second World War to reporting victory in terms of global strategy and the ultimately unconditional surrender of a hated enemy, lionized MacArthur and jollied on the U.N. forces toward Manchuria.

It looked like a simple Second World War solution: defeat and occupy the enemy-held North and reunify the country on U.N. (that is to say, our side's) terms. MacArthur was the popular and successful hero who would bring about this satisfactory end. When the president, as commander-in-chief, summoned his brilliant if insubordinate general to a meeting on a Pacific island, MacArthur greeted him insolently with a handshake, not a salute. The two men exchanged bare courtesies and little else. MacArthur went back with his confidence in his own power and judgment perfectly intact. The only enemy was communism, and it had to be stopped by courage and will and proof of the mightiness of American arms. There was not just the puppet Kim Il-Sung to break and humiliate, there was the dark barbarian waiting in the north, Mao Ze-dong.

Mao needed a warning. But he had a warning to issue himself. At the end of October Chinese troops joined with North Koreans and fiercely attacked the U.N. forces at just the moment at which their commanders believed that the blue berets would go home before winter. The Central Intelligence Agency made colorful estimates of the Chinese presence according to which 40,000 soldiers already committed, with 750,000 waiting on the Manchurian border, were overrunning U.N. outposts and trenches; these hordes of Chinese troops were all indistinguishably dressed in heavy padded coats and peaked caps with the little red star and were driven by their fanatical devotion to communism and their well-known Oriental disregard for life and death, their own or anyone else's. "How many Chinese make a horde?" a British journalist asked a U.S. spokesman, but wasn't coherently answered.

In the hobbit history of the Korean War, the Chinese struck on behalf of world communism against the class enemy and took the Americans by surprise. On the day after Thanksgiving, they attacked in a series of sharp probes, and the Americans ran. MacArthur ordered a scorched-earth policy as his troops retreated, and the B-29s put it into force. Reminiscences of the fighting are full of hand-to-hand, revolver, grenade, bayonet, and small arms combat, much of it in the dark, all of it underequipped and conducted in a bitter, inescapable cold that sank to what were for all British and most Americans

unexperienced depths. At minus twenty degrees Fahrenheit, men died in their
sleep and knocked off their frozen fingers as they stumbled blindly through the
dark. Oil froze, trucks would not start, and weapons jammed all the time. The
inefficiency and waste of the U.N. forces amazed the combatants themselves.
The only consolation was that the Chinese had almost no artillery. In December 1950, 100,000 men of the U.S. and South Korean armies were taken off
at Hamhung on the northwestern coast and sailed back to Pusan on the
southern tip. The only American unit that kept its great name for daring and
comradeship was the Marines. Behind the departing flotillas, the U.S. Air
Force bombed Hamhung into rubble, out of bitter vengefulness. MacArthur,
at seventy, had been defeated.

The weapon that would restore American supremacy, as everyone knew
and no one dared to say straight out, was the atomic bomb.

It is clear that it was at about this time, when hysterical coalitions in
Washington were clamoring against a Truman administration thought to be
unable to hold back the Soviets and thwart their goal of world communism,
a rogue MacArthur was set on avenging his first defeat, and Stalin himself, as
it later leaked out, was apprehensive about the immediate possibility of a war
breaking out between his still underfed, unrestored, stricken nation and this
rich and reckless superpower so suddenly astride the world, that the cold war
most nearly flashed to hot for the first time. George Kennan, harboring many
second thoughts about the doctrine of containment during these dark, insane
days, set himself the task of thinking about what such a war would mean. He
lamented the entire absence of any such rational inquiry into their own
objectives by those in power, who were responsible to the citizens they represented and to the future.

Kennan, indeed, saw the danger in terms of a theme central to this book.
He predicted that "wartime emotionalism" would seize his compatriots and,
renewing their commitment to unconditional surrender (which he was convinced the Soviets would never concede), that they would pursue punitive
nuclear destruction with no thought of how things would look the day after.
The familiar structure of feeling in relation to war, the enemy, victory, and
freedom was there to give form and meaning to the passion, terror, and
self-righteousness of the moment.[13]

These currents swirled through the capitals and governments of Britain
and the United States, and through their citizens also, at the end of November
1950.

On 30 November Truman—perhaps incautiously, and under heavy Republican criticism—said at a press conference that the United States would
"take whatever steps are necessary to meet the military situation." Pressed
further about the atomic weapon, he replied, "There has always been active
consideration of its use."

The world shook. Clement Attlee, the British prime minister and Truman's careful but constant ally, flew at once to Washington to express his nation's grave misgivings. He got a bit of reassurance, but not much. The British warned agitatedly against war with China, which as he urged, both reason and morality forbade. But the American chiefs of staff were blockading the straits of Taiwan and recommending air strikes on China. Senator McCarthy, as we shall see, was kicking up his hateful noise and threatening impeachment for the president, and MacArthur was on the rampage, choosing Christmas Eve to request that twenty-six atomic bombs be dropped on "retaliation" targets in China and North Korea.

In a vigorous little interview, reminiscing about those days, MacArthur reckoned that "I would have dropped between thirty and forty atomic bombs . . . strung across the neck of Manchuria." He would have followed up with an invasion of half a million Chinese Nationalist troops, borrowed from Chiang Kai-shek, and finally, would have "spread behind us a belt of radioactive cobalt, from the Sea of Japan to the Yellow Sea."[14]

It will not do to pay too much attention to the half-demented recollections of elderly and half-disgraced generals. As all the world now knows, MacArthur would eventually be sacked by Truman.

But in early 1951 the war was still going badly for the United Nations. The communists recaptured Pyongyang and pushed south. MacArthur had atomic bomb–loading apparatus ready in Japan by March. On 6 April the bombs were to be put in strictly military care. Only a phone call to Washington would be needed to send them off.

Truman rescinded the order. MacArthur had proposed a series of vast projects, half of which were never begun. There was his lunatic cobalt scheme. He was going to cut China off from Korea by unprecedentedly heavy bombing. He issued his own communiqués, contradicting those of the president, his commander-in-chief. "If we lose this war to communism in Asia," he wrote in a letter to a congressman who read it to to the House, "the fall of Europe is inevitable."[15] On 11 April Truman fired him.

MacArthur came home to a spectacular ticker-tape parade welcome. He gave an arrogant, self-justifying, and acclaimed address to Congress, and in testimony before Senate hearings his egocentricity never once yielded to grace. He was the recipient of a characteristic outpouring of support by Americans for one of their folk heroes: the great general, the big, vivid, handsome figure who cuts through bureaucratic vacillation, whose patriotism and courage are great and incontestable, who acts on his beliefs to kill the nation's enemies, a long way away.

V

But the war ground on. In his address to Congress after his dismissal Mac-
Arthur said:

> The war in Korea has almost destroyed that nation. I have never seen such
> devastation. I have seen, I guess, as much blood and disaster as any living
> man, and it just curdled my stomach the last time I was there. After I looked
> at that wreckage and those thousands of women and children and every-
> thing, I vomited. . . . If you go on indefinitely, you are perpetuating a
> slaughter such as I have never heard of in the history of mankind.[16]

Two years would pass before the war ended. The fighting veered up and
down the peninsula. The Chinese were heavily overstretching their lines of
supply, MacArthur's successor, Gen. Matthew Ridgway, fought a dogged
campaign, and territorial victories took each side across the 38th parallel. A
quarter of a million South Korean soldiers and the same number of Americans, with a sprinkling of British and token Australians, French, Belgians,
Dutch, Turks, South Africans, Canadians, Filipinos, Ethiopians, Thais, and
Greeks, faced the enemy: 500,000 Chinese and 200,000 North Koreans.
Behind the U.N. forces was massed the vast bombardment power of the
American fleet and air force, both of which could operate with virtually
complete immunity.

The Chinese were suffering heavy losses and were reluctant to go on doing
so. Their troops had little beyond the small arms they carried. They had few
tanks and no sleeping bags and trotted to war in canvas shoes. Parts of cold
war rhetoric then and now has been to call them "volunteers," the official
description used by the North to justify China's involvement in Korea. They
were no more volunteers than the conscripted defenders of our rights and
freedoms in the U.N. forces. Max Hastings, greatly to his credit, quotes one
twenty-three-year-old Chinese soldier: "We felt pretty confident because we
had just beaten the Kuomintang, with all their support from the Americans.
We expected to do the same to Syngman Rhee's people. We weren't very
wrong. They were a pushover compared with the Japanese."[17]

But the Chinese were ready to stop, the Soviets had proposed a ceasefire,
and the United Nations, after the United States had cooked up a vote in the
General Assembly condemning China as the "aggressor," was ready to pack
it in, although on strictly partial terms. While very heavy ground fighting
continued for little gain on either side of the 38th parallel, the endless negotiations began. These talks would be depicted by the cartoon pictures of the cold
war: the expressionless, ruthless Orientals in their drab uniforms setting im-

possible conditions, breaking agreements, walking out; the dogged, decent generals of the West trying patiently to save lives and freedoms.

While the talks continued, with rows and suspensions and both sides walking out in the usual theatrical manner, men went on getting killed, for a few hundred yards of cratered hillside, in a way strongly reminiscent of 1916; Porkchop Hill and Heartbreak Ridge joined Hill 60 as emblems of pointless heroism and slaughter. Britain's "glorious Gloucesters" were surrounded and wiped out along the Imjin River north of the parallel. Until the last stand, they held out with antique courage against the bugle charges of the Chinese infantry; half of the battalion disappeared into the prisoner-of-war camps, which posed the knottiest problem of the peace talks.

The talks began in a tent at Panmunjom on the 38th parallel on 10 July 1951; the armistice was signed on 27 July 1953. The United States had to discuss terms as military equals with two states, China and North Korea, that they refused to recognize diplomatically. The talks were a riot of propaganda for both sides: the assorted obduracies and acts of effrontery from the communists, reminiscent of Berlin, caught the headlines, while the cheerfully reasonable Western negotiators blithely ignored such momentous byplays as the signing of the peace treaty with Japan in September or the terrible bombing of the North at regular intervals throughout the rest of 1951 and most of 1952.

The major cities of the North were flattened by the American forces; as a little encouragement to negotiations, Gen. Mark Clark, who had replaced Ridgway when the latter's tour was done, ordered in May 1952 the bombing of the main dams, barrages, and power stations in the North. They were all hit; the tens of thousands of square miles of cultivated land below the dams were flooded clean of topsoil and left gleaming and sterile. In July the U.S. bombers went on the kind of air strike their commanders had always loved best since the lethal and useless 1,000-bomber raids of 1943 and 1944; they killed one-eighth of the remaining citizens of Pyongyang, dropped 10,000 liters of napalm in glass bottles, and flew 1,400 sorties over seventy-eight cities. The bombing of North Korea disgraced the great principles of the American nation, but the Americans made the movie *The Bridges of Toko-Ri* about it all, in which the commander, incredulous and overcome with admiration for his gallant airmen, asks, "Where do we find such men?"

While the Americans bombed and napalmed, the communists, less destructively, dug. They dug 155 miles of passable road from coast to coast; They dug and blasted and reinforced 3,427 miles of trench, 776 miles of tunnel. They stalled at the negotiations, their volunteers and their peasants endured the bombing, and by the end of 1952 were quite immovable.

For many months tiny degrees of leverage, prestige, and power were traded for lives in the negotiation hut. Soldiers were killed and civilians bombed while the arguments circled about the repatriation of prisoners—was it to be compulsory or voluntary? Voluntary repatriation, a shiny new idea dreamed up by the Americans, sounded straightforward. North Korean soldiers, they said, didn't want to go home. The victims of Yalta—the Russian cossacks who were forcibly repatriated and then shot at Stalin's command—were freely cited.

There is little consistency to the tales of how either side's prisoners of war were treated. What *is* certain is that the South Koreans frequently shot northerners out of hand, and that British and American soldiers wrote home to their congressmen and parliamentarians to say so. One of the best-known incidents was the report of the British journalist James Cameron, who filed his open attack on the Rhee regime, "An Appeal to the United Nations," in the British equivalment of *Life* magazine, *Picture Post*.

> I had seen Belsen, but this was worse. This terrible mob of men—convicted of nothing, un-tried, South Koreans in South Korea, suspected of being "unreliable." There were hundreds of them; they were skeletal, puppets of string, faces translucent grey, manacled to each other with chains, cringing in the classic Oriental attitude of subjection, the squatting fetal position, in piles of garbage. . . . Around this medievally gruesome marketplace were gathered a few knots of American soldiers photographing the scene with casual industry. . . . I took my indignation to the [U.N.] Commission, who said very civilly: "Most disturbing, yes; but remember these are Asian people, with different standards of behavior . . . all very difficult." It was supine and indefensible compromise. I boiled, and I do not boil easily.[18]

He nearly lost his job for the report, and soldiers were disciplined for writing in protest to congressmen and newspapers at the violation of those human rights and freedoms they supposed themselves, so far as anything could justify their presence in that awful country, to be fighting for.

The cant that surrounds the term "freedom of choice" has always been slimy, and sometimes emetic. In South Korea prisoners were tortured and bullied and tied down to be tattooed with anticommunist slogans so that freedom to choose became a decidedly abstract master value. In the North the freedom of prisoners was more abruptly curtailed by a bullet in the neck, usually for South Koreans, sometimes for Americans.

And then there was brainwashing, the first time the phrase turned up in the cold war lexicon. It was inscribed in popular rhetoric by the melodrama of the movie *The Manchurian Candidate*, to which we shall return. For now it is enough to say that there was so little evidence in that movie of anybody having enough brain to find and wash that fears of wholesale laundering of G.I. wits were a bit misplaced.

What is uncomfortably clear from the histories, however, is that certain groups of North Koreans and Chinese prisoners of war were indeed strongly motivated by their beliefs and retained a high military morale—such that, at Koje camp in May 1952, they held the camp commandant a hostage, discouraged waverers on their own side by tearing off their balls, and got an admission out of the Americans that they had at times tortured and killed prisoners of war. Brigadier General Dodd was released. The Americans denied their admission. A little later, the revolt was briskly put down with tanks and flamethrowers, and the Chinese prisoners conceded the point.

Across the border, amid the same reek of brimming latrines, the bitter cold, thin-walled huts, the stony ground, and the watch-houses of POW camps, the North Koreans conducted their strenuous and ludicrous classes in correct Communist thinking and left their POW audiences rigidified in boredom. A few broke down, and even fewer changed sides

While the talks either stalled or inched on and the bombs still fell, the U.N. forces dug in as deep as the Chinese, and both far deeper than anyone could ever have gone on the high water table of Flanders in 1916. Short of atomic warfare and many more deaths in combat, neither side was likely to budge. Dwight D. Eisenhower, the Republican candidate in the 1952 presidential election, came to Korea to plan the armistice. One million soldiers on each side faced one another just north of the parallel and kept up a low-level, harassing, patrolling sort of war, killing just enough men to keep the enemy on its toes.

Stalin died on 5 March 1953. The new regime under Malenkov pressed, from its always fairly distant position on the war, for peace. The dams were breached. The rice and quite a lot of people drowned below them. The grim, power-hungry, and self-righteous John Foster Dulles was now Eisenhower's homburg-hatted secretary of state. Jon Halliday and Bruce Cumings report that:

> In India Secretary of State Dulles told Premier Nehru, for onward transmission, that the USA was prepared to use the atomic bomb. Asked how he would feel if a peace agreement were reached at once, Dulles replied: "We'd be worried. I don't think we can get much out of a Korean settlement until we have shown—before all Asia—our clear superiority by giving the Chinese one hell of a licking."[19]

Rhee tried to scuttle the armistice agreement and refused to sign it. The Chinese and the North suddenly accelerated their battlefield efforts, and in the month before the end, both sides suffered over 20,000 casualties, most of them deaths. Shocked into stopping, the Chinese, North Korean, and U.N. generals signed a truce on 27 July 1953. South Korea sat it out.

VI

Halliday and Cumings estimate total deaths from the war at well over four million (in a country that had a population of thirty million). They guess that two million of these deaths were North Korean civilians and half a million were North Korean soldiers. Probably 750,000 Chinese soldiers died. In the South one million civilians and half a million soldiers were killed. The American forces lost 33,629 men, and in a diminuendo, the British lost 686.

At the end of the war the territorial settlement was much the same as at the beginning. But the dams lay gaping, the fertile land was drowned, the hills were seared and blasted. The major cities—Pyongyang, Wonsan, Chunchon, Inchon, Seoul—looked like the aftermath of Hiroshima. The Second World War had barely touched Korea. That omission had been corrected.

5

BIOGRAPHY II

The Diplomat—George Kennan

GEORGE Kennan was the author of the single most important document of the cold war, the so-called Long Telegram (and so it was: 8,000 words in all), which he sent from his sickbed in Moscow on George Washington's Birthday 1946. The next year it became the anonymous but soon attributed article by "X" in *Foreign Affairs* and was circulated to every desk of any size in the state departments, military headquarters, and foreign offices of the West.

It had, as I suggested in chapter 2, its limitations. Kennan dictated it at one of the lowest points of U.S.-Soviet relations in a bad year. He was fed up with Soviet suspiciousness, obstinacy, and pointless uncooperativeness, and he felt lousy into the bargain. A short time before he had jotted down some rules of thumb whose crispness and accuracy cut across the stately and untelegrammatic prose of the Long Telegram. "Don't act chummy with them," Kennan admonished himself and his staff. "Don't make fatuous gestures of goodwill. . . . Do not be afraid of unpleasantness . . . or of using heavy weapons for what seem to us to be minor matters," and more in the same vein.[1]

The rules of thumb were fine. The Long Telegram and the "X" article were fine so long as they were seen, as Kennan insisted, as "a guide to the technique of dealing with (I reiterate) the *Stalin* regime." Rereading his telegram, however, for his memoirs he is, as usual, hard on himself, but not wholly without justice: "I read it over today with horrified amusement. Much of it reads exactly like one of those primers put out by alarmed congressional committees or by the Daughters of the American Revolution, designed to arouse the citizenry to the dangers of the communist conspiracy."[2] Kennan was fed up with the Soviets; anyone would have been. He had seen a great deal of them over the past twenty years, and insofar as what he had seen had all been the product of the Bolshevik revolution, he had liked none of it. What he had learned of nineteenth-century Russia he liked, even loved, with the extraordinary intensity of feeling and that unremittingly self-critical nervousness that mark the man so deeply.

Kennan was born in 1904 in Madison, Wisconsin, where his father was a lawyer. His family had been eighteenth-century immigrants from Scotland, and then freeholder farmers for 100 years, never selling their labor (as Kennan has pointed out), living frugal, upright, Scottish Presbyterian lives.

Kennan went to a military academy as a boy. (His later idea of the ideal school for diplomats was not unlike such an academy, with solid doses of classical literature thrown in.) He went to Princeton, then joined the U.S. Foreign Service when that was not at all a fashionable or well-paid career. In 1929, after service in Riga on the Latvian coast during which he had picked up what he could about the revolution from a group of impoverished exiles who kept up their reading, their tea drinking, their shabby respectability, he was sent for specialist training as a diplomat in Soviet affairs in Berlin.

Sixty years later, he said to me, "The Russians are part Dostoyevski and part Chekhov. When I read Dostoyevski, I want to say to the characters, "Calm down, go home, have a bath." But they wouldn't listen, naturally. They won't even stop to eat. . . . The best of Chekhov is like the man himself, the one we find in his letters—kindly, modest, urbane, upright, all these things. And very funny and good-humored. I haven't myself written in this jocular, good-humored vein. But I could." Then, as Kennan has done so often, he reached for a trope with which to grasp the notion of national character. (He always believed that the careers of Ivan the Terrible and Peter the Great were as much keys to unlocking the Russian heart as those of Bakunin or Lenin.) "Russians, it seems to me, can be understood best if you go back a long way into their history and see them as a short people dominated by hordes on horseback. The crucial image is the man on horseback dominating the man on foot. There is the story of the Russian envoy meeting the emissary of another country while both are on horseback, and the whole extended parley consisting of an attempt by the Russian to make the other man dismount first."[3]

He went back to Riga from Berlin in 1931 and loved it. It is still a magical place: the high narrow houses of the old town, with gnarled and rusty iron railings curling beside their worn steps and the cobbled streets, twist up from the estuary shore until, breathless, you turn at the top and look down on the gray gulf under the low clouds. In 1933, when at last the United States recognized the new Soviet regime, Kennan took office in Moscow and watched the terror begin.

He hated the regime, as well he might. And then, after sitting through the German annexation while he was briefly in Prague, he grasped the contrasting nastiness of the Nazis when he was posted to Berlin early in 1939, only to be interned for six months when the United States entered the war after the attack on Pearl Harbor. In 1944, at the age of forty, he went back to Moscow under Averell Harriman's ambassadorship and took up a position that would remain at the center of the cold war storm winds as they blew at their iciest.

In spite of the striking command and eloquence of his summaries, he was always a slightly off-center figure. If his seniors and colleagues found his obliquity expressed with unfailing directness and probity, this was the first part of Kennan's paradox. He was the most important shaper of the most important foreign policy decisions that his country, superpower of the world, would take for fifty years. Yet he was solitary, fiercely self-deprecating, immediately doubtful of the decisions themselves, and rarely at one with anyone else in the State Department.

His two years in Moscow confirmed him in his view of the Stalinists. Things went from bad to worse, and he duly sent off his telegram. He went home disgruntled to Washington to find, to his consternation, that his paper, with its very general and diplomatic suggestion of ' containment" of the Soviet Union and its ambitions was fast hardening into military strategy and longterm commitment.

The Truman administration, to its credit, set up a new policy planning staff—just ten men, with Kennan at their head. His assignment was to plan U.S. policy for the world at large; his first momentous task was to implement the Marshall Plan, with a reminder on the desk in front of him from Marshall himself to "avoid trivia." Kennan's formidably intelligent staff had more work than a handful of men could possibly do toward the biggest economic reconstruction in history. Remembering the fierce unremittingness of those labors twenty years later, Kennan wrote, "So earnest and intense were the debates in our little body in those harried days and nights that I can recall one occasion, in late evening, when I, to recover my composure, left the room and walked, weeping, around the entire building."[4]

It is a mark of the man that he is so candid about the vehement depths of his feelings. Those feelings animate his memoirs, which reflect his formidable sensitivity to the lightest breath of animadversion, enmity, or rebuffal. It is surely a sensitivity almost impossible to bear for a diplomat (of the cold war, no less); Kennan's searing combination of honesty and touchiness landed him in social or actual exile from his kingdoms on several occasions.

At the same time, his determination to make plain his convictions with painful truthfulness secured for him his unique respect from his presidents and the American people. For the whole period of the cold war, everyone knew that George Kennan, speaking only for himself and not for his advancement or any hidden interests, would tell his audiences what he believed and that his arguments would have a blithe disregard for his listeners' preferences, the timeliness of his remarks, their propriety, or the way his opinions contradicted his own past thinking. Indeed, he is able to speak with equal impressiveness on both sides of the grandest issues of the epoch and is never other than substantial, deliberate, and possessed of a sweeping vision.

From the cold war's first coagulation when, Kennan reported, he "viewed the labors of the Potsdam Conference with unmitigated skepticism and despair," Kennan combined a deep, partisan feeling for the fervor and darkness of Russian culture and history with a settled conviction of the brutal gangsterdom of the Soviet leaders.[5] At this point in his career, around the time of the Berlin Blockade, Kennan was clear about his political purpose: to hold down Stalin and his dreadful minions. But when the old devil died, just after Kennan resigned the august office of ambassador to Moscow, everything changed.

If, at the end of the 1940s, no communist party (except the Yugoslav one) could be considered anything else than an instrument of Soviet power, by the end of the 1950s none (unless it be the Bulgarian and the Czech) could be considered to be such an instrument at all.

This development changed basically the assumptions underlying the concept of containment, as expressed in the X-article. Seen from the standpoint upon which that article rested, the Chinese-Soviet conflict was in itself the greatest single measure of containment that could be conceived. It not only invalidated the original concept of containment, it disposed in large measure of the very problem to which it was addressed.[6]

It was typical of Kennan that when he was slung out by the Soviets as ambassador in Moscow after less than a year, for accurately but injudiciously saying how frightful the quotidian oppressions of their government were, he at once resigned from the Foreign Service, with bitter self-reproach. He was always oblique to the machinations of power and always stood with his back turned to the inner circles of the power elite, even when his own plans were being carried out—though often long after he had changed his mind about what ought to be done. He made key contributions to the provisions of the

North Atlantic Treaty, for instance, but struggled to present the alliance as a temporary, not a forty-year, construction and to exclude those countries not grouped about the North Atlantic (Turkey, Greece, Italy) in order not to crowd Soviet sensibilities even further.

During the blockade Kennan was one of the group at the State Department waiting nightly for General Clay's telephone calls. He watched with repugnance his compatriots living in West Germany who "flaunted" (Kennan's word) their comforts in the face of the despair and starvation on all sides. It has seemed as though he cannot separate his real, suffering sensibility from his participation in the grand mendacities and amoral self-propulsion of states. He has insisted on *seeing* the starving Berliners and the cheery Moscow children whom the horrible Soviet cops chased away from his own friendly little boy.

These miseries make Kennan, well, a bit of a misery. He can take personal affront as hard as mass suffering, and he has often lacked the gift of gladness. But even with these contradictions, his lack of color, perhaps, and his strenuous, anxious responsibility for the whole bloody old globe, Kennan unmistakably has stood for the very best of certain strictly American values. It is these values that the present over-fifty generation in Europe has looked to the United States to defend, and that, to our shock and dismay, the United States has at times corrupted instead.

Kennan held these values within an older tradition of diplomacy. He said in 1989, "My generation of diplomats and the older men I looked up to, held to the idea of a gentleman. It's hard to use the term now, some people are so ready with the word elitist, but that's not it at all. *Gentleman* turns on matters of principle and conduct, not on what you wear or where you came from. New York doesn't have to be the city of origin, the provinces will do just as well." He approved strongly of the class of patrician Bostonians who were diplomats out of public-spiritedness and didn't have to count on State Department patronage.

Kennan's conservative intellectual ancestors were classical American republicans, not the old cynics of British conservatism. They were Alexander Hamilton, Thomas Jefferson, and John Quincy Adams. The values he steadily followed were more than anything those of his freeholder farmer ancestors, stiff perhaps in manner but also plain-speaking, law-abiding, Presbyterian, incorruptible. I like to think I can hear the manners of the Scottish enlightenment in Kennan's anecdotes: "For my sins, I was chairman of the committee drafting the Atlantic Treaty, and after one meeting a member of the committee from another country left behind, forgotten, his briefcase. My officials brought it to me and there it was, presumably full of secret papers. They asked, 'What shall we do?' 'Telephone him immediately,' I said, 'and tell him it's here. Give him my word of honor it hasn't been opened or tampered with.' "

By the same token, Kennan, against the great anticommunist tide of the 1950s, stood up calmly for his junior officer John Paton Davies, who was traduced on no evidence for his alleged favors toward China, betrayed by John Foster Dulles himself, harried by *Time* magazine, and disgraced into resignation.

In 1957, while a visiting professor at Balliol College, Oxford, Kennan was invited to give the Reith Lectures broadcast to the British nation by the BBC, named for its first and imperial director, John Reith. These annual lectures were in those days quite an occasion, and Kennan rose splendidly to it. In the process he upset the U.S. Defense and State departments and, indeed, the whole North Atlantic Treaty Organization. Yet in opposing vigorously the transformation of NATO into a long-standing nuclear alliance, in proposing that new efforts be made to reunify and demilitarize Germany and therefore to make some move toward healing wartime wounds that were still visibly raw, he pronounced in 1957 the lessons that most people did not learn until the Berlin Wall came down in 1989.

In the 1957 Reith Lectures—often written on the hoof, with emendations scribbled as the red broadcasting light went on in the studio—Kennan discussed the division of Germany and the place of nuclear weapons along the barbed wire. He pointed out that Khrushchev had only recently promised that the Warsaw Pact armies would leave East Germany and the satellite countries if NATO would promise to do the same; that the Polish foreign minister Adam Rapacki had proposed a nuclear-free zone in both halves of Germany and in Poland; and that the successful launching of *Sputnik*, in the week the Reith Lectures began, would ignite a terrified American rush to build new weapons unless a better peace was made at once.

Of course, *Sputnik* did just that. The Soviets were on their triumphant way to creating the intercontinental ballistic missile (ICBM), and the Americans would cash every check they could draw to catch up. Kennan, once again, said what was right at the moment, and not a soul would heed him. Remembering that moment, he said to me, who had heard that Reith lecture and first learned of Kennan when a new student at Cambridge just back from service in the Suez campaign, "When I suggested in those 1957 lectures how our dismal, wrong arrangements for postwar Europe might be dislodged, I was calumniated for it, made many enemies, was bitterly criticized, as I say in my memoirs, by men as sympathetic and intelligent as Raymond Aron, who said it was just too dangerous, we couldn't reunify Europe, we couldn't go back, it would be too precarious, too fragile. And by others like Dean Acheson and John Foster Dulles, who simply dismissed what I said out of hand.

"Now, thirty years later, those lectures seem to have greater resonance and find sympathetic echoes in the way in which the world is now being directed. Certainly if Gorbachev is successful. And they can't go back, they can't go back now. What Bush will do, I don't know. I look to my former colleagues to take him forward in some response to what Gorbachev has to say. They've got to make moves of that sort. It will continue to be a dangerous world, but the cold war is over."

In the lectures he pointed out the unreality of American insistence that a unified Germany be able to join NATO, a condition that, as he said in his memoirs, "in effect demanded a unilateral withdrawal of the Soviet Union from its positions in Central Europe without any compensation at all,"[7] and he shuddered at the vision of an arms race whose only conceivable end is the last day of the world, the day on which they are used. He spoke out against the tantalizing deceits of local or "theater" weapons, which were new in 1957 and promised to wipe out only two or three counties rather than two or three nations. In his final lecture he pointed out sharply the utter irrationality of a defense policy based on being the first side to use weapons that, once fired, could only bring on massive retaliation. A year later, also in a BBC lecture, Kennan rebuked the superpowers equally for supposing that their quarrels entitled them to threaten the world with extermination when they held it in trust, as all public servants should do, for future generations. To violate that trust was "simply wrong."

In and out of public life, Kennan went on speaking truths and repeating moral facts that the conductors of the tubby bus of the cold war simply ignored, in order to keep the old girl on the road In 1961 John Kennedy had the imagination to retrieve Kennan from the Institute for Advanced Study at Princeton and make him ambassador to the only non-Soviet but still Communist state in Europe, Yugoslavia, where Kennan took quite heartily to our old comrade-general of the partisans, Marshal Tito. Faithful to the letter and spirit of American principles, Kennan fought and lost a battle with a Senate that debarred aid to Yugoslavia and, in one entertaining incident, withdrew aircraft fully paid for by the Yugoslavs on the entirely erroneous ground that their country was a member of the Warsaw Pact. It seems that the senators just could not believe, even in 1962, that a self-proclaimed socialist state was not by definition in Moscow's pocket. In spite of his nobility of temper, Kennan is very brusque with such stupidity and ignorance.

American longshoremen's unions refused to load or unload Yugoslav ships in American ports; the Yugoslav consulate in Chicago was bombed; Congress occupied itself with retaliatory measures against countries, including Yugoslavia, whose ships visited Cuba; in short, the sterling quality of Ameri-

can anticommunism continued to be demonstrated daily in a dozen unpleasant ways.[8]

He left Yugoslavia the day after the dreadful Skopje earthquake (and,
typically, gave a pint of urgently needed blood during his packing), on 27 July
1963; he was fifty-nine.

All such sentences sound like valedictories. But Kennan was far from
finished with the cold war, and the American side of it was far from through
with him. In the years he spent at Princeton after 1963 he came to occupy a
unique place in the long argument of the world with itself about its best future.

One might say that Kennan has become to international politics what
T. S. Eliot had by the end of his life become to literature. He has striven for
a transcendental understanding of world struggle, not just a rationalization for
the American or Western point of view. He could not but *be* an American,
naturally, and he could not but have the oddities of character to which his
overstately, slightly marmoreal and chilly prose bears witness. There we can
glimpse his fierce family devotion, his recurrent depression, and his abomina-
ble touchiness in flashes that, one would guess, are involuntary. It is the prose
of a man for whom intimacy, even deep friendship, is cherished but difficult.

These qualities, however, are domestic details in the statuesque serenity,
instinct for justice, and searching high-mindedness of the public figure. His
efforts to argue not for the United States but for reason herself have brought
him obloquy as well as respect. For years he railed steadily against the arms
race; he was the first man of influence to press for huge reductions in the
stockpiles of nuclear weaponry, at a time when such ideas, under Reagan's
irresponsible presidency, looked, as they say, utopian. From no public platform except the eminence his own probity had given him, he spoke out against
the political folklore, superstition, and human silliness surrounding the awesome weapons. One by one he took the slogans—"nuclear blackmail," "Soviet domination," "Finlandization"—and showed the world, if it would only
watch, that "it is but the eye of childhood that fears a painted devil."

Over the years he has assumed without presumption his own remarkable
standing. He is not an intellectual on the French model, an Aron or a Sartre;
still less was he just a career diplomat, though that work brought in his only
income until he went to Princeton. In his memoirs, tired and in ill health, he
deprecates the incessant demands made on him to tell the public what he
thought. In England diplomats are sworn to silence and merely exude throttled platitudes. George Kennan, however, has spoken frequently to eager and
grateful audiences. He has worked out difficult, historically changeful ideas
slowly and without concealment. The ideas have been at times opaque, and
at other times contradictory, but they have never been servile, of time or

masters or anyone else, and they have always appealed to still noble ideals of public duty, diplomatic probity, personal faithfulness, and the canons and sentiments of his nation's great Declaration.

Kennan's significance is historic. In the excellent openness of American public debate, he built himself an unignorable lectern from which he summoned his compatriots—whom he has so often regarded with fear and loathing—to the best principles the happy chance of his upbringing had taught him. With unaffected moral authority, he could say to me, across a dining table, "I suppose loyalty is the only absolute virtue. I don't mean by that that everybody ought to practice loyalty; I mean that its demands are absolute until the loyalty to which one has given allegiance is broken by the institution to whom one gives it, having in some way flouted legality or rendered itself immoral, or whatever it may be.

"There is only history for us to learn from—I count literature as a way of meditating on history. The kind of history I favor is biographical because that alone is capable of reaffirming the value of the individual. This is a kind of history that's been out of fashion for a lot of years and now seems to be in revival. But praising biographical history doesn't mean individuals can do everything, or even very much. History defines what we can think (for example, the carpet bombing of 1943–45 shaped the thinking for 'mutual assured destruction').

"States differ, but they aren't individuals. States stand as agents but not as a congeries of principles. But, of course, they may act for better or worse ends. Evil, however, is never unitary—not a notion familiar to Americans."

6

FICTIONS I

Righteousness—The Magnificent Seven, The Manchurian Candidate, On the Waterfront, Animal Farm, 1984, and The Crucible

ROM 1946 until the Vietnam War really began to slide away from the Americans in 1963 or so, U.S. culture valued righteousness. For the Puritan fathers, righteousness was hard of attainment but once acquired, as *Pilgrim's Progress* told them, served as a breastplate against their adversary the devil, who walketh abroad, seeking whom he may devour. By its nature, righteousness turns at times into self-righteousness, a much less appetizing quality.

Yet even allowing for its nasty side, true righteousness remains a beautiful attribute and is easily recognizable. The righteous American moved calmly and bravely through the Second World War; those generals with their steady mouths and eyes—Clark, Ridgway, Marshall—convinced their soldiers and the American people that this was a just war in which dying was significant. Americans brought that same righteousness to the showdown with the hard-faced, coldly fanatical Russians. Whether exercising righteous anger or righteously putting to rights other nations' messes, Americans found plenty of opportunity during the cold war to experience the gratifying emotions that go with being in the right.

Righteousness has always been amply illustrated in the stories Americans tell themselves about themselves. Those stories taught that the American way of life, its fine independence and manly self-reliance, is the only meritorious way in a world of bad guys whether European or Asiatic, from left or right. The Fascists threatened over here, the Communists over there. Leave them to it, the national story said; we'll keep ourselves to ourselves.

After 7 December 1941, isolationism would no longer do. Duties and responsibilities crowded upon Uncle Sam, and the questions were flung at him from around the world. What was he going to do? Wouldn't he come to the aid of the poor, the weak, the defenseless, the oppressed? Wouldn't he put aside his principled peacefulness, the productive solitude of his farm, and sally out to do what was right, saving the freedom of poorer folk?

The questions crowd about the righteous man until he jumps up out of his chair, his boots heavy on the plank floor, and with a sudden movement of his arm cries out to his questioners to be quiet. He stares out of the low window, handsome, powerful, silent; his wife and children, his servants and paid men, watch him, the righteous man making up his mind. When he turns around, it will be made up, and he will be inflexible but gentle, impossible to turn from his purposes, just.

Of course it is right that righteousness should be directly connected to principles and procedures of justice; hanging onto justice gives righteousness the only chance it has of acting rightly. For the righteous man must be up and doing, and American politics puts action at the heart of efficacy. The righteous man and woman (John Proctor and Katie Elder) may have a touch of the tyrant about them, but they are the moving figures of American history.

The equivalent figure in Russian culture is different: he or she is a fearsomely uncompromising figure whose absolute standard is not right but truth, a saint who will keep faith with truth whatever destruction such behavior brings. To such a character—Dostoyevski portrays several, in the cold war Solzhenitsyn was one such, gentle Sakharov another, the two Mandelstams— personal survival is nothing, and compromise is literally unthinkable. Such people may be said to *be* rather than to do, but such being is like an intense cold flame that cannot be put out. It burns on; it does not change.

The righteous man puts things to rights. He gets things done. In the first freezing years of cold war, political doing was a patient, defensive business. The Berlin Blockade came as a relief; there was a job to be done, as was the Korean War, even more so for those well away from the action.

Hollywood captured the moral action of the cold war and turned its heart into simple, vivid parables. Time and again American filmmakers returned to their favorite form, the noble epic of the western, and its paraphrase the war movie, to tell the tale of righteous men opposed to wicked ones, of honest cowboys and heartless killers.

Just as the cold war began, wide-screen westerns came into their own.

Technology and total propaganda came together in the rich satisfactions of *The Magnificent Seven.*

II
_____

The Magnificent Seven (1960) is one of my favorite movies, as it is of the millions who have admired it. It was a formative film for many Englishmen of my generation; we saw it umpteen times in the cinema, long before it became a stocking filler in the videotape collections of our sons. If it looks, thirty years later, like a perfect parable for American hopes of themselves in their extended commitment to the cold war, this is not to impute to its great maker John Sturges, any beady-eyed intent. His success was due to his strong feeling for popular feeling, his vivid sense of where it flowed and how it could be idealized and thereby dramatized. And, of course, his inspiration was not politics but another film: the Japanese classic *The Seven Samurai* (1954), which he followed, according to his bright lights, faithfully.

The movie famously captures American feeling about all the wars impending at the time—that is, the small ones being fought somewhere else. A poor, one-street Mexican village is being plundered by a band of brigands led by Calvera, turned by the endearing clichés of Eli Wallach into the most likably roguish character in the movie. Advised by the village elder in his mountain eyrie, three of the peasants go to hire guns. They encounter a racial quarrel over whether a dead Indian laborer like themselves may be buried among the whites on Boot Hill. A powerful Slav in a spotless black shirt and jeans accompanied by a nervous, cheerful, streetwise city boy in a sweaty hat ride the hearse through the bigots to the syncopated majesty of a sort-of-Aaron-Copland soundtrack. Yul Brynner and Steve McQueen have replenished the rights of man.

Thereafter, the two gunmen recruit the rest of their battle squad. Each one comes not for money—for twenty dollars and a daily dollop of chili con carne—but for an unspoken cause, because they're on the run, because they are looking for a home; they come for friendship, for love of the sport, for love. The only one who comes because he can't believe it isn't all for the gold is briskly bumped off as punishment.

They stay. They drive off Calvera, but he comes back and takes them by surprise. He does not want to call down American revenge on his head by killing them, so he turns them loose.

But—James Coburn—"Nobody throws me my gun and tells me to ride. Nobody." The seven come back and rout Calvera. Four of them are killed in the action. Yul Brynner, his perfect shirt unmarked by dust or sweat, kills

Calvera-Wallach, who, as he dies, grates out, "You came back. . . . A man like you. . . . Why?" He is uncomprehending, but *we* know perfectly well why the burly, lissome hero comes back; he had given his word, he returns in defense of freedom, it is a duty to his name. To his name.

And yet we barely learn his name. They are all unknown soldiers who give knightly allegiance to the noble simplicities of the frontier and the Declaration of Independence. Each figure exudes a daunting adequacy; each is a fully achieved man, silent, un-self-questioning, completely trustworthy. (Even the one who loses his nerve for battle recovers it at the last moment.)

Such men are heroic ideals formed out of America's victory in the Second World War. They are ideologically applicable to any circumstances in which less adequate men deprived of their natural way of life may be brought to reclaim it. The seeds of "Vietnamization" were planted by the magnificent seven. When the seven return after their brief humiliation (due to treachery), those peasants who have spoken of knuckling under to Calvera and those who have turned cowardly recover their status as men by fighting and killing with domestic utensils—axes, chairs, reaping hooks, flails.

All the Americans are Hectors; none play Achilles or Thersites. They are knights, and knights do nothing except voyage and fight. The meaning they offer to ordinary householders like ourselves who watch and admire them is the transcendental old lie that a man is never more a man than when a warrior, but the greatest warrior is no mercenary. He is a free man and a democrat. When the occasion demands, we may hope to find him within ourselves.

It is a splendid film, gathering as it does the epic form of the Western into a heroic celebration of American values. As one would expect of an epic, it has no inner life. These are men of action. *The Manchurian Candidate* (1962) is as supremely confident in American ideals as *The Magnificent Seven* but complements the great epic by showing what can go wrong with those ideals when the inner life is made unsound or is invaded by political virus. It shows that righteousness must be of the right kind.

Laurence Harvey played the candidate of the title. He has been captured with several of his regiment in Korea, taken across the Manchurian border by the Chinese, and—in the scary phrase first used in reference to U.S. prisoners of war in Korea—brainwashed. His brain is washed of all natural desires and loyalties and is set like a timebomb to perform assassinations for the Communist cause. He is triggered out of conventional consciousness into the obedience of the automaton by a simple signal: the queen of diamonds, a symbol rich with the menace of the Tarot and the Orient, as well as diamonds themselves. When he sees the playing card, he blindly obeys the next voice to speak.

His control is his own mother, played by Angela Lansbury. (In a good joke by the unconscious, the two villains are played by an English actor and actress.) She expects a Manchurian candidate but does not know he will be, by a deliberately fiendish twist on the part of his Communist puppet-masters, her own son. Her communism is greater than her maternal love, however— indeed, each feeds on the other, as Michael Rogin emphasizes[1]—and she arranges for her son to assassinate the presidential candidate as he accepts the nomination at his party convention. Her husband, a McCarthyite fool and her unwitting creature, who by her expert manipulation of the caucuses has won the vice presidential ticket, will thereby assume the crown and she will be in a position to dominate the president on behalf of the enemy.

To test the completeness of Harvey's brainwashing, he is ordered to kill his fiancée, the only person who can shake him out of his catatonic coldness and clipped self-hate, together with her father, who is a politician of the best American stamp, honorable, intelligent, and fluently critical of the extremities of communism and chauvinism alike. Although Harvey is momentarily liberated by the lovely girl's suddenly appearing in a dress patterned with the queen of diamonds, his mother reassumes command, and he destroys the only two people he loves and the source of his redemption.

There is surely something in this, Rogin claims, of American matriarchy and the American mother's incestuous refusal to let her son go to the woman who will usurp her place. But the ideological point of the movie is more apparent: communism is the fatal usurper. It displaces family love and outrageously makes political belief stronger than natural affection and honest sex. It replaces the voluntary assent of the democratic voter with the choiceless obedience of the party robot. By invading domesticity, communism will conquer politics without anybody even noticing. The intelligence of a public composed of rational individuals will be perverted by the irrationalities of either right-wing hysteria or the cold calculations of left-wing manipulation.

In 1962, faced with the aftermath of McCarthy but also still believing in Soviet diabolism, moviegoers found this a congenial and all-American attitude. Behind it, I suppose, is Burke's "The price of freedom is eternal vigilance." Frank Sinatra, captured in Korea with Laurence Harvey but his own man (his own American) enough to ward off brainwashing, snaps the program in Harvey's brain and frees him. Once freed, Harvey is ablaze with hate for his mother, her treason, and her sickening desire for him. In a terrific coup de theatre, he first covers the putative president-to-be through his telescopic sights from his sniping position high on the roof of the convention hall, then suddenly switches, shooting his right-wing father and left-wing mother plumb through their foreheads.

The film grips like a strangler. If it gripped less, it might have been politically more astute. There are crassnesses, but they are carried off by the first-rate acting and screenplay and the dazzling surprises and cutting. The plot is incredible, the central place given to brainwashing absurd, the Communists' cunning quite ridiculous. But the visual effects match Hitchcock at his best: in one of the only believable bare-handed karate fights on film, Sinatra makes a dreadful effort to tear off the features of his Chinese opponent by digging his fingers behind the skin on the man's cheekbones. And the suspense, the stock, irresistible feature of all efforts to win the cold war on film, is painful.

The movie is a remarkable achievement in what may be called the propaganda of liberal totalitarianism. That is to say, it is *total* in its effect and its message. It warns that the enemy may achieve political victory by invading undefended and nonpolitical territory, family life and loves. The only defense is to politicize these lives and loves against the adversary. The film endorses fear of communism as rational and necessary, but it abjures hysteria. The high command is, as usual, complacently blind to the menace. Only Sinatra, with the love of a good woman to help him, saves democracy in the nick of time. We might any of us have had to do the same.

Taken together, *The Magnificent Seven* and *The Manchurian Candidate* mapped out political conduct and political self-awareness for cold war soldiers. When the call to defend freedom comes, then our men, the honorable gunslingers, are prompt to act.[2] When our domestic defenses are forgivably lowered in trustfulness, then they must be raised again by rational self-examination and, as in Laurence Harvey's example, meet and proper suicide.

III

In spite of Sinatra's patriotism and straightness, *The Manchurian Candidate* conveyed the simple truth that appearances may lie, and that cold war, unlike hot ones, depends on duplicity, even exercised by the righteous in the defense of freedom. It may corrupt those you love dearly, may even so invert trusted values that an informer is only doing what honor bids him do.

A cluster of Hollywood movies—*I Was a Communist for the FBI* (1951), *Kiss Me Deadly* (1955), *Jet Pilot* (1957), *Walk East on Beacon* (1952), *My Son John* (1952)—turned on the moral necessity of deceit for the sake of virtue, on lying for truth, and transferred these easy inversions, moreover, to the still tender taboos of sex, so that sexual intimacy itself was invaded by politics and the perfect safety of American domesticity politicized.

At first sight, many of these queasy political films were, as Gatsby once put it, only personal. In other words, they were out to vindicate a personal not a public righteousness. Elia Kazan's *On the Waterfront* (1954) is a very local cold war movie. Its locale was, of course, Hollywood, where Kazan, as Arthur Miller tells us, named the names of those he had known—and whom the House Un-American Activities Committee (HUAC) had also known, and known beforehand—to have been Communists and fellow travelers.[3] In *On the Waterfront* Kazan cast the young Marlon Brando in his marmoreal beauty as Jimmy Molloy, the free spirit brave enough to blow the gaff on the corrupt union boss. To the Communists, of course, Brando is a stool pigeon; to Kazan, precisely because Communists will make that judgment, Brando is standing up for more than truthfulness and against more fearful sins than venality. He stands up, Kazan hopes, for freedom as a virtue, rejecting corruption not because of the graft but because it makes the air stink. If you live like the union boss you cannot be a free man; if you cannot be a free man, you cannot be a good one.

The details of the film are extremely subtle, and the moral structure is crassly simple-minded. Its makers and supporters were left-liberals whose working-class sympathies and honest but inexpensive communism suddenly turned hot on them when the House committee began its roastings. Budd Schulberg did the script, Elia Kazan the direction, Sam Spiegel the production, Leonard Bernstein wrote the jerky, rhythmless music. Schulberg and Kazan had already made their confessions; the bullies of politics had bent the sissies of culture to their will. *On the Waterfront* was Kazan and Schulberg's apology and, in spite of their cowardice, their revenge.

Brando plays Jimmy Molloy, ex-pugilist brother of the henchman (the young Rod Steiger) to the rotten and power-mad union boss, played with thick lips, homburg hat, and astrakhan collar by Lee J. Cobb in his prime. The bosses kill an idealistic Irishman named, inevitably, Jimmy Doyle, who threatened to peach on their graft. Brando unwittingly set him up. The FBI are onto it all. Enter the beautiful and idealistic Eve Marie Saint, apostle of the working man's goodness and advocate for the repeal of his legitimate grievances. The grievances are, however, not against capital but against the union bosses.

Brando-Molloy, improbably roused to purity of motive and action by the girl's purity and stirred by a speech, conned straight from early Ibsen, made by the local Irish-Catholic priest, tells all in court. He is dreadfully beaten up by the union boss's thugs but revives to lead the workingmen triumphantly back to work, ignoring the abuse of Lee J. Cobb and under the admiring eyes of the rich dockside contractor.

The cold war-mongering of all this lies in the evident fear and hatred of organization among the working class. But Kazan is an old leftie; his true belief is in the spontaneous decency of workingmen. His true Americanism *and* his guilty self-justification of his confession to the House committee come out in his utter caricature of the trade unions (and this at a time when the AFL-CIO was at its most virulently antileftist). Kazan's real hero is the man whose conscience leads him to tell the truth about corruption whatever the costs. He is the unfrightening ghost of righteous and classical liberalism, and if he is caught, as Kazan was, standing up for the oppressed working man, he does so strictly in line with the Constitution.

This view comes out with painful and amusing sharpness during the film. Although the film takes the side of American law and order, all the real feeling runs the other way. Lee Cobb stands for conventional force and respectability—law and order. The FBI investigators are amiable good guys with no presence and less clout. They stand for idealized justice, the union boss for conventional order. Kazan restores his good name in a thousand lovely little touches—the rooftop life, the pigeon-keeping culture, the sweat-stained work, the delicious girl with a nature that turns to loveliness all it touches, even a beautiful pug with cauliflower eyebrows like Brando.

This is indeed the dream of American righteousness in the era of cold war, and Brando, Eve Marie Saint, and finally the stevedores themselves are the stars who look compassionately down and show the American people how to reconcile frontier values with working-class solidarity while keeping the Communist enemy resolutely at bay.

IV

The moral form of the cold war narrative, whether in real Korea or on the waterfront, is that conscience trumps solidarity. The inverse of that lesson is that mass organization *always* strips you of your humanity. Essential humanity, the source of freedom and happiness, thrives in the quirky, the private, the unplannable. Communism threatens humanity by its remorseless planning, its eradication of the solitary and the strange. The canonical text of this lesson is *1984*.

Strange that an Englishman—and such an English Englishman, reserved, obscure, difficult, uncongenial—should have written the key imaginative manifesto of the cold war. Would many people dispute that George Orwell's *Animal Farm* and *1984* are the two most important novels in English of the cold war? Not perhaps the best, but the best known and the most influential? As symptoms of the cold war, the two novels bespeak a widespread and early depth of distrust toward and disillusion with the very name of socialism and its state, across the English-speaking countries. As a contribution to the cold war, this "couple of horror comics" with which Orwell ran "shrieking into the arms of capitalist publishers," in the words of one Marxist critic,[4] not only did a great deal in Western consciousness to discredit socialist ideals in general and the Soviet Union in particular but also gave an extraordinary and potent imagery to those who would denounce socialism as totalitarian even in its mild British form and would make the whole doctrine sound as vile as the worst deeds of Stalin.

Both novels were jam for right-wingers. *Animal Farm* has become, after the murderous things done in the name of revolution in Russia and China, an allegory of what happens after *all* revolution. As a story, it is very well told. One thousand years of fabliaux and children's animal stories stand behind it, and Orwell marshals the reader's sympathies easily and vividly. Naturally, each characterization is as simple as a cartoon: the dim, honest cart horse Boxer, who was borrowed from David Low's political cartoon in which the cart horse stood for the trade union movement, "the cynical donkey, the cackling hens, the bleating sheep, the silly cows."[5] If this is the working class, it's no surprise that the pigs come out on top.

Of course, one always needs to remind Wall Street bankers, treasury officials, corporation presidents, and the like that Farmer Jones is their blood brother, and that at the end poor Clover cannot tell the difference between drunk and gluttonous farmer and drunk and gluttonous pig. But it's not much use. Once Snowball-Trotsky, trying to keep faith with *Animal Farm*, is run off the farm and into exile, everything goes the antileftist's way: the secret police, the new party elite, the old poverty and killing work, the turning upside-down of language, become the signs—the metonymies—not just of the Bolshevik revolution but of socialism itself. Richard Rorty points out that even by 1946 the available political lexicon was incapable of describing Stalinism.

> Efforts to see an important difference between Stalin and Hitler, and to continue analysing recent political history with the help of terms like "socialism," "capitalism," and "fascism," had become unwieldy and impracticable. In Kuhnian terms, so many anomalies had been piling up, requiring the addition of so many epicycles, that the overextended structure just needed a sharp kick at the right spot, the right kind of ridicule at the right moment.[6]

Orwell changed the language of political description. But then that new language was overtaken by his foes on the political right, who used it not only to congratulate themselves on their preferable method of exercising their own power but vigorously to denigrate as the practices of *1984* any political alternative that came along. Single-handedly, Orwell invented for the enemy a complete poetics of political invective.

It is a poetics of politics because it envisions an absolutely perfect system.

The totalitarianism of *1984* picks up the salient features of the Soviet system and idealizes them in such a way that we see in overwhelming close-up how they work, but also in such a way as to eradicate inefficiency, corruption, waste, and failure. The unforgettable apparatus of the ministries of Truth and Love, the Thought Police, the telescreen, the memory hole, Big Brother, and room 101, which contained the worst thing in the world, come together as the perfect structure of surveillance. The structure is given clarity and coherence by Newspeak, the circular political language in which it is impossible to distinguish truth from falsehood and impossible also to trace and record the moral values carried and transformed in that language. This is Kundera's and Kafka's dread theme: the politics of coerced forgetting.

Orwell's nightmare is indeed overwhelming. He lent piquancy to dreariness and made fearful conditions of life familiar. As soon as mass society had been recognized as such, American movies since Fritz Lang and novels since Sinclair Lewis had had the same nightmare about a world of collectivized anonymity and irresistible control of minds and spirits. Communism gave form and geography to that fear. Orwell showed what it would be like to live in such a way in England.

Many people at the time and since noticed how similar to the details of Ingsoc were the details of domestic life under Attlee's Labour government. Cold, shortages of fuel and food, dismal clothes, rationing, identity cards, bomb sites, and slums were the inevitable terms of life in London in 1947 when the economy failed to pick up and Britain tried to pay for both postwar reconstruction and its empire out of an empty checkbook.

In no time at all Orwell's iconography stood in for all state-directed policy, and Big Brother for any Communist dictator. There was a long gallery of models all across Europe. But Orwell's simple, vivid poetics went further than scenery and cast; his descriptions of the way the party machine organized its lying, of the contents of the illicit book, of Newspeak and indoctrination, have all become the stock in trade of schoolteachers alert to the ideological insinuations of socialism but entirely ingenuous about other doctrines and their devious little ways.

The persuasiveness of Orwell's poetics has left liberals cocksure about being in the right on the right ever since. Even as distinguished a liberal philosopher as Richard Rorty sees O'Brien, the character in *1984* who conducts the torture and breaking of Winston Smith, as Orwell himself saw him: the prototype intellectual of the totalitarian future, one for whom the old satisfactions of intellectual speculation, playfulness, and discipline—the classical forms of art and civilization—are focused and resolved by the keen pleasures of torture. Torture is the art of the state, and O'Brien the type of its artist. He is also how the future might go. He is wholly feasible.

What Rorty and Orwell alike underestimated was the sheer incompetence of the state, the certainty of corruption, and the social fact of rationality. Terror is the closing of the circle against rational objection and political argument.[7] To effect that closure you must have men like O'Brien. But as long as you do, as Stalin feared, they may go the other way, break open the memory holes, refuse to do as they are told.

The other answer to *1984* is the record of history. Winston Smith is a pretty poor thing, as one is oneself. He is abject in his grateful admiration of O'Brien. He is very frightened of rats and so can shriek "Do it to Julia" quite quickly. But there are countless tales of heroes and saints every bit as steely as O'Brien who outfaced other torturers and did not break down. They are the token that liberalism itself admires; less reassuringly, they are reminders that beliefs may indeed prove unassailable and that Stalinism may recur because some people are like Stalin.

Either way, *1984* was an overdetermined success of cold war. It taught mistrust of all grand systems of human optimism, but of socialism in particular, and it taught reflex distrust of power and the politicians who sought it. In this way, the book encouraged the mass withdrawal from public-spiritedness that sustained the stasis of cold war and nuclear standoff, and perpetuated the recurrent nightmare that history might come to an end and the long idyll of American prosperity with it. Its own style and the momentum of the moment made it available to every jack-in-office who wanted to make a political rally feel good on the cheap by denouncing the attentions of Big Brother. People who had never had a thought worth controlling, speaking themselves the unspeakable jargon of self-righteous politics, could wind on about thought control and the Thought Police and could accuse any old dissenter of Newspeak.

The trouble with the righteousness that shone so brightly from the breastplate of Americans in the first decade of cold war was that it so dazzled those who wore or saw it, they could not distinguish it from self-righteousness. Hence the one unmistakably great work of literature to come out of the period, Arthur Miller's *The Crucible* (1952), has been used to ratify individuality at all costs, and many of its deserved admirers have quite failed to see the harshness of its condemnation of all our bigotries (including those about our so very worthwhile individualities) that constitute our sense of ourselves.

The play's provenance is well known. When Kazan incriminated his friends as Communists to the House committee, Miller drove off to Salem to read its ancient archives and breathe life into the new cliché about witch-hunts.

The play counterposes the terrible self-righteousness of the witch-hunters, especially Hale and Danforth, the Massachusetts state officers, with the true righteousness of the martyrs. Miller forces that large number of Americans who know his play—from the theater, from the movies, and above all from reading it in high school—to go back to their most cherished myths about the homespun goodness and plain decencies of their Puritan fathers and to confront the real history of their hateful dogmatism, petty rivalries, and murderous superstitions. In a turn that is a mark of a bold dramatist, Miller even compromised the admirableness of his hero with the taint of adultery, original sin itself, and made John Proctor guilty as well as guiltless. And he concentrated the key moral struggle of his characters—between conscience and the social order, free speech and the preservation of belief and value, the very heart of the Constitution—in a wonderfully intense, bare, and biblical vernacular that is, as great poetry should be, at once ancient and modern, common and precise.

Miller's play was a detailed statement of those best American values that were under threat from both Moscow and Washington in 1952. Long after the disappearance of McCarthyism, *The Crucible* stands for a righteousness that fully deserves its victory over cold war.

PART II

The Balance of Terror

PART III

The Balance of Terror

7

The Rosenbergs and the Red Purge,
1950–1953

THE long, grim comedy of Berlin, the retreat of Chiang Kai-shek's men to Taiwan and the blockade along the straits, the Russian atomic bomb, and the near-total defeat in Korea after the first few months, the French on the run in Indochina, the Greek Communists only just flushed from the hills—with all this before him, it was not easy for Truman to see anything except the Red threat.

It would have taken a Roosevelt to overcome the feeling of a gathering storm. However well Marshall aid was going, however booming the U.S. economy and cocksure its world-scattered soldiers, the barbarian was waiting, biding his time. Supreme vigilance was called for, toughness, readiness, nerve.

In the early 1950s foreign policy became the most important domain of American political life and would remain so until the cold war ended. Truman, as he so disarmingly and frequently admitted, was no Roosevelt. He knew Washington, but he knew little of the world and could only try to follow in his master's footsteps. In trouble, he could only get tough, but he was no match for the toughs who wanted to wipe socialism out of the lexicon.

So with plenty out there in Europe or Asia to justify it, talking tough became the policy. Once NSC 68 and the Truman Doctrine were enunciated, once the state institutions were fully founded to circulate defense budgets and pass the projects for weapons research stretching decades ahead, along came the communicator ideologues to explain it all. These experts explained the balance of terror to presidents and to the public, and their audiences believed them. A total system easily described as totalitarian came into being in the United States, and the hideous stability of nuclear threats made good by local war became a way of life, within which a mute citizenry went about its private errands. To ensure that it could remain unpenetrated by alien ideas, the system began scrubbing the faintest tinges of pinkishness from its key offices: from the State Department, from weapons production facilities, and from the great story factories of the film industry.

It was a quite conscious exercise, put in hand by the men running Truman. But it was much encouraged by the world's most litigious state. The United States takes its laws seriously: The courtroom is the location for all the really serious American debates about the country's significance to itself. Hence the extraordinary prominence of the courtroom drama in the culture. During the Red purge, cultural imperatives and the national passion then running so high against the left naturally coursed together down judicial channels.

The Smith Act, in place since 1940, authorized the surveillance and enforced registration of all and any aliens; it had been used readily to initiate deportation procedures. Truman's attorney general from 1945, Tom Clark, was a Red-hating hard man indifferent to principles of justice when they clashed with the ethos of the country club or the maintenance of the president's serenity. As he blithely said early on in his bluff desecration of his country's great Constitution, "I ordered Mr. Eisler picked up because he had been making speeches that were derogatory to our way of life."[1] Clark was empowered to keep the attorney general's list of subversives; by the end of Truman's term, 300 people had been arrested for deportation.

It seems plausible to attribute to Truman motives for raising an extraordinary edifice of surveillance and oppression deriving as much from the hothouse of ideological neuroses in Congress as from any larger purpose of maintaining the world order. But David Caute's "great fear" could only take hold because of those half-crazy and collective neuroses.[2] For individuals, the vague fear of communism was connected with intensely personal fears of loss and menace: menace to the capital painfully built up against the family's own peasant or proletarian past; fear of the loss of position, of status, of Americanness.

These currents of feeling surged about Congress, and Truman was driven along with them. The House Committee on Un-American Activities, long the scourge of the old New Dealers, joined forces with the Senate Internal Security Committee and came up with the loyalty order. A compulsory oath to be taken by teachers and civil service workers, it was deprecated by the president but fervently supported by a whole political class in thrall to the national hysterics.

Committee generated committee, each stauncher than the last, each binding dissent into silence: the Senate's Permanent Committee on Investigations, its Loyalty Review Board, its Subversive Activities Control Board.

The energy of fervent anticommunism surged through the exorbitant anti-Soviet rearmament budget; it united a Republican Senate under a Democratic president. Given the savagery of the anticommunist feeling in the United States and the slamming of the bolts of the iron curtain all over Europe, one can only be relieved that the collective hysteria that boiled up and over between 1947 and 1958 was no worse. Lives were broken, names dishonored and defiled, and a very few people died. But if we jog our memories with the help of Solzhenitsyn's memoirs of the terrible gulag, then America's demented recourse to law to identify and extirpate moral error may look egregious, but it was rarely this agonizing to defendants:

> But the most awful thing they can do with you is this: undress you from the waist down, place you on your back on the floor, pull your legs apart, seat assistants on them (from the glorious corps of sergeants) who also hold down your arms; and then the interrogator (and women interrogators have not shrunk from this) stands between your legs and with the toe of his boot (or of her shoe) gradually, steadily, and with ever greater pressure crushes against the floor those organs which once made you a man. He looks into your eyes and repeats and repeats his questions or the betrayal he is urging on you. If he does not press down too quickly or just a shade too powerfully, you still have fifteen seconds left in which to scream that you will confess to everything, that you are ready to see arrested all twenty of those people he's been demanding of you, or that you will slander in the newspapers everything you hold holy. . . .
>
> And may you be judged by God, but not by people.[3]

II

The American system had, however, its solitary and awful chamber of electrocution.

Julius Rosenberg was the youngest son of fifteen children, born in 1919 to Polish-Jewish immigrants. Ethel, his wife, was three years older and from the same stock. At nineteen she led a walkout of women clerical workers and got fired; Julius, a dud electrician, had barely graduated from City College of New York and was an eager enlister in the Young Communist League. They were earnest, working-class leftists. They sold the *Daily Worker* door to door and lived poor, shabby, respectable lives in Brooklyn with their two sons in a rundown house with a little porch.

Ethel's brother, David Greenglass, knew a chemist in Philadelphia named Harry Gold, who was named as a runner of secrets by Klaus Fuchs. Fuchs, in turn was an important German atomic scientist exported to Britain in the 1930s who gave secrets to the Soviets and was then sent to jail for fourteen years. His hated name drowned the Rosenbergs by association and sent seismic waves of fear through the country.

In truth, the whole rank-smelling affair sent a shock through the liberal classes of all the self-styled Western democracies. It is a shock not yet subsided. But while others no doubt died from persecution, the Rosenbergs were the only two Americans to be executed on their own soil and by the state for alleged treason during the cold war. Even in a state so ready as the United States to bend the law to its ideological will, so tiny a death roll bespeaks some regard for individual rights, and some squeamishness over pressing hysteria into the service of military power.

Toward the end of the Second World War, Julius got a Signal Corps job supervising—at a not very senior level and with not a bad salary—the manufacture of the new electronic valves that would change the world and its weapons. The army sacked him, however, as soon as it learned he had been a Communist. When the Rosenbergs were put on trial in March 1951, Irving Saypol, then federal prosecutor of indicted Communists, pressed the charge of *conspiracy* to spy because, in the loose law of the time, a conspiracy charge admitted hearsay evidence.

The hearsay evidence of Mrs. Greenglass clinched the execution of the Rosenbergs. She said that in 1944 Julius told her that he wanted to assist the Soviets by telling them what he could about the new electronic science. Like Fuchs, Rosenberg was quoted as having said that he hoped his help would prevent the mystery bomb from ever being used by one of the great powers against the other. Of course, in 1944 there simply was no bomb, and Los Alamos under General Groves was as security-tight as it could be. But David Greenglass, pleading guilty in the hope of clemency and with a perjury charge behind him, must have persuaded the jury of the truth of his story that Julius showed him a sketch of the atom bomb and that he learned that Rosenberg had secretly received a decoration from the Soviets.

Evidence-conscious as we all now are, we have to say solemnly at this juncture that we cannot know for sure that Greenglass's story was all tarradiddle. But Greenglass was a known perjurer who was desperate to climb out of trouble; he was also an extremely low-level lens-grinder for the Los Alamos project. Julius Rosenberg was an honest sap. At the trial Rosenberg doggedly, ingenuously said that he felt the Soviets had made life better for the underdog, had restored the fabric of the country, and had helped destroy the "Hitler beast" who destroyed Jewry.

Then a second perjurer, Max Elitcher, testified that the Rosenbergs had collected a packet of film from a Communist party man he knew named Morton Sobell. Sobell, on this testimony alone, was sentenced to a truly Soviet nineteen years in Alcatraz. In his admirable history of the anticommunist hysteria, *Naming Names*,[4] Victor Navasky notes the readiness of the American Jewish Committee to come out against all these Communist Jews, and to declare again and again its All-American patriotism. The Rosenbergs bore their obloquy, too.

The Jewish judge in the case, Irving Kaufman, handed down a death sentence that, delivered with an intemperately self-indulgent virulence exceptional even among judges, startled and gratified the most vengeful Red-haters and god-fearers. Kaufman suggested that the Rosenbergs' obscure actions had directly caused the death of 50,000 of our boys in Korea.

No civilian had ever been sentenced to death for treason in the United States. The Rosenbergs were not on trial for treason; they had been indicted for conspiracy to spy. The spirit of the sentence was caught by Eisenhower's attorney general when, on 19 June 1953, he purportedly told his chief executive, after the Supreme Court had revoked a stay of execution by six votes to three, "Mr. President, those people have got to fry."

When the Rosenbergs were arrested, their small sons were left pretty well without care. They were moved from an inadequate aunt to an orphanage, and from time to time to the protection of the Rosenbergs' shapeless, generous, clumsy lawyer, Emanuel Bloch. Much later they were adopted by a kindly academic household. They would eventually return to attention in E. L. Doctorow's fictional version of their story, *The Book of Daniel* (1971).

The extent of the Rosenbergs' guilt has been bitterly contested. Although Kaufman's claims were fatuous and irresponsible, Ronald Radosh and Joyce Milton conclude that the pair were pretty certainly mixed up in espionage of some kind. But it can hardly be doubted from a position higher up on the hill of history that the death sentence was an expression of mob hysteria in which justice had no part. Doctorow thinks so.

Electricity flows in circuits. If the circuit is open or incomplete electricity cannot flow. In electrocution the circuit is closed or completed by the human body. My father's lips were sucked up between his teeth as the hood came down over his face, every last tremor of his energy gathered in supreme effort not to cry out. The hood is black leather and is offered in respect for the right of privacy at the moment of death. However, it is also possible that the hood is placed on the head to spare the witnesses the effect to the musculature and coloration of the face, and the effect to the organs of the face, the tongue, the eyes, of two thousand five hundred volts of electricity. My father's hands gripped the wooden chair-arms till it seemed as if they could squeeze them to sawdust. The chair would kill him but at this moment it was his only support. The executioner took his place behind a protective wall in a kind of cove. On the wall in this cove was a large handled forklike switch. The switch is thrown from an up position to a down position. The executioner looked through a glass panel, and observed the warden observing my father. Waiting a moment too long the warden turned to the glass panel and nodded. The executioner threw the switch. My father smashed into his straps as if hit by a train. He snapped back and forth, cracking like a whip. The leather straps groaned and creaked. Smoke rose from my father's head. A hideous smell compounded of burning flesh, excrement and urine filled the death chamber. Most of the witnesses had turned away. A pool of urine collected on the cement floor under the chair.

When the current was turned off my father's rigid body suddenly slumped in the chair, and it perhaps occurred to the witnesses that what they had taken for the shuddering spasming movements of his life for God knows how many seconds was instead a portrait of electric current, normally invisible, moving through a field of resistance.[5]

Perhaps the state, if not civil society, would have been more tolerant if it had not been for Klaus Fuchs.

III

Klaus Fuchs was the very big fish whose confession led to Gold who led to Greenglass who led, officially, to the Rosenbergs. Klaus Fuchs was born in 1911, the child of a devout and upright Quaker, clever, fervent, and solitary, suffused, as such a man was likely to be in Germany in the late 1920s, by socialist idealism.[6]

Known as a Communist party member, Fuchs was quickly marked down by the Nazis and instructed by the party to leave the country. With his family's support, this very reserved, passionate, and silent young man, with a great gift for theoretical physics, did so.

He shone at his subject. Neville Mott took him on at Bristol, then as now a university physics department second only to Cambridge. Fuchs remained a theological Communist, devastatingly ignorant of the practice of politics but magnanimous in his innocence. When he was roughly interned as an alien after the outbreak of war and transported to Canada with hundreds of other refugees from Hitler in a rusty, bucketing old freighter, he said long afterward, "I felt no bitterness at the internment, because I could understand that it was necessary and that at that time England could not spare good people to look after the internees, but it did deprive me of the chance of learning more about the real character of the British people."[7]

He believed the party line on the treason trials in the Soviet Union in 1937. When the British, in need of good physicists, dug him out of internment in 1941 and set him to work on the atomic bomb project, Fuchs at once looked around for a Soviet connection.

He found one quickly enough. From 1942 to 1945 he relayed information to Moscow by way of his quiet, competent "control," Sonia Beurton. His justification for this action was twofold, and no doubt shared by many of the idealistic leftists in Britain, some of whom became liaison officers for SOE like Frank Thompson, and some of whom were scientists like Rudolph Peierls and Peter Blackett who believed not only in socialism but also in the openly collegial nature of science. Fuchs held that the weapon on which he was working was so dreadful that no one nation should own it exclusively; that in any case communism was the name of a future in which there would be no war and all nations would learn to cooperate; and that science transcended national interests and of its nature compelled its practitioners to share all that they knew, whatever the politicians said. In 1945 these were not ignoble beliefs.

By 1950 he had, as he said, his doubts. He may have been an innocent, but what he had seen of Soviet methods brought him to the judgment that he had done wrong. He still believed that socialism would redeem the world, but not yet. He resolved to confess his espionage to the British intelligence authorities.

The secrets he had sent to the Soviet Union were serious stuff. Fuchs had been, successively, Peierls's research assistant at the University of Birmingham on the British atomic bomb project, code-named "Tube Alloys," in 1942; from 1943 to 1945 Peierls's deputy at Los Alamos in the implosion dynamics group, answerable directly to Oppenheimer; finally, head of theoretical physics at the British Atomic Energy Commission at Harwell near Oxford in 1950. At regular intervals throughout these years he was sending descriptions and analyses of the very latest nuclear research in the United States and then in Britain to the Soviet Union.

There is still perfect inconclusiveness about how much his disclosures mattered. According to his official confession, as edited by the FBI, Fuchs at first disclosed only what he was working on himself in 1943, but his contact seemed in some respects to know of more advanced work than he did. He subsequently and fully described the design of a plutonium (atom) bomb but did not include instructions for its manufacture because he did not know the engineering and industrial processes involved. As his misgivings grew between 1947 and 1949, he provided more limited fissile state mathematics but withheld such crucial formulas as the efficiency calculations. He was, incidentally, struck by how much the Soviets already knew; he was certain that when their first successful testing took place in 1949, they could not have had time to use what he had told them. (It is worth adding that his British interrogator was convinced that Fuchs said this out of surprise and not in an effort at self-exculpation.)

One charge, widely made and still believed, is false. Fuchs could not have given the Soviets any help with the projected super bomb, in which nuclear fission would be used to trigger nuclear *fusion,* turn hydrogen to helium, and release vastly more destructive energy than those superannuated siblings Little Boy and Fat Man put together. John von Neumann at Princeton reported in 1950 that the mathematics would not work. It was another year before Edward Teller cracked open the puzzle. But the fact that Fuchs could not have sent the Soviets hydrogen bomb secrets did nothing to prevent the notoriety of the case being used to justify all possible haste on the hydrogen bomb project—to get ahead of the Soviets and, more or less, to send the Rosenbergs to their death.

IV

The Fuchs case, which was zealously played down in Britain because it made the never-very-bright intelligence authorities who had vetted and cleared him *six times* look such monkeys, brought out the worst kind of anti-intellectualism in American culture. It made all physical scientists, Nobel Prize winners or not, fair game for abuse, interrogation, and arbitrary suspension in connection with national security. Meanwhile, all scientists became the more liable for their well-known scholarly habits of Pyrrhonism, atheism, criticism, socialism, and God only knew what else.

The anti-intellectuals' greatest catch in the United States was J. Robert Oppenheimer, formerly director of the Los Alamos laboratory during the Manhattan Project and by 1950 chairman of Truman's atomic energy advisory committee and director of the Institute for Advanced Study at Princeton. Oppenheimer, like many of his scientific generation, was a genial supporter of leftist causes and had collaborated in the advisory document commissioned by Dean Acheson that preceded the Baruch plan to shape atomic energy knowledge. He was a gregarious man of exceptional character and greatness of spirit. Freeman Dyson was a friend and colleague of Oppenheimer's from 1948 to 1965.

Of course, the appalling J. Edgar Hoover and the FBI had been after Oppenheimer since Los Alamos. He had defended scientists who were security suspects, he had told the air force that its nuclear strategic plans were negligent, dangerous, and absurd, as indeed they were, and he had crossed Adm. Lewis Strauss, a very big and self-important noise on the Atomic Energy Commission (AEC). By 1953 Strauss was the AEC chairman; from the moment of the first Soviet atomic bomb in 1949, he had been devoted to a policy of the more nuclear weapons the merrier.

Under pressure as a security risk, Oppenheimer resigned from the AEC. Documents that had been cleared for his retention were taken away from Princeton. His wife had been a Communist in the thirties; a list of his ex-communist juniors and appointments was compiled. The most distinguished scientist in the United States had his phone tapped, his mail read, and his lawyer followed. The bloodthirsty senators prepared to pounce on so eminent a name.

Oppenheimer testified to the house committee that he had "been an idiot" in his views, and in his clumsy protection of one or two friends, although like most academics under interrogation he never incriminated other people. Great names lent their names to support his. But one scientific name, already imposing in 1953 and destined to become celebrated if not notorious, spoke against Oppenheimer and became irrevocably stained by doing so. Edward Teller, begetter of the hydrogen bomb and for years king of Mount Livermore's golden laboratory, testified before the inquiry: "I thoroughly disagreed with him in numerous issues and his actions frankly appeared to me confused and complicated. To this extent I feel that I would like to see the vital interests of this country in hands which I understand better, and therefore trust more."[8]

Teller spoke at once truthfully and disingenuously. But with his declaration, made as a Hungarian who feared Stalin as the Jewish Oppenheimer feared Hitler, Teller lost his good name and retained only his eminence. Admiral Strauss ruled that Oppenheimer had been guilty of lies, evasions, and imprudence, and the doyen of the first atomic bomb retired from public life.

V

Stardom is so central to the political economy of the United States that the history of the anticommunist hysteria offers itself naturally as a series of stories about starring parts, each played in the courtroom: Fuchs, the Rosenbergs, Oppenheimer, Alger Hiss, Owen Lattimore, Lillian Hellman, Ring Lardner, Arthur Miller.

Alger Hiss, like the Rosenbergs, still stands in an obscure position between innocence and guilt in the case history of American law. He was an assured Harvard lawyer who had been a senior civil servant in several offices of the State Department, a member of Roosevelt's entourage at Yalta, and head of the Carnegie Endowment for International Peace. He was an Acheson-style figure: Savile Row suits, Boston accent, New Deal formation, tall and upright. But the torrents of class hatred running below the surface of mainstream anticommunism plucked him down arbitrarily, sucked him under, and drowned him.

A sort of Quasimodo was after Hiss, the villain in a new political part: the self-confessed Communist informing on the secret of his comrades in the name of apostasy and a free America. Whittaker Chambers, frowsty, corpulent, bespeckled with food and cigar ashes, was all that the audience wished him to be. Speaking with a florid, kitsch eloquence, he named Hiss to the House Committee on Un-American Activities. He led two of the committee's heavies to his back garden a few miles south of Washington and, reaching triumphantly into a pumpkin gourd, held aloft the microfilm passed to him, he avowed, by Hiss for forwarding to the Communist party.

Hiss denied the espionage charge before a grand jury in 1948. At the end of the year he was indicted for perjury. Chambers became more and more volubly circumstantial. There had been fishy business with a typewriter. A confessed Communist, Julian Wadleigh, testified that it might well have been he who passed the microfilms to Chambers, but all his evidence did was endorse Chambers's steamy picture of a State Department seething with spies.

After two trials and much sanctimonious comment about New Deal liars and crooks by the young Richard Nixon, a member of the prosecution team, Hiss was sentenced to five years. Twenty-odd years later, the excellent Victor Navasky tells us drily, the three microfilms withheld from the trial for security reasons were prized out of the district attorney's office; one was blank, and the other two carried high-tech information about naval fire extinguishers.

VI

Hollywood was the biggest bust of all during the ten or so years of full-blooded Red purge. It was inevitable that so rich, influential, remorselessly maudlin, and self-referential an industry should have written itself a juicy part.

Ring Lardner was as good a screenplay writer as ever went to Hollywood. He was part of a political climate that held sunnily in Hollywood from the late thirties until the beginning of the cold war: the best liberal values of the New Deal happily combined with wartime anti-Fascism and a decent, unquestioning, low-key sort of progressive patriotism. In 1948 its adherents included the writers Lillian Hellman, Clifford Odets, and Albert Maltz, the directors John Huston and Elia Kazan, and the actors Edward G. Robinson, Humphrey Bogart, Katharine Hepburn, and Gene Kelly.

It all began in 1947. As was no doubt appropriate, it was Hollywood that turned the Red purge into living theater and mythologized political choice, creating a narrative that more than anything else has given form and meaning to our epoch.

The House committee's arraignment of the Hollywood left memorably dramatized the anticommunist decade. The committee issued subpoenas to those it knew had some sort of radical past, part, or production on their résumés. The dramatic tension lay in discovering whether the accused would name friends, colleagues, and associates. The thrill in the loins (Arthur Miller was surely right to give the witch-hunt in *The Crucible* such a smell of sex) could be felt either way: when Lillian Hellman refused to name her associates, or when, cravenly, Elia Kazan did.

The peak of the drama, to which the suspense led, was the moment of self-cleansing avowal: the defendant naming names. To name was to be exculpated; to refuse to name was to prove yourself guilty. The First and Fifth amendments, the great pillars of the most carefully written constitution in the world, were no defense. To invoke them was to deepen the miasma of eager suspicion around you, palpable as a smell.

The Hollywood hearings had some splendid parts. J. Parnell Thomas the raucously bullying chairman of the committee, a man of perfect and apoplectic obduracy, whose notion of the grave decorum of the law was compromised by a Wild West view of its possibilities that only Hollywood could have imagined. Thomas had armed guards on tiptoe at whom he bellowed orders to bustle out recalcitrant witnesses whenever his peppery temperament boiled over.

The president of the Screen Actors Guild, Ronald Reagan—the leading player of the final station of the cold war—with his own inimitable combination of sincerity and hypocrisy, incriminated his own trade union members, who had entrusted him with representative office on their behalf. Perhaps my favorite witness is Walt Disney, who, David Caute reports, bore witness that the Cartoonists Guild was trying to lead Minnie and Mickey Mouse and the faithful Pluto into the socialist paradise.[9]

The Hollywood Ten, the first such cohort in postwar American courtrooms, were duly sent to jail. They included Ring Lardner, Albert Maltz (*The Pride of the Marines* [1945]), and Edward Dmytryk (*Hitler's Children* [1943], *Crossfire* [1947]). In a delicious grace note, Lardner met the choleric Parnell Thomas in jail, where he was doing a bit of time for embezzlement of House funds. They did not speak.

But the Hollywood Ten were not the last victims of this hysterical campaign. The House committee then reached beyond Hollywood to Broadway, where it found the only writer of genius who would be dragged into its sewage and have the odor of obloquy cling to him. He cleansed himself with the best astringent: Arthur Miller wrote *The Crucible*, his great work of art, the only one to come out of a shameful moment.

The House committee returned to Hollywood in 1951. This time it went straight for individual targets. To be named by a witness was to be blacklisted and excluded from work for a long time, ten or more years, sometimes for life. The inquiries were noisily encouraged by the poetically named Motion Picture Alliance for the Preservation of American Ideals, with John Wayne as its chair, backed by nice guys of the trade like Clark Gable and Gary Cooper. Ronald Reagan introduced a loyalty oath for the Screen Actors Guild.

The new inquiries pulled in a rich haul. Navasky quotes Sterling Hayden's long self-disclosure to his analyst, a document of unremitting comicality even when we recall that Hayden was not at all the worst of them; he was only trying to hang onto his access to his kids and preserve some vestiges of self-respect as, still, a parlor pink. The transcript from his couch makes a hilarious read:

> "Doc, I can't go through with it. Since the subpoena two weeks back I've tried and tried to convince myself. They know I was a Party member—they don't want information, they want to put on a show, and I'm the star. They've already agreed to go over the questions with me in advance. It's a rigged show: radio and TV and the papers. I'm damned no matter what I do. Co-operate and I'm a stool pigeon. Shut my mouth and I'm a pariah."
>
> "I suggest, Mr. Hayden"—the analyst's sober voice—"that you try and relax—just lie down. . . . Now then, may I remind you there's really not much difference, so far as you yourself are concerned, between talking to the FBI in private and taking the stand in Washington. You have already informed, after all. You have excellent counsel, you know, and the chances are that the public will—in time, perhaps—regard you as an exemplary man, who once made a mistake."[10]

But as Hayden and Lee J. Cobb and Edward G. Robinson and Clifford Odets caved in, the fingers of suspicion pointed across the continent and, falling on the shoulders of Elia Kazan, reached forward toward Arthur Miller in New York. Miller's prose is quite up to the melodrama of the moment.

> I saw the civilities of public life deftly stripped from the body politic like the wings of insects or birds by maniac children, and great and noble citizens branded traitors, without a sign of real disgust from any quarter. The unwritten codes of toleration were apparently to be observed no longer. . . . If under the pressures to go to the right I moved even further left for a time, it is explicable, if it is at all, as a willful act of self-abandonment and

defiance of my new-won standing in the world. Respectable conformity was the killer of the dream.[11]

But Kazan, maker of the films *A Streetcar Named Desire* (1951) and *Viva Zapata!* (1952)—the last named not the most reassuring work for the touchy souls on the House committee—was very frightened. He refused to name names. Then he rang Miller.

Kazan had produced the most important versions of Miller's *All My Sons* (1947) and *Death of a Salesman* (1949). He was a Greek immigrant who had made it to the top according to America's classic protocol: by unstinting hard work, genius, and luck. He paid his way through Yale and Williams, joined the Communist party and Lee Strasberg's Method school, directed winners on Broadway, and wrote a laudatory play about Bulgaria's lethal commander-in-exile, Georgi Dimitrov.

He was jam for the committee. He phoned Miller in April 1952. He had said no to the committee, but they had put the screws on; so now he was going to say yes, and he wanted Miller to agree that that was just fine.

Kazan was going to sing a full-throated song, and confirm a dozen names. (In the hearings, before a committee laden with prior reports from the FBI, confirmation was the crux of a confession. This is the deep trope of *The Crucible*.) As Miller said later, Kazan was a genius of the theater. Miller thought that it was—as the self-pitying rhetoric of the successful always puts it—unthinkable that he should be broken by a refusal to tell those ghouls what they knew already, and "at the height of his creative powers" never make a film again. Kazan, however, if he needed to, would have ratted Miller out as well.

Kazan sang all right, and Miller drove straight from the meeting to Salem, in steady gray rain, to begin work on *The Crucible*. Miller, in his plays as in his life, was the ambiguous Turgenev of the whole business. As Isaiah Berlin says of that great original,

> The situation that he diagnosed in novel after novel, the painful predicament of the believers in liberal western values . . . is today familiar everywhere. So, too, is his own oscillating, uncertain position, his horror of reactionaries, his fear of the barbarous radicals, mingled with a passionate desire to be understood and approved of by the ardent young. . . . The figure of the well-meaning, troubled, self-questioning liberal, witness to the complex truth, which as a literary type, Turgenev virtually created in his own image, has today become universal. These are the men who, when the battle grows too hot, tend to stop their ears to the terrible din, or attempt to promote armistices, save lives, avert chaos.[12]

Miller combined the scrupulous inefficacy and moving eloquence of his poetic ancestor. It is fitting enough that one can now read *The Crucible* as that most traditional of American myths: the honest man holding out against the graft and corruption of the big system. It is fitting because Miller himself most wished to recall his much loved and abused America to this best version of itself, before it was lost to view.

His later plays record his helplessness and dismay as it disappears over the edge of the falls.

VII

Sen. Joseph Raymond McCarthy is the all-American gangster-bully of this part of our story, as everybody knows. But he joined the action late and made less of a contribution to anticommunist hysteria than Truman or Tom Clark or even the prim *New York Times*. It is clear that he was more interested in his own publicity than in fulfilling a historic mission. But he gave his name to the era of the informer, however much his eruption onto the stage was late and opportunistic.

When McCarthy began early in 1950, he reared monstrously up holding "a piece of paper with the names of 205 Communists presently employed at all levels in the State Department."[13] He had bulldozed his way into the Senate on a Red-hating ticket in the 1946 election.

He was the simplest version of a memorable kind of American public man: the gangster-populist of the saloon bar and Tammany Hall, the hard-drinking, big-hearted, loud-mouthed fixers of late nineteenth–century politics. He adored Teddy Roosevelt and abominated Teddy's namesake. He loved dollars and gambling and women, flashy women with amazing hair and loads of money. He sweated and fought dirty. He made fantastic allegations on no evidence at all and lied serenely about the evidence he did have. He had been a colorful, reckless, boozy war hero, too, and that was part of his appeal, his irrepressibility and hard-to-hate swank. He bullied and lied, but he touched a nerve in Americans and ran the thumbnail of his barroom eloquence down the nerve, so that people quivered a bit, and backed him.

He went too far in the end, but what the hell, he had a good run for his money. He got himself a committee, the Senate Permanent Subcommittee on Investigations, and two nasty hangers-on, Roy Cohn and his dim partner David Schine, who raked through the muck for him and came up with the dirt about infidelity and homosexuality that made Miller's combination of sex and communism, the forbidden, dark declivities, ring truer than ever.

McCarthy did not, according to Richard Rovere, care much about communism.[14] He cared about the public noise and the coarse public drama, about picking a fight with intellectuals and Bostonians and pacifists and all the spineless crew who would render American manhood goosey and gutless. "Let me assure you that regardless of how high-pitched becomes the squeaking and screaming of left-wing, bleeding heart, phony liberals, this battle is going to go on."[15]

By 1952 he was so engorged with his own audacity that he called the mighty, taciturn, and gentlemanly George Marshall, author of the plan itself, a traitor. Acheson and Lattimore and their like had long been vilified, but with Marshall he went too far. Eisenhower, always cautious and sometimes craven, trod warily around McCarthy until he was elected president and did not stand by Marshall, his great mentor. Well into his first term, Eisenhower was still prepared to deny free speech to the extent of proposing that Communists be denied their citizenship.

McCarthy clattered on his destructive way. He ruined careers by the mere mention of a name. People lost their jobs or went under when he claimed that they had Communist associations. Between 1953 and 1955 the State Department received an astonishing 273 resignations.

When he brought his charges against the U.S. Army, however, the Army paused, and then broke him. McCarthy illegally suborned spies in the military, claiming that secret sources and confidential evidence had revealed to him that even in the Pentagon there were leftist helots and incipient traitors. In a last, extraordinary display of bluster and not unendearing braggadacio, he performed like an outlaw of the Wild West, flouting protocol insolently, his face pitted and shadowed by his black bristles, his stubby fingers punching the air. But the Army, quite rightly, wouldn't have it. It made discreet representations to the Senate, and the Senate, in the gradual American way, put McCarthy on a leash. He shrank and dried out, and before long, he died.

VIII

If Korea is a forgotten war, the Red purge is its forgotten propaganda policy at home. Yet it was an ugly business and struck deep roots. The lies and madness, the cowardice and mad rage, the hypocrisy, retaught American society all the old lessons about commies and roused the ghosts of long-dead Reds like Eugene Debs to terrify God's Americans once again.

There is still so much sentimental reminiscence about the bustling, perky, and honest Truman, and about Eisenhower and the quiet eight years of his presidency. But Truman launched the loyalty program and, like Mark Anthony, ordered his stooges to get on with the work of oppression while pretending to know nothing about it, in fact, being inclined toward tolerance himself ("Fool Lepidus"). The old tale goes that McCarthy ran the purge; but it was well established by 1950, and he only brought it the notoriety of famous names, gave it the stardom of Lattimore, Oppenheimer, and Linus Pauling and the slogan "twenty years of treason in the state department." Eisenhower kept quietly on with the purging process, curbing McCarthy's mouth over George Marshall but still cutting deep into the radical veins of the culture and letting them bleed dry.

The eradication of the Wobblies had pretty well stopped socialism as a political force in the United States before 1930; the short World War II alliance with the Soviet Union revived it. With the will of two presidents behind the effort, socialism was again excised from American culture after 1946. The intelligentsia was subdued, and its pupils, the schoolteachers, after being questioned and sacked when necessary (Caute records 321 teacher firings in New York City alone), were silenced utterly.

The unions were even more comprehensively stripped of the socialist vocabulary that in Europe had been their native speech. They readily ac-quiesced in the removal of their own tongues; Walter Reuther of the United Automobile Workers (UAW), the most powerful union leader in the country, entirely cooperated in Truman's project. The Taft-Hartley Labor Act (1947) made party membership illegal for union officials. Recalcitrant locals were brought to heel by withholding the rich defense contracts, and the United Electrical, Radio and Machine Workers (UE) (Julius Rosenberg's union) was harassed by the FBI, subpoenaed by the House committee, and split by a rival union set up by the General Electric bosses until, enfeebled and leaderless, it capitulated.

It was a precise campaign, at once national and local. David Caute tells bitter tales of employees dismissed on trumped-up charges of security lapses in the mammoth defense industries, of noncommissioned officers stopping UE members doing their job on similar excuses, of an orchestrated customer complaint movement to firms with large UE labor forces. Bright young libertarian sparks in the Democratic Senate, like John F. Kennedy and Hubert Humphrey, won their spurs by showing anticommunist zeal on the House Committee on Education and Labor and the House Committee on Labor Relations.

The enormous apparatus of the state, accurately applied without remission to the pressure nodes of working-class dissent and self-organization, eradi-cated socialism as anything but a swear word and an academic allegiance in the United States from 1947 to the present day.

So there is more to the Red purge than Joe McCarthy, J. Parnell Thomas, and a passing hysteria understandably set off as a response to Stalinism. The combination of popular passion, rigging of the legislature, and the hysterical complicity of the press turned the state into the liberal-capitalist version of totalitarianism.

U.S. totalitarianism was much less deadly than Stalinism, but more efficient. Its instrument was the law, not the secret police; its interrogations took place in courtrooms, not cells; its victims became unemployed, not dead. But it lied, bullied, and hectored until it had released a strong tide of feeling that swept away all opponents and set all its creatures swimming in the same direction.

A long American quiet in the cold war supervened after Korea until Cuba; the row was all in Europe between 1953 and 1961. But *la grande peur* was not quieted until Richard Nixon fell. And it is not dead yet, not by a long way. Anticommunist fever is carried inertly in the blood until a nightmare quickens it, and it stirs, and sends a country mad with its heavy poison.

8

EVENTS V

Budapest, Warsaw, Suez, 1956

THE Yalta fault line twisted along the frontiers of the Soviet satellites, crossed the Balkans and an ambiguous corner of the Mediterranean, and reappeared along the length of the rock-strewn deserts of what the Western powers who arranged futures for other people had always referred to as the Middle East. In the days before radical Islam, the cracks ran across Israel, naturally, and then followed the oil pipelines across to the Persian Gulf.

In 1956 the fault rumbled in Poland, in Hungary, and on the barren, disputed, and holy sands of Sinai.

In February the twentieth congress of the Soviet Communist party had conferred the succession on Nikita Khrushchev, who, as we shall see, had his points. He said himself of Lavrenti Beria, Stalin's police chief, that "he was a really awful man, a beast, to whom nothing was sacred. . . . Not only was there nothing Communist about him, he was without the slightest trace of human decency."[1] Khrushchev had his decency. He was set upon "de-Staliniization," upon denouncing what he called "the cult of personality" as a crime against the human race.

There were contradictory pressures on him; although he was first secretary of the party, his position was not certain, and his rivals, even if less murderous than in the dark days, were well able to gang up on him. Edward Crankshaw reckons that his enemies thought Khrushchev would be finished merely by speaking against Stalin.[2] In the event, he utterly denounced his dead master and, emerging triumphant from the subsequent tumult, would be credited with the mildly liberal policies of post-Stalinism.

Khrushchev made his first, devastating report on Stalin at a secret session of the twentieth congress. It was the first congress held in three years. The new leaders had to break with the fearful past, and to get all the opprobrium unloaded onto Stalin's corpse they had to do the deed quickly. There was no knowing how disclosure would turn out.

But the white light of Khrushchev's revelations could not be kept secret. The *New York Times* carried a summary in mid-March 1956. Allen Dulles, head of the CIA, had a copy smuggled out of Poland by the end of the month.

A few years later Khrushchev spoke genially about his attack on Stalin:

It was supposed to have been secret, but in fact it was far from being secret. We took measures to make sure that copies of it circulated to the fraternal Communist Parties so that they could familiarize themselves with it. That's how the Polish Party received a copy. At the time of the Twentieth Party Congress the Secretary of the Polish Central Committee, Comrade Bierut, died. There was great turmoil after his death, and our document fell into the hands of some Polish comrades who were hostile toward the Soviet Union. They used my speech for their own purposes and made copies of it. I was told that it was being sold for very little. So Khrushchev's speech delivered in closed session to the Twentieth Party Congress wasn't appraised as being worth much. Intelligence agents from every country in the world could buy it cheap on the open market.[3]

The Poles, in character, told all their senior members. Antonín Novotný in Czechoslovakia and Mátyás Rákosi in Hungary were aghast; they saw plainly enough their own likenesses limned in the polemic penned by the Presidium and embellished by the First Secretary. So they paraphrased it gingerly, and in their countries, the cult of personality turned out to be less a crime and more a deviation from the lessons of Lenin.

Khrushchev undoubtedly weakened the hideous strength of Stalin's structures of power. He allowed official Soviet policy toward Tito to warm up a bit; he was content enough that the people's democracies along the frontiers of the Soviet Union should work out their own economic if not political salvation.

But he remained the revolution's man. Edward Crankshaw speaks of him as being "naturally a violent man (his violence would still break out even when he had become an international statesman) and violence was the mood of the times."[4] Khrushchev knew that things couldn't go on as they had under Stalin.

But he also knew that less than forty years after civil war and famine, merely ten after world war, famine again, and (say) twenty million dead, the job of putting the Soviet Union back together was barely thinkable. Haste and violence could easily masquerade as decisiveness and, in Khrushchev's vocabulary, be readily justified. He was, after all, Stalin's man as well.

II

Khrushchev reanimated the discourse of state socialism with a vigorous new triangle of concepts—"cult of personality," "rehabilitation," and "peaceful coexistence." The exploration of these ideas certainly made for intellectually easier times, even if there still were not enough overcoats to go around. Taken with the startling visibility of the new Soviet leader, who dashed from one Soviet capital to another to bully, cajole, conjure, and joke with the comrades of the empire, it is no surprise that two of the satellites—Hungary and Poland—took these signs as wonders and started out on a different path to socialism.

Rehabilitation included Yugoslavia, and if Khrushchev wanted new collaboration with Tito, then anything was possible, reasoned the progressives in Budapest and Warsaw. They sought to follow Tito's pioneering path. While remaining faithful to the precepts of revolutionary socialism, as well as acknowledging the mouth-filling laws of historical materialism, these progressives hoped to use a bit of their local knowledge and recover some of their own traditions.

In Hungary the party leader, Rákosi, felt the new wind blow cold; he had dug himself into a dutifully Stalinist position, and change was impossible without looking foolish. Once a totalitarian Communist leader looks a fool, he is finished. The softer totalitarianisms of the West allow fools to stay out in front.

Rákosi's competitor was his vice-chairman, Imre Nagy, who had long and openly criticized his leader's ludicrously dogmatic insistence on heavy industrialization of a small country with very little iron ore or coal. After Stalin's death, when Nagy became almost Rákosi's equal, he set in train radically liberalizing measures all through Hungarian society: in production, certainly, but also in the press and the academies, in freeing political prisoners, and in abandoning collectivism. Nagy went too far; in 1955, in a routinely sickening display of duplicity, Rákosi threw Nagy out.

But the damage had been done. This was more than a headline struggle between the grim, gray Stalinist Rákosi and the idealist, socialist, and cultivated intellectual Nagy. Hungary has always cultivated an active intelligentsia, which took Nagy's part, even after his explusion. As Rákosi tried to steer his way between the contradictions of faith in Stalin and obedience to Khrushchev (who detested him), he tried rather too casually to exculpate himself for the execution of an earlier rival, Laszlo Rajk, whose name was inconveniently up for rehabilitation in Moscow.

The Hungarian intelligentsia was outraged. Rákosi tried to hold them off by giving official status to a wide coalition of reformist groups called the Patriotic Popular Front which had massed behind Nagy. When at last his supine parliament began to gang up on him, he resigned.

That was, however, only the beginning. As always in totalitarian regimes of this sort, characterized as they are by intense ideological self-consciousness, once a crack appears in the monolith of ideas, the water gushes through. In 1956 Hungarian and Polish intellectuals were swept along by a vision of socialism with a human face, with a program for the empowerment of the silent masses, with—as Kierkegaard put it—"the purity of heart to will the one good thing" into existence, a just and truthful state.

In the heady current of revisionism in Hungary, new journals and newspapers called in detail for reform. Rákosi tried to put down the argument but did not have the Kremlin's backing for a new wave of arrests among Nagy's circle of supporters. At once dithering and instinctively repressive, Rákosi was bluntly told by Moscow to resign in July.

His replacement, Ernest Gero, was not much better. Intellectual criticism of and popular resentment at the continued policy of forced industrialization were never placated, and Nagy, a staunch socialist, remained the unofficial legislator of public opinion. Opinion, indeed, was his only weapon, together with the reputation he shared with many men who later became pure tyrants—Dimitrov, Hoxha, Kim Il-Sung—for having been a brave anti-Fascist when it counted. His declaration of faith made to the party's central committee on 4 October still sounds a chord to all those who still believe, after 1989, that the best values of socialism have a future, may yet, indeed, *be* that future. "My place is in the party where I have spent nearly forty years, and in whose ranks I have fought to the best of my ability with gun in hand . . . or by means of the spoken word or the pen, for the cause of the people, the country and socialism."[5]

Dissent mounted. On 6 October there were public funerals for the rehabilitated victims of the state (that is, Rákosi's victims): the hapless "eliminations," that horribly precise abstraction for the act of firing a bullet at close range into the base of the skull. Huge crowds walked peaceably through Budapest behind Nagy. Two weeks later, while party leaders were away anxiously toadying in Moscow (always an ill-advised move for leaders in trouble—the palace putsch takes place when the prince is off on a jaunt), the Petofi Circle, a group of intellectuals named after a hero-poet of 1848, published a modest manifesto.

Their students promptly published an immodest one. Without asking anybody, they called for the withdrawal of the Soviet troops, multiparty elections, revolution. They were fired by the images of 1848, like many unpolitical people before and since. They wallpapered the city with their demands; they painted its slogans, in those pre-aerosol days, in big, red, dripping letters on blank white walls. The rivulets of paint ran like blood.

The students were, heroically, looking for trouble, as their brave predecessors did in 1848, and trouble was indeed what they found. Without such troublemakers the crowd would never have broken onto the stage of history at all, and all our freedoms would be the less, as would our death tolls.

In his novel *News from Nowhere* (1891), William Morris points out quite levelly that a liberalizing government trying to mitigate oppression and respond to popular and usually vague aspirations for something better must be ruthless and decisive. If it dithers, it will fall. Gero, back from Moscow, dithered. The central committee elected Nagy prime minister but left Gero as party chief. It also, almost as an afterthought, asked for help from the Soviet troops in policing the streets and put its own conscripts on alert.

The heavy-treaded tires of Soviet armored cars warbled on the cobbles. Nagy was half in charge but caught between the Kremlin and the people, with authority over neither. He hovered cautiously between the two, and the Kremlin's traveling salesmen, Mikoyan and Suslov, came and went, after installing Janos Kadar as Nagy's deputy.

But the Budapest streets were still restless places filled with shouting students. Beyond the capital, worker-soviets took possession of factories, and even local governments, according to Bolshevism's classic protocols. They put up with Nagy, but the mood of the capital and the nation had gone far beyond the spirit of his genteel revisions. Footloose lads were making pikes again with which to break a few police heads; the graffiti on the walls became bolder; a wild excitement coursed through the people's blood, and the students wanted more and more.

The Soviet soldiers waited with their Hungarian comrades for firmer orders. When they regrouped in the suburbs, it looked like withdrawal. Nagy sighed with relief and was at once deafened by those many voices belaboring him to do away with one-party rule, with communism, with the Russians, with everything. Kadar kept his own company. Nagy consulted neither the police nor the army about keeping the peace; in truth he was being rushed he knew not where by the flood of insurrection. Hungary seceded from the two-year-old Warsaw Pact.

The Soviets were baffled. Khrushchev certainly was not going to pull out his troops, but he did not want to practice crude coercion only months after

Figure 8.1   A citizen of Budapest burning a poster of Stalin, 1956. (The Hulton Picture Company)

his twentieth congress speech. The city of Budapest was on a tremendous jag. Probably the most famous bit of newsreel shows some intrepid young men climbing out on the pediment of the main government administration building and slowly prizing the Soviet star that crowned it free of its frame. The tin star, big as a man, leaned forward, swung out, toppled, and fell, turning and wheeling against the facade of the building until it crashed heavily on the street below.

Khrushchev went to Belgrade to consult with Tito; he was anxious not to damage the new rapprochement with the eminent deviationist. To his surprise, Tito agreed that the Soviets would have to put down the Hungarian insurrection. The Soviet interpretation of events was not, of course, that the Hungarians had risen spontaneously against foreign occupation but, following dogma, that counterrevolutionaries were subverting socialism from within. In his reminiscences, Khrushchev uses all the routine cant of Soviet political theory. He apparently believed some of what he says and acted in the kind of faith allowed for by an all-inclusive ideology that fuels blindness and unreason. All leaders do it, perhaps must do it.

In the cold, dark dawn of 4 November 1956, the Soviet tanks began their terrible, screeching trundle in from the suburbs of Budapest. The Hungarian generals who were known to support Nagy were arrested. The tanks swept away the feeble barricades the students and their supporters had put up against the troops. They fired their heavy 12.7mm machine guns against the tiny weapons of the rebels and killed more than a few, but they used the T-55's big turret gun only a very few times, against the hopeless, gallant fighters bolted behind the gates of a handful of factories in Budapest and the other big towns and against the miners at Pecs. The workers and students fought with

Figure 8.2   A secret policeman shot by his compatriots in Budapest, 1956. (The Hulton Picture Company)

that courage against all odds learned from a century of guerrilla defeats, but they never had a chance. The tanks, the absolute symbol of Soviet power from the beginning to the end of the cold war, were unstoppable. The resisters had no leaders, no organization, and no strategy: they believed that their high hopes and dauntless idealism would see them through.

But the peasants went on working in the fields. A few Hungarian soldiers, not many, joined the insurrection. The students and the factory workers were shot down or rounded up in no time, and a handful of doomed athletes scrambled across the rubble to snipe uselessly at the armor. Caught by the cameras of Western journalists, they would be changed utterly into the deathless, beautiful image of the patriot-partisan, holding back an army with a pack of grenades.

Marshal Ivan Konev, supreme commander of the Warsaw Pact armies, told Khrushchev that he would need only three days to crush the counterrevolution, and he was, as a good soldier should be, deadly accurate. The last hope of the rebellion was the capital's radio, the symbolic center of any attempt to overthrow a state. Some nameless student-romantic continued his broadcast to the West as the gunshots and the sound of boots came closer and closer; he appealed desperately for help in the name of the freedoms for which he gave his life. The listening West did not move.

Nagy was exiled, disgraced, and, in 1958, executed. Kadar took over, a safe, prudent man untainted by association with Rákosi or Nagy. He received a great deal in aid from Khrushchev and started his country on its long, wary search for a socialism it could call its own.

There is a minor but not unimpressive epilogue. In 1957, only about six months after the shooting stopped, Khrushchev went on one of his infamous trips to visit Hungary, and in particular the mining region of Pecs where the miners had fought the Soviet troops so fiercely. Commanded by ex-partisans and carrying only small arms, they had used the mining plant, the shafts, and the slag heaps like true guerrillas, fighting, wounding, and melting into the blackness of their weird landscape, so intractable to the iron monsters. They had lost, of course, and surrendered, been shot, or gone to prison. But it had been a bonny fight. It took nerve, even so, for Khrushchev to challenge them with this apparent sincerity:

Because I was a former miner myself, I felt I could take a tough line with the coal miners. I said I was ashamed of my brother miners who hadn't raised either their voices or their fists against the counter-revolution. They hadn't taken an active part in the mutiny, but they hadn't put up any resistance either. They had let themselves become demoralized and apathetic. When I finished speaking, the miners said they were sorry. They

repented for having committed a serious political blunder, and they promised that they would do everything they could not to let such a thing happen
ever again.[6]

Khrushchev meant it; perhaps some of the miners meant it. In any event, the
miners got a quieter thirty years out of Kadar than they would ever have got
from either Rákosi or Nagy.

III

Hungary might have gone so differently. If Nagy had had more authority or
more sense; if he had had some kind of power base or had made friends with
the army or the police; if he had persuaded the firebrands to settle for less. The
other side of possibility was the course events took in Poland at the same time.

Those events also began in the ruthless resettlement of 1945. At Potsdam
in July, immediately after the surrender of Germany, the Soviet leaders agreed
with Attlee and Truman, Bevin and Byrnes, that in return for their withdrawal of reparation demands from a starving and devastated Germany,
Poland's borders would take in the formerly eastern frontiers of Germany as
far west as the famous line formed by the Oder and Neisse rivers.

The form the Polish government would take had already been agreed to
in a rather fishy deal in Moscow between Molotov and Harry Hopkins, an
emissary from Truman without portfolio but with the credit of being an old
chum of Roosevelt's. It would be a broad coalition of leftist and peasant
parties, the non-Fascist right having been left to its bloody fate by the Red
Army in 1944 and usefully eradicated.

Poland settled uneasily into its subaltern position. Its application in the fall
of 1945 for large-scale aid from the World Bank, invented at Bretton Woods
the year before, was blocked by the United States. Stalin committed the
security of his new empire to a crash program of heavy industrialization. In
Poland, as elsewhere, there were still nationalist partisans and anticommunists
fighting underground. To win Soviet aid and stabilize his regime, the liberal
prime minister Stanislaw Mikolajczyk fell in with the Communist parliamentary majority, wiped out the partisans, and having done his bit, vanished in
the election roundly rigged for a Communist landslide in 1947.

Even then, the Communist government expressed warm interest in the
Marshall Plan in the summer of 1947. But in September, playing host to the
new Cominform (Communist Information Bureau), it was briskly instructed
by Andrey Zhdanov, the Soviet representative, that imperialism was on the
march from Washington. Subsequently, a grimmer new militarization of

politics turned the attractive designation "people's democracy" into a Thing neither democratic nor belonging to the people.

Poland was the only one of the "people's democracies" to keep the victims of this lying and murderous bit of Stalinism cut of the hangman's hands. Wladyslaw Gomulka, the party's first secretary, was merely imprisoned. The drearily orthodox and obedient Boleslaw Bierut, the Polish party leader, put down dissent more flatly than before, and fell in with a deal to overwork his labor force, then to underprice their goods in sales to the Soviet Union. In a not very fair return, the Soviets helped Poland set up the Lenin shipyards in Gdansk (once the port of Danzig), where first Gomulka and then, after 1980, Polish communism itself were to go under.

After the twentieth congress of the party, Bierut died—of shock, as many savage Polish jokes would have it at the time. But he and his granite-jawed henchmen had had a bumpy year. The Polish intelligentsia—necessarily socialists, many of them idealistic graduates of the anti-Fascist underground imbued with the stirring spirit of hopefulness abroad in Europe after 1945— came out openly against dreary Stalinist orthodoxy in 1955. Perhaps the two most honored names in this unofficial opposition were the poet Czeslaw Milosz, Nobel Prize winner in 1980 and already an exile in 1955, and Leszek Kolakowski, the brilliant young Marxist historiographer who, in 1957, at the age of twenty-six, launched an incisive, bitter, and global attack on the horrible distortion of socialism and its values that governed Poland. His attack appeared in the dauntless journal of the Polish intelligentsia, *Nowa Kultura*. The same year Kolakowski wrote his declaration for the times, "Responsibility and History" (also published in *Nowa Kultura*) and spoke up for a socialism of conscience; Milosz answered him—ironically, from Paris—in "The Spirit of History":

> *You, in whom cause and effect are joined,*
> *Drew us from the depth as you draw a wave*
> *For one instant of limitless transformation.*
> *You revealed to us the pain of this age.*
> *So that we could ascend to the height*
> *Where your hand commands the instrument.*
> *Spare us, do not punish us. Grave are our offenses,*
> *We tended to forget the power of your laws.*
> *Save us from ignorance, accept our devotion.*[7]

When Bierut fell so conveniently dead, Khrushchev himself came to Poland not to bury him but to appoint a good Khrushchevite to instigate gingerly reform. He picked dull, sound, but unobtrusively liberal Edward Ochab, and the members of the central committee naturally found that they could bring themselves to vote him in.

Ochab was a loyal workhorse. He was also honest and his own man. He sensed the popular feelings surging toward liberalization: there had been silent demonstrations at Bierut's funeral to wish him ill beyond the grave. At the same time, Ochab was clearly a dependable Communist, he was not power-mad, the army trusted him, and he was capable of intelligent compromise. He released Gomulka from house arrest, issued an amnesty for 30,000 political prisoners, and sacked a bunch of Stalinists, including the attorney general (inspiring much dancing in the streets). He joined the rehabilitation game and sincerely encouraged the prime minister and his parliament to be open to criticism and reform.

At this time, in another essay called "History and Hope," Kolakowski wrote, "The world of values is not an imaginary sky over the real world of existence, but also a part of it, a part . . . that is rooted in the material conditions of life."[8] In a way that has been unfamiliar in university life in Britain and the United States for a generation, the students and writers of the Polish 1956 lived out this commitment to identifying the facts of life with the desires of the dream. The Polish writers' union published its manifesto; *Nowa Kultura, Nomka Polska, Po Prostu,* and a litter of other magazines gave voice to more and more demands and dissent, and on 28 June insurrection followed a strike in the industrial town of Poznan. The townspeople turned out into the streets; in the same blazingly brave and reckless spirit of their Hungarian comrades, students and workers commandeered the local radio station, sang Catholic hymns to the crowds and the airwaves, and opened the prison gates. The minister for industry sent in the tanks. Polish tanks.

Fifty-four people were killed in Poznan (10,000 or more would be killed in Hungary). Those were the only corpses at military hands in Poland that year.

There was a public inquiry, and Gomulka, free but without an office, spoke up for the oppressed, poverty-stricken workers: "They had had enough." He got a job back.

Gomulka had wide popular support, some modest but sincerely held principles, and a respectable intelligence. He kept the pressure up for change. The Poznan rebels got light sentences in September, and Kolakowski published his heterodox "Intellectuals and the Communist Movement" in the intelligentsia's house magazine. The Catholics were on the move, a million on one Sunday to march and pray with the exemplary dignity and discipline that would hold for another thirty-odd years. Karol Wojtyla was one of the officiating priests, learning the hard lessons of church and state. There were seminars in the factories and calls for worker-soviets. Gomulka was lionized as the only leader of the day.

The minister of defense, Konstantin Rokossovsky, was a Kremlin stooge and kept his allies fully informed, as Ochab and Gomulka well knew, by phone each night. His only reliable military allies in Poland, however, were Soviet generals. They plotted a coup in fine old Conradian style and drew up a list of the 700 people they wanted behind bars. Members of one of the new factory committees, in a bold and undiscovered office theft, snaffled the list. As many people as had telephones were called and went into hiding. The rest were warned in a thrilling, last-minute sprint from door to door and a head-long drive from suburb to suburb. They missed their Soviet pursuers by a whisker.

Gomulka knew how precarious things were. But he had the authority and organization that poor Nagy in Hungary lacked. Members of the Zeran factory committee were his runners, in bashed-up, rusty Mercedeses left behind by the Nazis eleven years before.

The eighth plenum—the policy and executive meeting—of the central committee was scheduled for 19 October. That morning, the Zeran factory published a leaflet laboriously duplicated in thousands of copies and rushed by hand and bike and train and old car around the largest Polish cities. It read: "We strongly affirm that we have committed ourselves to the power of the people as a matter of life or death. We contend with all who believe that our democratization represents a first stage on the way back to bourgeois democracy."[9]

The central committee met against a national background of extraordinary excitement. The first item on its agenda was the formal return of Gomulka to committee membership, and the second, his election to the post he had held in 1948, first secretary. As the committee members took their seats at 10:00 A.M. in the Belvedere Palace, a Soviet airplane asked permission to land at Warsaw's airport. Edward Ochab was advised that a Soviet delegation wanted to join the plenum. Consternation.

With impressive coolness, one of the reformist members proposed that the airplane be kept circling the airport while Gomulka and three other liberals were reelected to the committee. Carried unanimously.

It was Rokossovsky who had asked his masters to turn up so abruptly. But he too was under watch. The Polish generals confided steelily to their minister that if Soviet troops marched in, the Polish army would fight them if it had to. Rokossovsky's plot was blocked. At the last minute he phoned his Kremlin allies and asked them in but warned them that if the Red Army followed, there would be open war.

Khrushchev, drinking vodka in gulps as his aircraft circled Warsaw, was violently angry when it landed. Red in the face, swearing volubly, he refused to acknowledge any of the reception party on the tarmac. The big black cars swept off to the Belvedere. Khrushchev dominated the meeting at first. He banged the table in his usual way, accusing the country of being in crisis and the committee of the standard roll call of counterrevolutionary tendencies. The committee members stood up to him. When Khrushchev threatened them with Soviet troops, Ochab retorted that the Poles would fight.

Anastas Mikoyan deftly moved to adjourn for lunch, and Khrushchev trotted darkly over to his embassy. Rokossovsky followed. It was true, he reported, that the Polish army, probably the nation, would join in battle against invasion. When he returned, the committee members stood together. Khrushchev softened his manner and turned up the charm that was undeniably part of his ruthless nature. Gomulka, in his turn, displayed both calm resolution and a reassuringly compromising attitude. He reminded the Russian of Polish unity and its sovereignty, and he agreed that the Warsaw Pact agreement giving garrison rights to the Red Army was inviolable.

He kept his word. When settled the next day in the office of first secretary, Gomulka set out a new program of liberalization to which all his colleagues, once Rokossovsky was thrown out, gave assent. The economic reforms, which included a decent price for sales of coal and steel to the Soviet Union, held for only a decade. The political reforms, which included rights of worship in the Catholic Church as long as the Church kept away from state politics, held for long enough to be instrumental in overturning the politics of state socialism in Poland in the 1980s.

IV

The Western nations stuck tightly to the Yalta and Potsdam agreements, which marked out the spheres and limits of mutual influence in Europe. Therefore, when that heart-rending appeal for help came brokenly over the airwaves as the last hero of Radio Free Budapest stayed at his post, they did nothing, nothing at all.

No doubt there was little they could have done. Full military support for the Hungarian rebels would have been criminally dangerous and in defiance of international law. The Poles, at least officially, would have had none of it anyway. They held onto sovereign control by the skin of their teeth. No American president, least of all Eisenhower, could have risked significant help with tanks to a small country bound by treaty to the Soviet Union. In any case, the United States had allies bent on a little shooting war of their own, and alliance obligations to stop them behaving in too completely lunatic a way. While history played out as tragedy in Hungary, it played at the same moment in the Middle East as farce. But it is a commonplace of the theater that farces have plenty of blood in them.

On 26 July 1956 Gen. Gamal Abdel Nasser, an Egyptian army colonel who had recently become president of his country, nationalized the Suez Canal. Once upon a time every child in France and Britain was taught that the Frenchman Ferdinand de Lesseps planned the construction of the canal, that the British owned, as was only proper, the lion's share of the shares, and that it was ours till kingdom come, or at least for ninety-seven years. In the days before long-running freighters could make the haul around Africa's Cape of Good Hope at ten or twenty times the load of the old ships, the canal was the most important fixed link in the world; Western Europe's trade depended on it.

Nasser had come to power a turn or two after General Neguib had persuaded the corpulent and lubricious King Farouk I to retire to Monte Carlo. He was one of a group of idealistic professional soldiers who sought to modernize their country according to the terms of a genteel socialism and a dignified version of Islam. They were yet another measure of the end of empire for the old Europeans. In 1954—the same year the remnants of the French Foreign Legion were rounded up and humiliated by a coalition of Communist and nationalist groups, the Viet Minh, at Dien Bien Phu—Nasser signed a treaty with Britain arranging for its complete military withdrawal.

Nasser's program for the country was classically progressive: industrialization, bureaucratization, and, in spite of traditional Arab hatred for Israel, careful neutrality in his foreign policy. The Third World—not at that time a synonym for backwardness and poverty—was beginning to declare itself, with India's Jawaharlal Nehru as its uncrowned king and Nasser as its very capable new recruit. He had done one deal with Czechoslovakia in September 1955 for a handsome cache of weaponry (all of it Soviet, naturally), which caused a right old commotion; he promptly did another in December of the same year with Britain and the United States for the gigantic Aswan Dam project, which was to upset so comprehensively the ecology of the Nile.

Nasser was, in other words, a resourceful balancer of the great powers that surrounded Egypt. On meeting him, the British Conservative prime minister who succeeded Churchill in power, Anthony Eden, noted that "Nasser was a fine man physically" and "a great improvement on the Pashas." This was high praise of an Egyptian from the British Foreign Office. The senior official in charge of Middle East affairs had previously described Egyptians in a memorandum as "an unreliable and improvident people"; Churchill's adviser advised colleagues "to reflect soberly" if they were "tempted to believe that we can ever place much trust in the Egyptians." When Brian Robertson, hero of the Berlin Blockade, became military negotiator in Egypt, he agreed with

Eden. "Nasser is a man of moods. . . . But he is no fool. . . . It will be in our own interests to support the present regime." Nasser, however, was up against Eden, a much moodier man, one who was to see in his new enemy another Hitler, another Mussolini, and a suitable case for the assassination treatment.

Eight months after the United States and Britain agreed to put up the cash for the Aswan Dam, they took it away again. Congress had no interest in Nasser and wouldn't vote the money; Britain's attitude was caught by the *Sunday Express* (15 July 1956): "Does anyone in his right mind really *want* to give a 5 million present of British taxpayers' money to Colonel Nasser? If the Dam does not go up he may fall down. But why should we help to maintain him in power?"

Nasser was hurt by the rebuff, but dignified. And he had his contingency plan. It was seamless. Legally, the canal company was incontestably Egyptian; Nasser said that the 1888 international convention governing the use of the canal would be faithfully observed and the shareholders would be compensated. He nationalized the canal.

The report of his announcement was read on the ticker tape in the Foreign Office and carried around the corner to 10 Downing Street where, it so happened, Eden was wearing his full fig as a Knight of the Garter in order to camp it up in front of his old friend King Faisal II of Iraq, who was there to dinner. The news caught Eden in knee-breeches and in the queasily intemperate state brought on, so the medical reports go, by a bile duct leaking into his digestion.

It left him, well, bilious. As his hot suspicion of Nasser had grown in the previous months, he had one evening phoned his under secretary of state, the upright and excellent Anthony Nutting, Eton- and Trinity-educated war hero whose principles over Suez were to lose him his job, many friends, and the prime minister's preferment, but not his honor. Nutting had written a long memorandum suggesting not only how the shock of Nasser's growing hostility to Britain might be cushioned, but also how "to demonstrate to the Arab world that we wished to see justice done for their cause."

In the middle of dinner . . . I was called to the telephone. "It's me," said a voice which I recognised as the Prime Minister's. If his esoteric self-introduction was meant to conceal his identity from the Savoy Hotel switchboard, our subsequent conversation could hardly have done more to defeat his purpose.

"What's all this poppycock you've sent me?" he shouted. "I don't agree with a single word of it."

I replied that it was an attempt to look ahead and to rationalise our position in the Middle East, so as to avoid in the future the kind of blow to our prestige that we had just suffered. . . .

"But what's all this nonsense about isolating Nasser or 'neutralising' him, as you call it? I want him destroyed, can't you understand? I want him removed, and if you and the Foreign Office don't agree, then you'd better come to the Cabinet and explain why."

I tried to calm him by saying that, before deciding to destroy Nasser, it might be wise to look for some alternative who would not be still more hostile to us. At the moment there did not appear to be any alternative, hostile or friendly. And the only result of removing Nasser would be anarchy in Egypt.

"But I don't want an alternative," Eden shouted at me. "And I don't give a damn if there's anarchy and chaos in Egypt."[10]

After his resignation over Suez, Nutting published his celebrated account only after observing the ten-year silence rule. But some years after that, asked by Christopher Hitchens if there were any juicy details he had left out, "he sat thinking for a bit; at length he said, 'Well, I am prepared to tell you that Eden didn't say he wanted Nasser destroyed. He said very emphatically that he wanted him killed. And, in fact, the idea of killing him was put to some senior officials.' "[11]

It is hard, in this biographically inclined history, not to dwell on Eden. He had then such a handsome past as well as face: he had resigned at the right moment and in the right way over Munich; his wartime diplomacy, especially with the Americans, had been resolute and judicious; he had done the negotiations that got the French tidily out of Indochina; he had just succeeded to Churchill's throne as Prime Minister. Only later did his callousness and treachery over the Cossacks, sent back to be shot by Stalin, emerge. But the most delicious tale from 1956 is told as an aside in Leonard Mosley's biography of the Dulles brothers and their mother.

Eden was an upper-crust diplomat who possessed great charm, a gift for languages (eight, including Arabic), and a perfect conviction about the continuing might of his nation, together with his right to use it whenever he chose. The dangerousness of such a man in the world of 1956 as the postwar order settled into place and it became increasingly clear that Great Britain's place in that order was much diminished, is as clear as a fuse in a dynamite factory.

Eden burst into tears when he learned in November that the Americans had called in his overdraft. He tore up a report on the legality of Nasser's nationalization and threw the bits in the lawyer's face. Earlier in the year, he had asked the celebrated military strategist Basil Liddell Hart, by then great with honors in the field, to help his chiefs of staff prepare an invasion plan for Suez. Liddell Hart complied but had his first, second, third, and fourth drafts curtly sent back for improvements by Eden, who fancied himself a military tactician. Extremely annoyed, Liddell Hart waited a day or two, and then sent back, unemended, the first draft. It was fine by Eden, who summoned him and said:

> "Captain Liddell Hart, here I am at a critical moment in Britain's history, arranging matters which may mean the life and death of the British Empire. And what happens? I ask you to do a simple military chore for me, and it takes you five attempts—plus my vigilance amid all my worries—before you get it right."
>
> "But sir," Liddell Hart said, "it hasn't taken five attempts. That version, which you now say is just what you wanted, is the original version."
>
> There was a moment's silence. Eden's handsome face went first pale and then red. He looked across at the long, languid shape of Captain Liddell Hart, clad in a smart off-white summer outfit, then he reached out a hand, grasped one of the heavy, old-fashioned Downing Street inkwells, and flung it at his visitor. Another silence. Liddell Hart looked down at the sickly blue stain spreading across his immaculate linen suiting, uncoiled himself, picked up a government-issue wastepaper basket, and jammed it over the prime minister's head before slowly walking out of the room.[12]

At the beginning, in August 1956, there had been no conspiracy—or at least, no agreement to lie to other nations. The British and the French discussed together how to recover the canal by invasion. They could invade Alexandria, which would be a nice World War II sort of objective—taking out trains and airfields and radio stations—but not what they were supposed to want. Or they could invade small, mucky, and inaccessible Port Said at the top of the canal. Either way they would need 80,000 men, among whom, in Britain's conscripted two-year army of goofy boys, would be a lot of inexperienced soldiers.

Eden tried to square the Americans with a very watered-down account of his plan, but Eisenhower hated it. He was fighting his reelection campaign in a tone so soporific people could not hear enough anticommunism in it and he was having to speak up. He certainly was not going to endorse his allies in waging an unnecessary war. So he sent John Foster Dulles to London to slow things up.

Dulles was not at all averse to a spot of warfare; had he not been photographed for the world in a trench on the 38th parallel in Korea, looking shrewdly north and wearing a black homburg? He mollified Eden. "A way has to be found to make Nasser disgorge what he has attempted to swallow. . . . We must make a genuine effort to bring world opinion to favour international control of the canal. . . . It should be possible to create a world opinion so adverse to Nasser that he would be isolated."[13] He talked about a military operation, if there had to be one, and Eden was cheered up. He disliked Dulles as much as most people did and did not understand Dulles's bulky and seal-like slipperiness. Eden himself was a gentlemanly conciliator-turned-bully, and he did not recognize the new type of diplomat embodied in Dulles compounded of self-righteousness, a flat-footed Catholicism, Republican business obduracy, rocklike egoism, and power-worship.

The French having sent legates to Eden, on 14 October a new plot began. The French had suddenly delivered a large consignment of fighter planes to Israel. They now proposed a put-up job: The Israelis would attack Egypt across the Sinai Peninsula. The French and the British, knowing all about it beforehand, would allow them to capture the desert, which the Israelis wanted for their own protection anyway. The French and the British would then order both countries to cease hostilities. When they refused—or when, at least, the Israelis by arrangement refused—the French and the British would move in to separate the armies and secure the safety of the waterway for the benefit of free world trade (meantime seizing the two main canal posts, Suez and Port Said, for the happier monopolizing of the passage).

Eden was very excited after the French left. He shouted at anyone who crossed him. He recalled his foreign secretary from New York, then on the very edge of a respectable settlement with the Egyptians. All over Britain, nineteen-year-old soldiers were ordered to respray their motor transport and gun carriages in sand yellow, and the army began its laborious transformation into a squeaking machine of war. Nutting, alone in his office above Parliament Square after last being shouted at by Eden, certain that he must resign and thereby end a political career that everyone had predicted would reach Downing Street, was left to this bleak and accurate prediction.

> We might never regain our reputation in the Middle East, and our friends, such as Nuri in Iraq, King Idris in Libya and King Hussein in Jordan, together with pro-Western rulers like President Chamoun and King Saud, might be engulfed and overthrown in a violent anti-British and anti-West reaction. Finally, we should confirm the deep-seated suspicion of many Arabs that we had created Israel, not as a home or refuge for suffering and persecuted Jewish humanity, but to serve as a launching platform for a Western re-entry into the Arab world and a military base, organised and financed by Western governments and Western money, to promote Western "imperialist and colonialist" designs.[14]

Dulles tried to set up the Suez Canal Users' Association, which was intended as a sort of internationalization of the canal. The Egyptians, reacting to the implication that they could not run their own canal, would have none of it. When the British and the French organized a deliberate pile-up of maritime traffic at each end of the canal in an infantile attempt to prove the incompetence of the Egyptian management, the canal pilots worked around the clock and cleared the waterway quicker than ever before.

Very slowly, an invasion force assembled at Southampton and Portsmouth. Assorted plans were devised and discarded. The soldiers finished their painting. The French and the Israelis sharpened things up. In Jordan, a new government signed military agreements with Egypt and Syria. The Israelis booked their Sinai invasion for October 29th. The three British parachute battalions, green as grass, were told that they would halve their jumping height to combat level: 400 feet. They became pensive. This sounded much more dangerous than falling out of old Hastings aircraft at Weston-on-the-Green. After all, they had only joined for the beret.

Britain and France, however, were once more retrieving their imperial purple by force, with a little help from a small country with more at stake than prestige and an absolute ruthlessness about holding onto land, even the barren and stolen Sinai.

The Israelis attacked, as arranged, on 29 October. The French and the British issued their phony ultimatum. Each of the belligerents was to withdraw to a ten-mile margin from the canal, while the fighting was actually going on one hundred miles to the east. The instructions to the Egyptians were to withdraw over one hundred miles from within their own country; couched in the same terms of a ten-mile margin, the same instructions to the Israelis gave them latitude to advance correspondingly until they were ten miles from the canal.

Eden and his foreign secretary Selwyn Lloyd then went off to lie to Parliament about the collusion with the Israelis. For the first time, Britain cast its veto at the U.N. Security Council, and—to Soviet glee—against the American ceasefire. There was uproar in the House of Commons, and a vote of censure was moved by the Labour party. Eden and Lloyd lied, and lied again, to Eisenhower as well, and Eisenhower, with presidential polling due on 6 November, was very angry.

For capitalism was still rolling on, in spite of Suez and Budapest. Sterling was pouring out of London like dirty water from the bath as the finance houses took fright at Eden's antics and sold off their pounds at any price. To buy them back, Harold Macmillan, the chancellor who got Eden's job a few months later, spent his much-needed gold and dollar reserves. When he and Eden asked Eisenhower for a loan to plug the now torrential money leaks, the President turned them down flat, and Eden wept for very frustration. Unless the British did as they were told by the Americans, they would slide to the edge of bankruptcy.

The humiliation was irrevocable. For on 1 November, as prearranged, the

Royal Air Force bombed the Egyptian Air Force with great precision and effectiveness. As they flew in, the innocent Egyptians in the control tower asked if the plainly British aircraft needed help. "Yes, please," came the reply, so the Canberras and Valiants were beamed in to destroy 260 planes sitting outside their hangars, without hitting any civilians.[15]

On Sunday 4 November Ambassador Yuri Andropov in his embassy ordered the Soviet tanks to start killing in Budapest. At the next dawn, the British 16th Parachute Brigade peeled out of the squadrons of elderly Valettas and tubby Hastings and floated briskly down onto their dropping zones. While they did so, portmanteaus filled with leaflets detonated softly above the heads of Egyptian civilians going to work in the capital. The fluttering pieces of paper exhorted Egyptians to rise up and overthrow Nasser, otherwise known as Gamalov, the dupe of the Soviets and a false prophet to his people.

Ten miles away from the canal, the easily victorious Israeli troops stopped in a straight line and waited. The French airborne troops, accustomed to treating the locals in Vietnam badly, took few prisoners. The British paratroops exacted surrender from the town of Port Said.

In the morning of 5 November Nasser revoked the just-agreed ceasefire, the seaborne landing came ashore with 30,000 French and British soldiers, and a lot of civilians were killed. The secretary-general of the United Nations, Dag Hammarskjöld, frenziedly telephoned all the capitals involved. The Soviet president, Khrushchev's sidekick Bulganin, sent letters to Eden and France's premier, Guy Mollet, threatening them vaguely with nuclear missiles he did not have; a companion letter went to Eisenhower suggesting they join forces to stop the war. The British and French armies were told by their governments to stop fighting twenty-six hours after they had started.

Eisenhower laid out what the British and the French were to do: to get out and hand over to U.N. troops. Eden squealed and tried to hang on, reinforcing his troops all the while. The British Tories became delirious, one cabinet minister claiming that Nasser had been put up to his tricks by the Soviets. Eisenhower insisted; otherwise he would leave sterling to its collapse. Eden caved in.

Nothing had been achieved. Everything was lost. Eden tried to cover up with the help of his cabinet, but he failed. Ill and broken, he resigned early in 1957, cannily going off for a suntan in Jamaica. All that Nutting foretold had come to pass. The canal was blocked with forty-seven shiploads of concrete. Britain was shamed in the Arab world. The Soviets had a little bit of space in the Middle East. The Israelis, disgusted at being ditched by the Europeans, turned to the Americans and have been paid by them ever since.

V

1956 is both a battle honour and a disgrace in the history of the cold war. Socialism as a term of art and hope for the purportedly socialist nation-states was probably killed off. By extension, the concept was badly wounded further west in Europe as well.

The year had begun with Khrushchev's secret speech. Stalin was dead, thank God, and at last his ghost could have a stake impaled through its heart. So what stirred so vigorously in Poland and Hungary in 1956 was the spirit that had been abroad in Europe in 1944: the hope for a better future for its children, one over which invading armies do not roll indifferently, and in which a decent government of one's own people keeps an everyday kind of faith with the grand abstracts of justice and equality. When that spirit was again killed with the cruelty usual to despots, and men and boys were shot in the back as they ran before the tanks, communism was stopped just as dead as a force for good. Communism became equal to power without value. The only opposition possible was a helpless nobility.

1956 rendered socialism immobile. The balance of terror had not wavered, despite Bulganin's empty scabbard-rattling over Suez: Soviet brutality had been sufficiently matched by the double-crossing arrogance of the British and the French.

The Americans had discovered Khrushchev to be, in the last analysis, a chip off the old Stalin block. The only opposition possible, as so many poets and intellectuals from Poland and Hungary discovered, was morality without a politics. Since there was nowhere to put their politics, the bare necessity became exile.

The murders of 1956 drove all but grim Communists out of the parties of Western Europe. They had borne the revelations of Stalin's slaughter, but he was three years dead. This new cruelty was too much. As they left their party in thousands and in silence, the two liberal democracies sprang their evil-smelling plot on Cairo and pointlessly killed a still unverified number of Egyptians in order to keep their rheumatic fingers on the cash. Power without value there, too; no comfort for any honest man or woman looking for a home to put politics in, nor East nor West.

VI

On 16 July 1989 the remains of Imre Nagy were reburied in Budapest. The Committee for Historical Justice, an impromptu group of citizens, had patiently accomplished its task. They had obtained official permission to exhume Nagy's body where it had been thrown uncoffined in a prison yard grave. They removed his skeleton bone by bone and with archaeological care. The jaw had become detached from the skull. The rough boots in which he had been hanged were still on his feet. His daughter stood watching the exhumation with folded hands. She did not weep.

The whole city turned out in silence to watch the funeral hearse pass through its streets. Thirty-three years later, 1956 was over.

It was a bad year for hope. In 1956 the United States learned of the irredeemability of its enemy, and of the untrustworthy incompetence of its two purportedly great-power allies. Both discoveries pushed the Americans, not unwillingly, nearer to the center of the world stage. They had shoved Britain aside over these months; whatever tub-thumping British prime ministers later indulged in, the old country was hereafter to be held firmly as dogsbody to the United States, a role in which she complied entirely.

At the same time, Khrushchev's bloody intransigence had impressed the Americans. They dug in deeper in Europe, fattened up their air force in Japan, Korea, and Guam, and prepared themselves to sit things out for as long as the Soviets liked. German reunification? Nuclear-free Europe? Forget it. The temperature of the cold war dropped below freezing again.

9

The Berlin Wall and the Cuban Missile Crisis, 1961–1962

THE Berlin Wall, torn down in that wonderful moment of popular exuberance in November 1989, stood for twenty-eight years. More than any other material object in Europe—more, even, than the giant cockatrice silhouette of missiles on their transporters creeping about the landscape—the Berlin Wall symbolized the cold war. It was a novelist's symbol.

> He had known Berlin when it was the world capital of the cold war, when every crossing point from East to West had the tenseness of a major surgical operation. He remembered how on nights like these, clusters of Berlin policemen and Allied soldiers used to gather under the arc lights, stamping their feet, cursing the cold, fidgeting their rifles from shoulder to shoulder, puffing clouds of frosted breath into each other's faces. He remembered how the tanks waited, growling to keep their engines warm, their gun barrels picking targets on the other side, feigning strength. He remembered the sudden wail of the alarm klaxons and the dash to the Bernauerstrasse or wherever the latest escape attempt might be. He remembered the brigade ladders going up; the orders to shoot back; the orders not to; the dead, some of them agents. But after tonight, he knew that he would remember it only like this: so dark you wanted to take a torch with you into the street, so still you could have heard the cocking of a rifle from across the river.[1]

It was less an international danger than a quite incredible expression of the enemy's blank obduracy. The wall fixed the division of the city, thereby signifying the division of the world. It froze East Berlin stiff at a sunny moment in August 1961, and on the other side, the plenitude of Western consumer imagery and physical delights shimmered and bulged just out of reach of the expressionless, envious poor who watched from behind the wire. To the world, the wall was a plate-glass window between a prison and a Blooming-dale's; it was a categorical distinction between the two kinds of political economy that had made their promises in 1945, the labor culture and the consumer culture.

II

In June 1961 Nikita Khrushchev met the new American president, John F. Kennedy, in Vienna and warned him that Berlin would again be a cause of trouble between the great powers. Khrushchev knew that as long as the anomaly of West Berlin remained so visible and thriving in the middle of East Germany, he could never calm Walter Ulbricht's rule over the Democratic Republic and draw the people's democracies together into a harmonious and sympathetic union of European socialist republics. Khrushchev was his dreadful predecessor's pupil. Like almost all those who wield vast power, he saw no irony in that fact; also like most leaders, he lacked imagination. He was not short, however, on either vision or subtlety. He gambled a lot, but never his shirt.

Kennedy took his own comparably enormous egoism with greater solemnity. Not long after his Viennese summit, his secretary Evelyn Lincoln has recorded, she picked up a *pensée* that had fallen to the floor below Kennedy's desk. It expressed this revolting sampler: "I know there is a God—and I see a storm coming; if He has a place for me, I believe I am ready."[2]

Kennedy, in spite of his famous ego, had a remorseless boyishness and his wit; certainly he was charming as well as prudent. But his thousand days were among the stormiest of the cold war, and it is not unfaithful to the facts to say the storms were blown from the cherubic mouths of the two great-power leaders, twenty-three years apart in age, whose only similarity was their utter representativeness of their national characters.

Their clash in Berlin was overdetermined. By the end of June 1961 the steady stream of refugees leaving the Democratic Republic for work, better pay, and new lives in the West had swelled to over 4,000 per week. Most of them came through Berlin simply because passage from one sector to another was so much easier there. People could work in either half of the city and live

in the other half. It was necessary to have official documents and identity
cards, of course, but they were not too difficult to obtain, and in any case,
leaving the Soviet zone for another was not a crime, not even after that zone
became a nominally sovereign state bound tightly into Stalin's orbit.

During the blazing days and hot nights of July Khrushchev began to test
his opponent's nerve at the node of the cold war in Berlin. As the exodus of
refugees, particularly the collectivized farmers, continued, there were marked
increases in Soviet troop movements around East Berlin, in the garrison, and
in secret agent activity. At the same time, First Secretary Khrushchev beat a
militant drum in all his speeches about his plans for the Democratic Republic,
especially a proposed peace treaty that would flout the four-power agreements
on the future of Germany by excluding the other powers from the new
settlement as they had excluded him from the making of the Federal Republic.

Kennedy asked Acheson, who was no longer in office but served as a sort
of chamberlain at the court of Camelot, to draft a Berlin contingency plan.
Acheson rattled the President by recommending institution of the draft for
several hundred thousand men, including reservists, and setting up an emergency armored corps that would be ready at a moment's notice to blast its way
up the autobahn if the Soviets closed it again. Khrushchev promptly added
30 percent to the Soviet defense budget.

The farmers continued to walk off the flat East German farmland, leaving
the crops to rot. The rumor flew that Ulbricht would seal off East Berlin from
the rest of the city. The agitation spread to the wealthier citizens of the western
half of the city, who began to move their furniture back into the Federal
Republic. Kennedy turned down Acheson's plan but turned up his own
military budget by $6 billion over the Eisenhower estimates. His national
security advisers—Dean Rusk, Walt Rostow, Robert McNamara, Maxwell
Taylor, Edward R. Murrow, a galère of the glittering people—were all prepared to endorse actions based on the psychotic pseudo-nationalism that
dominated the culture. It would last until the memory of Vietnam was assuaged.

Kennedy went on national television to talk to the people in their favorite
kitsch metaphors:

> The world is not deceived by the Communist attempt to label Berlin a
> hot-bed of war. There is peace in Berlin today. The source of world trouble
> and tension is Moscow, not Berlin. And if war begins, it will have begun in
> Moscow and not Berlin.
>
> For the choice of peace or war is largely theirs, not ours. It is the Soviets
> who have stirred up this crisis. It is they who are trying to force a change.
> It is they who have opposed free elections. It is they who have rejected an

all-German peace treaty, and the rulings of international law. And as Americans know from our history on our own old frontier, gun battles are caused by outlaws, and not by officers of the peace.[3]

Khrushchev, however, was not so much looking for a fight as trying to keep the cowhands from walking off the ranch. Three and a half million men had left the Democratic Republic since 1946; 20 percent of the remaining population were pensioners; one in ten of the country's doctors had defected during 1960. The Soviet Union's technological victories since *Sputnik* in 1957 had shaken its rival, but its economy was on the rocks.

The solution to the disappearance of the German labor force down the drain was simple to a degree: put a plug in. On the quiet, sultry Sunday of 13 August 1961, the East Germans put into effect a large-scale sealing of the capital's zonal frontiers that was astonishing in its scope and efficiency. It had clearly been some time in the planning. Although rumors had been rife about some such eventuality, the city had had to live off rumors for fifteen years; they were its news and the source of its nervous energy. In any case, the socialist mayor, Willy Brandt, one of the heroes of the blockade and, indeed, of the cold war, had discounted them. The East, he said, needed Berlin as a valve for the release of the steam of discontent.

Nonetheless, soon after midnight on 13 August, the soldiers and police of the eastern sector started to roll huge coils of dannaert wire across the main thoroughfares. It is a measure of the way the world economy was already going that much of the wire was made in England.

The folklore of that time is rich in escape stories. Innumerable houses, gardens, allotments, railway sidings, the backs of gasworks, and the edges of canals gave onto what was now to become the most tightly patrolled frontier of the iron curtain. Seventeen-year-old Ursula Heinemann wriggled under the taut wire and through the coils laid across a courtyard, tearing her jeans and her behind, leaving her mother, to reclaim her job as a waitress at the Plaza. Emil Goltz left his wife Anna and ten-year-old daughter Beate, in the hope of getting them across later, and escaped by tucking himself above the massive axles of the Moscow-Paris express, hanging above the terrible wheels until he reached the zoo.

It was a few days before those escaping were shot at. A *Vopo* himself jumped over the two feet of wire barring the road to the West, as Willy Brandt was watching, and thereby became an emblem reprinted in every European newspaper. Another couple, kissing passionately at the wire, used that public intimacy to conceal their wirecutters; when the job was done, they dashed across.

The work of sealing off the retrograde half of the city went stonily forward.

The high, nineteenth-century apartment blocks on Heidelberger, Harzer, and Bernauer strassen, with their tall, pedimented windows and solid architraves across the tall doors, were steadily blinded by being bricked in. The chicken-wire barricades were steadily replaced by the wall itself. The subway lines stopped, and the tunnel was thickly blocked. Watchtowers rose beside the canals and the Spree River, and swimmers were shot from the catwalk under the arc lamps and sank into the black water.

The three Allied powers did not, indeed, could not, do much. Ed Murrow sent a report from the wall saying that West Berliners' morale was on the floor, but Kennedy remained detached. On 19 August he sent his vice president, Lyndon Johnson, to cheer up the Berliners with a New York motorcade, and he appointed the hero of the Berlin Blockade, Gen. Lucius Clay, to the Berlin embassy.

Johnson had a whale of a time, Clay was feted and garlanded, and half a million Berliners turned out to cheer them. And the wall still grew in mass and menace. When Clay heard that the East Germans were up to the old Soviet tricks—stopping American and British vehicles for searches and pointless quizzing of their papers while on lawful passages through each of the four sectors of the city—he turned up at the Friedrichstrasse checkpoint, the infamous Checkpoint Charlie, and went past the shocked *Vopos* with a unit of fully armed infantry. The East Germans retorted with humorous humorlessness by broadcasting a report about drunken American officials out with girls.

Clay, burning on his short, decisive fuse, decided to bring the opposition out into the open. He sent ten tanks to point their long barrels straight down Friedrichstrasse. The Soviets brought up ten T-34s to stare back down the barrels. Clay had made his military point diplomatically. The zones were still governed under four-power agreements, each power having its agreed rights of access.

Khrushchev takes up his own story. After hearing from Marshal Konev, whom Clay had in effect flushed out to take responsibility for this little showdown,

> I proposed that we turn our tanks around, pull them back from the border, and have them take their places in the side streets. Then we would wait and see what happened next. I assured my comrades that as soon as we pulled back our tanks, the Americans would pull back theirs. They had taken the initiative in moving up to the border in the first place, and therefore they would, so to say, have been in a difficult moral position if we forced them to turn their backs on the barrels of our cannons. Therefore we decided that at this point we should take the initiative ourselves and give the Americans an opportunity to pull back from the border once the threat of our tanks had

been removed. My comrades agreed with me. I said I thought that the Americans would pull back their tanks within twenty minutes after we had removed ours. This was about how long it would take their tank commander to report our move and to get orders from higher up of what to do.

Konev ordered our tanks to pull back from the border. He reported that just as I had expected, it did take only twenty minutes for the Americans to respond.[4]

It was an exact rehearsal for the staring match to be staged in the Caribbean fourteen months later.

III

The infantile maneuvering at Checkpoint Charlie was linked by both Kennedy and Khrushchev to the difficulties each statesman was having with Cuba.

Fidel Castro, it will be recalled, had been swept to power in Cuba in 1959, only two years before. He had emerged from the Sierra with his guerrilla of the 26th of July Movement to overturn the unedifying Batista by great good luck, extraordinary dash and bravery, and the overwhelming assent of the urban poor. Communism now had its redoubt only ninety miles from Key West. In 1960, in the casually cruel way of great states, Eisenhower cut off the Cuban sugar trade quota in the hope of starving communism out.

The Cuban bourgeoisie, a decadent and seedy bunch, exiled themselves with their dollars to Miami and began that endless cacophony that has deafened and diverted presidents from then until now. John Kennedy, inexperienced, hesitant and cocksure, arrogant and timid, first backed, and then backed away from, the first attempt of these buccaneers to overthrow Castro and reclaim the *patria* for drugs, prostitution, and frozen sugared daiquiris.

The rebels were to land 1,500 men at Cochinos Bay—the Bay of Pigs—and they were to be helped with a lot of clandestine CIA funds. Castro got wind of it. When the U.S. planes camouflaged with Cuban markings came, like the RAF in Egypt, to bomb Castro's air force on the ground, his planes were gone. When 1,500 badly armed and half-trained cavaliers struggled ashore from defective landing craft on 16 April 1961, they found good new roads that no one had told them about and a defending enemy capable of bringing up 20,000 men thoroughly equipped by the Soviets with armor and artillery.[5]

Kennedy withdrew the second and crucial air strike. The U.S. Navy came to evacuate the few who could get away and then vanished before collecting them. Castro's men rounded up the humiliated band—they suffered few casualties—and put them, justly enough, in a very nasty prison. Kennedy and his toadies rushed to cover things up and to reissue his equivocations as judiciousness. In an impressive orchestration of falsehood, he wrote a memo specifying the journalists to be, in his simple words, "briefed and brainwashed."[6] They were the *Times,* Lippmann and Alsop—he pulled prior reports of the invasion from a compliant *New Republic,* and he set up Adlai Stevenson as a patsy to tell journalists innocent lies.

So Khrushchev had Cuba on his mind in Berlin; and so did Kennedy. A small sugar island with florid dreams of becoming a socialist commonwealth was the location for the most dangerous moment in the history of the world.

After the Bay of Pigs invasion, the Soviet Union increased arms shipments to Cuba. From the end of August 1962 onward, as the Berlin wire was replaced by the Berlin Wall, reports and aerial photographs coming in from Cuba increasingly suggested that Soviet missiles of indeterminate strength were being installed on the island.

The Soviet ambassador to the United States promised the president's brother, Robert Kennedy, that no *offensive* weapons would be placed in Cuba, only weapons of the kind the Cubans themselves had asked for to help them keep out the Miami freebooters. The two sides then launched an exchange of high-minded, vaguely threatening press statements, until Congress was brought up short by Senator Keating, who claimed to have the entirely dependable information that the Soviet government was building nuclear missile ramps in Cuba, pointed at the United States.

The CIA had shrewdly been doing its camera work well away from the missile stations, in case its U-2 airplanes were hit. Once it received its orders, however, the CIA came up with ample evidence that there were indeed medium- and intermediate-range launching platforms already being constructed on the western end of the island.

The next morning, Tuesday, 16 October 1962, Kennedy appointed an executive committee—"Excomm," consisting of Rusk (though he was mostly absent), McNamara, his brother Bobby, Maxwell Taylor, and Nitze, among others—to range over and review all the possibilities. Acheson came and was vinegary about the lack of procedure. The committee canvassed the following possibilities: doing nothing; a deal with Khrushchev over Jupiter missiles in Turkey; blockading; air strike; invasion.

It is said that it was in this committee that the useful cliché category "dove" and "hawk" originated. The doves asked, on what grounds could military action be justified? The Soviets were themselves ringed by awesome missiles, Cuba was a tiny Soviet ally and no business of the United States, and in any case, a sudden air strike—a "surgical" strike, in hawkish jargon, "Pearl Harbor in reverse," according to dove thinking—would be dangerous, bloody, and by no means guaranteed of success. The hawks noted that the Berlin Wall

Figure 9.1   Castro and Khrushchev in Cuba, 1964. (The Hulton Picture Company)

was up, and Khrushchev was threatening to annex the city by blockade again. The two strategies were obviously connected; any such annexation must be blocked at once.

The committee continued to argue through Wednesday and Thursday, and gradually the idea of a blockade—to keep the Soviet supply ships out— gained the advantage. It was still abominably risky—what if Khrushchev ordered his ships to flout it? What would the Europeans do, so much nearer the Soviet missiles, so much more exposed a target? Macmillan and de Gaulle could be squared, but their citizenry might be out in the streets, refusing to oblige the United States with support. A blockade it was to be, however; it would be more gradual, less direct, more (the new cant) "flexible."

The same afternoon, the best deadpan liar of the epoch, Andrey Gromyko, the Soviet foreign minister, spent two hours with Kennedy without mentioning the missiles, not knowing that the President knew and not being told either, ending brazenly with complaints about the recent congressional authority given to call up reservists in time of emergency. He also said that the Berlin anomaly must be tidied up, and that the Soviet Union was going ahead with a new one-sided treaty with the Democratic Republic to that effect.

Kennedy heard him out silently, but Gromyko's effrontery had made him very angry, as well it might. Garry Wills cites Kennedy's apothegms during the crisis made in stage asides, intended for requotation, to James Wechsler and Arthur Schlesinger: "If Khrushchev wants to rub my nose in the dirt, it's all over," and, "That son-of-a-bitch Khrushchev won't pay any attention to words. He has to see you move."[7]

In a not unfamiliar spasm of selective moralizing, however, Kennedy was angered by Gromyko's cheek at the same time as he himself lied to the public about the Cuban threat, the missiles' range, and the sheer unimportance of the Soviet weapons. The executive committee itself agreed that the missiles did not alter the balance of nuclear power. In *Thirteen Days*, Robert Kennedy willfully exaggerated their strike capacity, claiming they could "kill eighty million Americans." Their maximum reach was 1,000 miles; Cuba would have been a radioactive dustheap before one such missile struck Miami.[8]

Looking back now, not a generation later, nothing, nothing, could justify the risks those assured courtiers of Camelot took with the future of the world. Millions may still nonchalantly say that Kennedy was right, in the right, and in any case, victorious. But there were awful dangers. The crisis brought to light how erratic the flow of information to Excomm was: it didn't know that the CIA was still running covert operations on the island, nor that the Defense Department thought the Turkish missiles belonged to the Turks.

Khrushchev wanted to forestall another Cuban invasion. He thought the risks were under control. But the world could easily have been detonated over a handful of missiles and a president's or a first secretary's resentment, each impelled by vast power and an egoism to match.

Kennedy was locked into taking action, in thrall to his own style: first the blockade; then, if necessary, the air strike. Acheson, acerbic about what he saw as the inadequacy of the operation, was dispatched to Macmillan and de Gaulle in order, as de Gaulle drily said, "to inform rather than consult them." "But," the old nationalist went on, "I think that under the circumstances President Kennedy had no other choice."[9] "No other choice." Each leader insisted he had no choice, until Khrushchev chose the unchoosable and was humiliated. Meanwhile, Kennedy went on television again to make the most blood-freezing speech of the cold war.

Our policy has been one of patience and restraint, as befits a peaceful and powerful nation, which leads a worldwide alliance. We have been determined not to be diverted from our central concerns by mere irritants and fanatics. But now further action is required—and it is under way; and these actions may only be the beginning. We will not prematurely or unnecessarily risk the costs of worldwide nuclear war in which even the fruits of victory would be ashes in our mouth—but neither will we shrink from that risk at any time it must be faced.[10]

He announced a meeting of the Organization of American States (OAS) and encouraged the Cubans to rise up and throw off their leaders, calling them, in ritual cold war invective, "puppets and agents of an international conspiracy," which must have rung a bell in Moscow. Kennedy appealed to Khrushchev to call off his adventure. Finally, he announced that the United States would implement a "quarantine," which, he explained, was less belligerent than a blockade: all ships bound for Cuba would be stopped and searched for weapons. The quarantine went into effect on Wednesday, 24 October, 500 miles out in the free waters of the international Atlantic. As a Soviet convoy sailed toward the line, nineteen American warships waited for it.

The Soviet cargo carriers were due to touch the invisible line across the moving Atlantic soon after 10:00 A.M. The Strategic Air Command nuclear bombers were scrambled and set to prowl the skies with their irrevocable loads fully primed. That Wednesday, the executive committee met early, and the President and his brother met earlier still.

I talked with the President for a few moments before we went in to our regular meeting. He said, "It looks really mean, doesn't it? But then, really there was no other choice. If they get this mean on this one in our part of the world, what will they do on the next?" "I just don't think there was any choice," I said, "and not only that, if you hadn't acted, you would have been impeached." The President thought for a moment and said, "That's what I think—I would have been impeached."[11]

They went into the meeting and were presented with literally miles of aerial photography documenting the launching platforms, the warhead bunkers, and the Soviet and Cuban backs and faces bent to the job.

The news was brought to the quiet room that a Soviet army submarine had taken up a position shepherding the Soviet naval vessels the *Gagarin* and the *Komiles*. McNamara told the meeting that the USS *Essex* would instruct the sub to surface and identify itself. If it refused, depth charges would be dropped from helicopter.

The navies seemed certain to engage. Robert Kennedy recalled his brother saying, "We must expect that they will close down Berlin—make the final preparations for that," but all orders had now been made as far as the men in that room were concerned.[12] As in future models of the moves along the algorithm to nuclear war, everything was now in the hands of commanders out on the open sea.

At 10:25 A.M. the Soviet ships stopped dead in the water. They rolled heavily for a while in the mid-Atlantic waves and then cumbrously turned to set off for home. In John Updike's novel *Couples* (1968) two of the characters are on their way to play golf at that moment.

> They agreed not to cancel. "As good a way to go as any," Roger had said over the phone. Stern occasions suited him. As Piet drove north to the course, the Bay View, he heard on the radio that the first Russian ship was approaching the blockade. They teed off into an utterly clear afternoon and between shots glanced at the sky for the Russian bombers. Chicago and Detroit would go first and probably there would be shouts from the clubhouse when the bulletins began coming in. There was almost nobody else on the course. It felt like the great rolling green deck of a ship, sunshine glinting on the turning foliage. As Americans they had enjoyed their nation's luxurious ride and now they shared the privilege of going down with her. Roger, with his tight angry swing, concentrating with knit brows on every shot, finished the day under ninety. Piet had played less well. He had been too happy. He played best, swung easiest, with a hangover or a cold. He had been distracted by the heavensent glisten of things—of fairway grass and fallen leaves and leaning flags—seen against the onyx immanence of death, against the vivid transparence of the sky in which planes might materialize. Swinging, he gave thanks that, a month earlier, he had ceased to be faithful to Angela and had slept with Georgene.[13]

That fateful morning Walter Lippmann, the most august and heeded voice in the national press, had reproached the president in the *Washington Post* for not having tried enough diplomacy, for not having faced the Soviet Foreign Minister with Gromyko's barefaced concealments and thereby, perhaps, giving Khrushchev space to save *his* face. But Kennedy was to drive Khrushchev under in his determination to concede nothing.

On Thursday morning Khrushchev sent an unofficial letter, followed immediately by one that seemed to be more the work of the central committee and its clerical staff. The first letter, however, was different in tone and eloquence; spoke in a suitable prose of the matter of life and death before them. To Acheson and company, the scornful soldier-gentlemen, and their mimics in the posh press like James Reston of the *New York Times*, the first letter was—in Acheson's words—"confused, almost maudlin"; he claimed unbelievably that Kennedy's overcautious response was bailed out by "Khrush-chev's befuddlement and loss of nerve." Reading the letter again at this distance, it expresses rather well the quite unprecedented combination of accident, interest, doctrine, and will that by the late evening of Thursday, 25 October was only barely under human control.

Early in this letter—whose standing has never been clear but which, surely quite rightly, Kennedy chose to take as having diplomatic precedence over the second, more formal one—Khrushchev wrote that the missiles in Cuba would never be used to attack the United States and were being installed strictly for defensive purposes. "You can be calm in this regard, that we are of sound mind and understand perfectly well that if we attack you, you will respond the same way. But you too will receive the same that you hurl against us. And I think that you also understand this. . . . We want to compete peacefully, not by military means."[14]

The passage at least raises the possibility that when Gromyko made Kennedy so angry by saying that any weapons sent to Cuba were for strictly defensive purposes, he meant exactly what he said.

But Khrushchev went on, in his maudlin way:

> If assurances were given that the President of the United States would not participate in an attack on Cuba and the blockade lifted, then the question of the removal or the destruction of the missile sites in Cuba would then be an entirely different question. Armaments bring only disasters. When one accumulates them, this damages the economy, and if one puts them to use, then they destroy people on both sides. Consequently, only a madman can believe that armaments are the principal means in the life of society. No, they are an enforced loss of human energy, and what is more are for the destruction of man himself. If people do not show wisdom, then in the final analysis they will come to a clash, like blind moles, and then reciprocal extermination will begin.

There glimmers in that passage a light of hope for a less lethal world; generously responded to, it might also have illumined a path toward the clearing in which a cold war armistice could have been read. But Kennedy had Cochinos Bay just behind him and congressional elections just in front. Visions of a safer world would have to wait. Besides, a contest was on, and he, a handsome, attractive, rich, and young-enough American, was going to win.

Khrushchev's letter offered a settlement: withdrawal of the Cuban missiles in exchange for a promise not to invade Cuba. The official letter also asked Kennedy to trade in the Jupiter missiles in Turkey, missiles the President knew

Retreat

Figure 9.2   Khrushchev's retreat from Cuba, 1962

to be so clumsy they were already obsolete; in fact, he had already asked for a report about their removal.

Nevertheless, he was going to give away as damn little as possible. Kennedy made his only concession in a subparagraph; after insisting on his own conditions, he diminished his responsibility for menacing Cuba by aligning himself with all the other nations of the hemisphere, few of whom had thought to bother Castro very much.

> We, on our part . . . agree—upon the establishment of adequate arrangements through the United Nations to ensure the carrying out and continuation of these commitments—(a) to remove promptly the quarantine measures now in effect and (b) to give assurances against an invasion of Cuba. I am confident that other nations of the Western Hemisphere would be prepared to do likewise.[15]

Khrushchev broadcast his reply in a statement on Moscow Radio, early on Sunday morning. The missiles were to be dismantled, crated up, and sent back to the Soviet Union. Cuba tried to hang onto a squadron or two of bombers, but by then Kennedy knew he could take all the tricks. He insisted that the bombers also be sent back, and that the United States be allowed to count them as they went. Khrushchev bowed to this humiliation as well.

IV

In 1969 Dean Acheson published his version of the Cuban Missile Crisis in *Esquire,* an organ not at that date noted for its contributions to foreign affairs. He said he thought the President had pointlessly procrastinated, had lacked resolution, and, as Acheson told him just afterward, had been "phenomenally lucky." (I particularly relish his reference to Khrushchev's letter as "rambling off again on the horror and folly of nuclear war"—the authentic voice of the tough realist.)[16]

Well, one not altogether lucky consequence of Khrushchev's extremely public humiliation was his being ousted from power soon afterward. It is also almost certain that another consequence was the authorization by the central committee of an enormous increase in the manufacture and research and development of nuclear weaponry after October 1962. And it can plausibly be claimed that American foreign policy makers, congratulating themselves on their own coolness and hardheadedness, then went forward to Tonkin Bay and the Tet offensive, blindly certain that they could always get their way.

Sitting on the northwest rim of Europe, on an old island that bristled with

American nuclear weaponry as well as a few bits of its own for four decades, it is hard to deal anything but harshly with American conduct during those thirteen days in October 1962. The Kennedy courtiers insist that, in Roger Hilsman's words, the Cuban Missile Crisis was "a foreign policy victory of historical proportions."[17] Hilsman was a senior Kennedy official whose judgment is less easily hooted at than Arthur Schlesinger's embarrassingly pop-eyed adulation in *A Thousand Days:* "It was this combination of toughness and restraint, of will, nerve and wisdom, so brilliantly controlled, so matchlessly calibrated, that dazzled the world."[18] But Schlesinger's conclusion was almost the same as Hilsman's, that "the thirteen days gave the world—even the Soviet Union—a sense of American determination and responsibility in the use of power which, if sustained, might indeed become a turning point in the history of relations between east and west."[19] The corner Kennedy and his merry men reckoned they had turned was one after which *their* will and *their* decisions—benign and liberal ones, naturally—would be running the show. Lyndon Johnson, no global thinker even on his best days but a decent reformer, was broken by the same assumption—and destroyed a country making it—but had at least the guts to admit his error.

The more we learn about 1962, and the better our historical vantage point on it, the more perilous a time it seems, and the greater the momentum its events gave to the systems of destruction.

I. F. Stone, the great dissident journalist, writing four years after the event, pressed the grand arbitrary of the crisis: what if Khrushchev had not backed down? Stone contends that the issue in the Cuban crisis was only ostensibly American prestige; more deeply, it was Kennedy's own standing at stake on which he gambled the safety of the world. He had dithered at Cochinos Bay, and had had a poor first year in Congress. The Republicans would have pilloried him if the Soviet missiles had been left untouched in Cuba, and then they would have won a House majority. Yet a temporary setback to the Democrats does not now, and did not then, seem much of a pretext for hemispheric incineration.

Once the Kennedy brothers had got hold of a clear line of argument, the momentum of their own decisiveness gave conviction to their feelings. This explains the grotesque overstatement of John Kennedy's speech to the nation at seven o'clock on the first Sunday evening, 21 October:

> The path we have chosen for the present is full of hazards, as all paths are—but it is the one most consistent with our character and courage as a nation and our commitments around the world. The cost of freedom is always high—but Americans have always paid it. And one path we shall never choose, and that is the path of surrender or submission.[20]

His courage and commitment appalled one senior ally. Two years later President de Gaulle led his nation out of NATO in a stately huff, complaining that France's commitments were not those of the United States; instead, France was committed to avoiding the consequences of American recklessness on its behalf.

The president was not helped by the White House's awful atmosphere of toadydom, with which only Acheson broke, in his high-handed, damn-and-blast-it-all sort of way—and Acheson held no office and, in any case, wanted to bomb the missile platforms instantly. The atmosphere was quite uncritically caught in the CBS movie in which Martin Sheen played Kennedy with a Boston accent as harsh as an arcwelder. When the news comes through that the Soviet ships have stopped, all the Excomm members applaud and Sheen turns away, silently moved. The fawning took the standard of talking tough. Curtis Le May, a senior general, zealously wanted to bomb Cuba because the Soviets would not react; even when the milder step of blockading was agreed upon, he said, "Nobody in the White House wanted to be soft. . . . Everybody wanted to show they were just as daring and bold as everybody else."[21]

Only Adlai Stevenson suggested conventional negotiations, and the poor chap was quickly sat upon. Indeed, the whole affair was notable for the absence of any of the ordinary diplomacy or use of the institutions by which nations are supposed to wage war and decide to destroy or not to destroy one another. Kennedy published Khrushchev's private letter, having used it quite improperly as a formal proposal. He ignored the United Nations. He ignored the sensible offer of trading useless missiles in Turkey for highly effective ones in Cuba. He was left by the great liberal institutions—Congress and the grave sages of the fourth estate, James Reston, Joseph Alsop, and Alistair Cooke—to do just what he pleased. It is hard to escape the conclusion that all these men were more terrified of taking responsibility and looking chicken than of rational recoil from such a danger. It is usual, the poet says, for a man of courage to respect fear; these ordinary men disregarded ordinary, wholesome fears in order to achieve a transient and a trivial victory.

Kennedy lacked the experience, the generosity of feeling, and even the class tradition, to do any better. He mobilized the institutions he knew and could handle—the official and opaque mendacity of state utterances, the press, and television. He set off a sequence of events, timing them so that they would reach crisis and therefore consequence. He sacrificed thought and the obliquities of debate to the arbitrary energy of *events*. The events came out his way and locked his successors and his nation onto course for Vietnam and 1968.

10

The European—Willy Brandt

THE fulcrum of the era and its balance of terror stood poised in Germany: the crucial weights were placed in Berlin. When they trembled, weapons were cocked in Korea, Cuba, and at the rim of the world.

Willy Brandt guarded those weights and felt the trembling of the balance of terror for the length of his political life. In old age, at the end of a career conducted full in the center of the history he was sometimes accused of experiencing with too great an enthusiasm and awareness, that history delivered to its patient sentinel a victory of common sense and compromise: the everyday liberty and decent equality for which he had striven since his youth.

It was a long journey across the century. It began in Lubeck in 1913 where Brandt was born the illegitimate son of a working-class and actively socialist mother, a woman of rare energy, some cultivation, and a keen intelligence. He learned early the powerful lessons of European socialism, its commitment to wresting material and political comforts and freedoms from a ruling class that put down working people all the way.

As a very young man he was prominent in the German revolutionary party of the left. He was convinced, as Fascism swelled monstrously toward the apogee of its power, that only insurrection by the working class would defeat it, but he underestimated the Fascists' dedication to wiping out all opposition whatever. When that zeal became unpleasantly apparent in 1933, the twenty-year-old Brandt was dispatched by his party to safety in Norway.

In Norway he learned to admire the virtues of the indigenous Labor party, its patient gradualism and its keen, effective conscience. He had taken the name Willy Brandt as a political disguise in Germany, and because of his alias and his formal illegitimacy, the Nazis had no record of him. He still worked energetically for his party and its international connections but, being virtually stateless, identified himself as much with Norway as Germany. His mission as liaison officer to the workers' army during the Spanish civil war and his dismay at the Communist militia's idealistic ineptitude, as well as at the cynical maneuvering of the Stalinists, also formed the great breadth and depth of Brandt's Europeanness, which was as rare in 1936 as it is plain to us now.

He dodged the Nazis when they invaded Norway and found refuge in Sweden, where he emerged as the best representative of the German exiles from the Third Reich. Barely thirty, strikingly handsome and dashing, irresistible (all his life) to women, he spoke for the humane values his country had given itself to befouling. At a famous conference of junior branches of the Socialist International in Stockholm in 1943, he sketched out a postwar settlement that would have diminished the virulent nationalisms laying waste to the Continent: the United Nations of Europe he envisioned, with certain common laws, would ensure social welfare, workers' rights, and a free confederation of democracies.

So when Brandt returned to Germany as a Norwegian press attaché at the Nuremberg trials, he brought with him the vision of concessionary, polite, and negotiatory politics that he would hold to all his life. That vision led to variously accurate charges against him of deviousness and indecision; nonetheless, it provided not only the model for that captious entity, the European community, but also the only hope for a wider world.

Was this a conscious project? Historians who have asked him, as I did, about the theory and principles of his politics are put deftly aside. "My political principle is that principles are bound to conflict. You will only stop them conflicting violently if you let them rest side by side. If this leads to contradiction, too bad."[1]

Brandt came back to Germany in 1947. One year later he gave up Norwegian citizenship and in February 1948 he started work in Berlin as a liaison officer for the Social Democratic party (SPD), postwar Germany's labor party and the only party that survived the Nazi putsch with its moral credit intact.

The Berlin Blockade effectually turned him into the mayor's unofficial deputy. Ernst Reuter was the brave and overwhelmingly popular mayor; he expressed and, in the parliaments of Germany's recently mortal enemies, fought for his townspeople's fervent wish to stay out of the Communists' brutal hands. Brandt knew the Communists and their little ways by heart. His became almost as well known a face as Reuter's during the long year of blockade.

By the end of the blockade, Brandt knew what he wanted from his party and for his country. He was a party man all through. But he also saw that his party must change itself, must drop the old simplicities of working-class preference and speak to the citizens of a quite new class structure that was emerging first in West Germany but presaging the forms of European society to come.

In 1990 Brandt claimed that he anticipated this long revolution; he has never been loath to appropriate history to his credit. "What I helped the SPD to become is what, after 1981, Mitterand made of the French Socialist party and now you, too, in Britain are making of the Labour party."

But this was more than a party matter. Brandt's vision was of a federal membership and mutuality that would hold all Europeans in a common affection. Berlin was the ground upon which such an experience could be foreshadowed.

In 1957 he was elected mayor, crushing the old-style leftists of the SPD under Franz Neumann. Then, as the young, handsome, brilliant candidate of his transfigured party, he ran for chancellor against the ancient tortoise Konrad Adenauer, then eighty-five. Adenauer conducted a notoriously dirty campaign, impugning Brandt's years in exile, lying about his softness on communism, and digging up his bastardy. These appeals to the dark suppressions and ambivalence about their past of the German people did enough damage to deny Brandt the office.

But Adenauer's day was over, and in spite of losing the election, Brandt was soon the hero of Europe and the toast of Washington. When the Berlin Wall began to go up, Brandt took the lead in urging Kennedy, who was as charmed by Brandt as Brandt was by Kennedy, to support the city and in persuading his baffled fellow-citizens to vent their anger and incredulity without provoking the *Vopos* to shoot at them. "I remember feeling quite cheerful and confident when I went down to see the American tanks looking down the barrels of the Russian tanks. I thought it would end tidily. They wouldn't take the wall down. That was not to be expected for years. (Actually it came down sooner than I thought.) But we would negotiate. We would find ways—slow ways, doubtless—to make it transparent. And then superfluous."

No one can say that Brandt fails to look on the bright side. His method has always been to bend with the wind (but not to be uprooted), to move nimbly and lissomely between positions in which others might embed themselves. Living in Berlin made last-ditch stands seem entirely unattractive; Berlin *was* the last ditch, and Berliners would either live or die in it. Brandt stood on the side of life every time.

His liveliness and reputation caught up with him from time to time. He was attractive to, and attracted by, many women. He was liable to chronic depressions, which at times he drank himself into and out of. The depressions were surely an expression of the lived difficulties of keeping up hope and charm and good cheer in the face of the old entrenchments of Catholic and Communist parties in his sundered land.

At this distance, he is wry about the slipperiness of his political method. "When President Kennedy telephoned me to tell me his decisions, I bowed my head—metaphorically, you understand?—and said with great respect that the decision was entirely his, and that I could only support him in it, whatever it was." Brandt's eyes twinkle. "Then Kennedy said on American television, 'Even the Mayor of *Berlin*. . . .' " But Kennedy paid off his debt to Brandt in 1963 in his celebrated "Ich bin auch ein Berliner" speech at the Brandenburg Gate.

In 1966 Brandt almost died of a mysterious respiratory attack and this, too, I believe, was an expression of the extraordinary intensity with which he joined body to soul, public to private life, in his political endeavors. Those endeavors *required* that division and ambiguity become synonyms for his principles. Consequently, that same year he led the SPD into coalition with its traditional enemy, the Christian Democrats, and became foreign secretary.

This was the official beginning of the *Ostpolitik*—Eastern politics—for which his life and his historical vision had prepared him. Over the next eight years, until his sudden (and temporary) disgrace in 1974, he applied his unique powers of conciliation, sensitivity, deliberate evasiveness, and high-toned duplicity to the future unification of Germany, and of Europe. "We Germans must not forget our history. But we can also not continually utter confessions of guilt. . . . We must intensely, responsibly ask for our right to self-determination, to national self-realization and with this make our contribution to the healing of the wounds of Europe's center."[2]

In 1969, by very narrow margins, Brandt led his party triumphantly to power and himself to the chancellorate. He was backed by an intelligentsia with Günter Grass and Heinrich Böll in its vanguard, and also by the mainstream groups of the radical student movement of 1968. As chancellor, maybe he was the ghostly forerunner of a long-distant president of the United States of Europe.

Brandt brought off the four deals that clinched *Ostpolitik*—with Moscow and Leonid Brezhnev, with Poland, with East Germany, and with the four powers in Berlin. Formally consolidating the status quo as a peaceable arrangement for all concerned, the treaty kept open future prospects of unification and cooperation. In East Germany he was given the welcome of his life by rapturous crowds hailing him, despite the set teeth of their official leaders, as the greatest German of the day. Brandt, as always, was the picture of modesty, but he loved it. Anybody would have.

In 1971 he was awarded the Nobel Peace Prize. There was no worthier prizewinner in the epoch.

On 6 May 1974 he resigned in a sudden flurry of opprobium because a confidential clerk of his was discovered to be an East German espionage officer. His party was probably helpless to save him; at any rate it didn't try. There were rumors of love affairs and of other, more compromising detours. "I resigned," he said stiffly, "out of the experiences of the office, out of my respect for the unwritten rules of democracy, and in order to preserve my personal and political integrity."

Earlier than any other person, he began to close the cold war epoch, at least in Europe. As he would later say to me, perhaps he pressed for too little, or "maybe the early preparation was what was needed to make things happen quickly fifteen years later. . . . I had judged that the foreign policy talks about German unity, the so-called 2-plus-4 talks, would take several years. I'm happy to say I've been wrong."

After his resignation, he naturally assumed the singular role that had shadowed him all his life, of, so to say, First European-in-Waiting. He had always been quick to sense new political growth and sentiment. A supporter of NATO since its birth, he nonetheless stretched out his hand in a kind of blessing on the German peace movement and the Green party when they surged to prominence in the early 1980s. Some people called him two-faced. Only two? These days, a two-faced politician is a halfway honest one.

And when the cold war ended, the old socialist was the first to say, with characteristic generosity, "It is very important not to disappoint the hopes associated with the current arms negotiations, i.e., that at least part of the peace dividend will benefit the countries of the South. My proposal is that one-third of the net savings on arms expenditure should be set aside to be used for international tasks by reliable multilateral institutions. . . . We would then jointly be able to make a major contribution toward improving life on this planet. Our one world. . . ."

I I

The Scientist—Freeman Dyson

D IPLOMACY, politics, and science. The three realms of the cold war have become, since the advent of nuclear weapons and ideological specialization, more and more exclusive. The divisions of labor run deepest between just those people who most need to understand one another's language. Scientists are summoned, as Oppenheimer and Teller were, to the deliberations of politicians and diplomats, but they are not expected to decide anything. They are called as expert witnesses but are not asked to make general applications of their expert knowledge. The culture's stereotypical scientist is marvelously ingenious but scatterbrained in everyday life. As Churchill noted of Niels Bohr, scientists have wild hair blowing about their heads, a certain sign of unkempt absentmindedness.

Yet the three institutions, diplomacy, politics, science, stand side by side in the great offices of state. Power opens its doors, not quite equally, to each.

Freeman Dyson was a familiar figure passing from the great institutes of scientific inquiry through the portals of power for over forty years. For most of that time he was a neighbor of George Kennan's at the Institute for Advanced Study at Princeton. Dyson told me a short cold-war story that illumines the Institute's celestial role in the American polity.

"The finest moment in the history of the Institute for Advanced Study in Princeton where I work came in the 1950s when we appointed the mathematician Chandler Davis a member of the institute. Chandler was then sitting in jail because he refused to rat on his friends when questioned by the House Un-American Activities Committee. He was convicted of contempt of Congress and sent to jail for a year. He came straight out of jail to Princeton and continued doing mathematics. That is a good example of science as subversion. Chandler is now a distinguished professor at the University of Toronto and is still actively engaged in helping people in jail to get out.

"Another good example of science as subversion is Andrey Sakharov in Russia. Chandler Davis and Sakharov belong to an old tradition in science that goes all the way back to the rebels Franklin and Priestley in the eighteenth century, to Galileo and Giordano Bruno in the seventeenth. If science ceases to be a rebellion against authority, then it does not deserve the talents of our brightest kids. I was lucky to be introduced to science as a rebellion against Latin and football. We should try to introduce our children to science as a rebellion against poverty and ugliness and militarism and economic injustice."[1]

Dyson came to science as a way of defying school. He was born in England in 1923 and was sent at the age of eight to one of those horrible little boarding schools in which the wretched little boys of the English upper classes were incarcerated in those days. In such schools, science was the unrespectable, nonclassical refuge of the clever and the intractable. Dyson was both and wore science as his badge. When he went on to Winchester, he was able to flourish more openly as a scientist. He also worshiped, with the fervor such schools nurtured, the color, dash, and gregarious brilliance of Frank Thompson, big, noisy, accomplished, and four years older.

Like Thompson, Dyson was at odds with his social place. Thompson became a socialist and went to join, as Dyson put it in his memoir, the "worldwide struggle of Socialism against Fascism."[2] Dyson, being a remarkable mathematician, was conscripted to count and spent his two years of war studying, for the sake of its greater efficiency, the statistics of Bomber Command. Those statistics convincingly proved that carpet bombing in no way affected the course of the war beyond pointlessly killing bomber crews. One particularly painful fact that Dyson and another youngster discovered was that many more aircrews would be able to bail out if the Lancaster bomber, by then the biggest bomb-freighter of the RAF, had an escape hatch only two inches wider, like that of the U.S. bombers. No senior officer would listen to the two twenty-one-year-olds for a moment. Their work was ignored, and Dyson turned his back forever on the British staff-officer caste and its disregard for reason and humane fact.

In August 1960, I came back to America from Europe with my wife and eleven-month-old daughter. After thirty-six hours of traveling, we arrived at the old New York bus terminal and bought a carton of fresh milk for baby Dorothy. But we still had a problem. Dorothy could not drink out of the carton and her milk bottle was rancid in the August heat. I went to a public drinking fountain and began rinsing the bottle. Two tall crew-cut American soldiers stood nearby and watched my ineffectual efforts to get the bottle clean. Finally, one of them said, "Say, that's not the right way to clean a nipple. Let me show you how to do it. I happen to be an expert." He took the bottle and cleaned it out with the thoroughness and precision of a well-trained nurse. I thanked him and thought, Well, this is America and I am glad to be home. So was born the dream that the military establishment of my adopted country might serve a humanly comprehensible and ethically acceptable purpose.[3]

It is an affecting parable. Although Dyson writes very fluently and openly, he speaks very slowly and with great reserve. He is small and elfin and more than a bit shy; his large, impressive, elfin ears are just like Spock's. However diffident he is himself, Dyson believes that warmth and personal spontaneity are critical qualities in those whose work is the popular exposition of science. He is as innocently free with maxims about how to ensure the peace of the world as were some of his celebrated predecessors. Speaking from England, the official adviser on science to several governments, the imposing Lord Solly Zuckerman, was unkind about these propensities: "Dyson is a romantic— even a fantasist. This is strange for a man whose brilliance as a mathematician and physicist was manifest by the time he had reached his early twenties. . . . He supports his prescriptions with sweeping, but often dubious, generalities, and makes assertions that are sometimes outrageous."[4]

But it has never been strange for the best scientists to use their great prestige, their special knowledge, and their intelligence to warn the world about where it is going and suggest what it might do. The real thing strange is the very pure combination in such men and women of innocence and attentiveness. They have exceptional standards of truthfulness and accuracy. Those in the science trade agree on who the people are who are really good at it. Someone regarded as a good scientist, as Dyson certainly is, will be listened to by his or her peers. The scientific culture teaches that a good scientific case is persuasive of its nature; such a lesson is largely denied in political circles.

In spite of being wrong and being rebuffed, Dyson still believes—as, indeed, Kennan does—that the internationalism of science remains as good an example as there is of common human endeavor on behalf of the future. Dyson went to the United States in 1946 because that was where all the best science was, and because, "when I came here, the U.S. seemed more socialist than England in Attlee's day—more opportunity, free education for G.I.s, even cheap public housing."

He was first picked out for his brilliance by the great Hans Bethe at Cornell, whose graduate student Dyson had been while still a fellow of Trinity College at Cambridge. He spent a year at Princeton's Institute for Advanced Study in 1948 before being appointed to Cornell for a couple of years. Then Oppenheimer took him back to Princeton permanently. It was Dyson and some other irresistibly bright young men who would transform the postwar physics of the "middle realm" (the wide space between galactic and particle theory), roam across America with Richard Feynman and Julian Schwinger, befriend both Teller and Oppenheimer, and live at the very center of one of the most creative periods known to science.

After his recruitment by Oppenheimer, Dyson was, like all men of his generation conscripted into the cold war. Oppenheimer would try to be both scientist and politician and would come to grief. The ambiguity of the scientist's moral position, pulled by nationalism in politics one way and by internationalism in science the other, brings out the innate contradiction of scientific knowledge. Dyson also lived with this contradiction, though he believed at first that the scientist-in-politics could forswear the seduction of power, even when he was himself caught up in the mad exhilaration released by the exercise of pure will, which *is* science in successful action. Oppenheimer, a Jew, worked on the atom bomb in order to defeat Hitler; Teller, a Hungarian, worked on the hydrogen bomb in order to subdue Stalin. In 1956 Dyson rejoined Edward Teller in San Diego to work on safe and peaceable applications of nuclear power.

General Atomic started life in a disused school. It was a typically American compound built with both public and private money, although not a lot of either, and it had on the spot three or four of the best young physicists in the world. They worked furiously and merrily toward the design of a foolproof nuclear reactor, and they brought it off. But at that time, in the late fifties, it proved impossible to keep peacetime and military nuclear research apart. The launch of *Sputnik* had projected the arms race into space; Werner von Braun, Hitler's V-2 genius, soon had an American satellite orbiting alongside the Soviet vehicle. Dyson was pulled into General Atomic's Orion project, which was looking for a more ambitious, and cheaper, space rocket program than a trip to the useless rock and dust of the moon.

There's something agreeably batty about Dyson's space speculations, about his cheerful cost comparison of William Bradford's establishment of the Plymouth colony in 1620 with what it might take to set up a space colony with its own little greenhouses on an asteroid by the year 2000. Such wildness invigorates science, but it also indicates a certain airiness about Dyson's career.

Abruptly, the Orion project was ditched after a year as a civilian space program, and its developments were sucked into the nuclear weapons race by the federal government, Dyson with them.

The Orion rockets were exceptionally mucky things, adding more than their share of radioactive dust to the planet's atmosphere. Dyson spent a brief time at Teller's Livermore Laboratory to work on the rocket's cleanliness problem and became caught up in the work on the new fission-free nuclear weapons. These weapons, which would go off with a hell of an explosion, would shake neutrons out and contaminate a queasy population, but would not poison the planet for years with fallout, could be made from a much smaller stock of fuel than the fissile weapons, and would be usable, the tacticians began to think, in infantry warfare. They would enable the United States to break free from the doctrine of massive retaliation and to pursue what very briefly appeared to be the safer, less apocalyptic strategy of "limited nuclear war."

Dyson stood up for Teller against Hans Bethe when the great physicist tried once again to persuade his government to stop testing nuclear weapons. Dyson even joined the cold warriors in the pages of *Foreign Affairs* to disagree with his mentor. The quarterly was glad to have this new star on their contents page; he was joining the ranks of the hardliners and militarists and was welcome. He became so politically respectable that he was appointed scientific adviser to Kennedy's new Arms Control and Disarmament Agency (ACDA), which was set up in 1961.

ACDA's birth coincided with the launching of more and filthier bomb tests by both superpowers than ever before. The spanking new agency's officers were all made to feel important by being placed on the mailing list for diplomatic telegrams. Everything seemed fine until Dyson turned his deft mathematical skills to a diagram plotting all the tests.

> As soon as the diagram was finished, the situation became clear. The curve of cumulative bomb totals was an almost exact exponential, all the way from 1945 to 1962, with a doubling time of three years. A simple explanation suggested itself for this doubling every three years. It takes roughly three years to plan and carry out a bomb test. Suppose that every completed bomb test raises two new questions which have to be answered by two new bomb tests three years later. Then the exponential curve is explained. Having discovered this profound truth about bomb tests, I was ready to draw the consequences. Some questions have to remain unanswered. At some point we have to stop. That evening I accepted for the first time the inevitability of a test ban.[5]

It was a start. It wasn't much. When Kennedy and Khrushchev signed their test ban treaty after Cuba, Dyson judged it to be hardly more than a distraction. Dyson was, however, captured by Khrushchev himself. In 1945, before the cold war froze up all such channels, Dyson had expected to go and work in the Soviet Union, his idealism about the new social experiment of the Soviets still fired by memories of Frank Thompson. His own great gifts had made the acquisition of several languages easy for him, and Russian was among them.

> I read every utterance of Khrushchev that I could lay my hands on. Khrushchev I found invaluable. Unlike other official Russians, he spoke from the heart. No hack speech-writer would have dared to write for him the things he said: often inconsistent, often bombastic, surprisingly often human and personal. I had a strong sense that this was a unique moment in history, when a man so open and so whimsical was in power in Russia. If we did not start quickly to negotiate with him about basic issues in a language he could understand, the opportunity might be gone forever.[6]

Dyson's soaking himself in Khrushchev's writings led him to strongly heretical views about the Cuban Missile Crisis. He saw it as bluff. He believed that Khrushchev was impressively declaring solidarity with his first Hispanic-American ally, but no more. Dyson also thought that the American demolition of the Soviet missile bluff by public display of aerial photography forced the Soviet Union to replace its fictitious missile force at home with a real, vastly expensive one, and that the United States, by banging on about missile gaps and missile advantage, had only redoubled the ever-suspicious Soviets' weapon-building. When he voiced his views on behalf of the ACDA to a Senate committee, he added, after a prompt from Senator William Fulbright, that he thought the bombastic Khrushchev's threat against the United States that he had come out with so infelicitously at the United Nations, "We shall bury you," was no more than the good Russian's boast that Mother Russia would outlive her rival—not an order to his air force.

Dyson was having second thoughts about his recruitment into the cold war; slowly he began to think and write on behalf of a safer homeland. "It wasn't a question of politics—or not in the usual sense. It was a matter of human survival. After all, whatever our present differences, they will hardly matter at all in five hundred years."

In 1962 he joined the Pugwash Conferences, an international meeting of scientists that began at the home of a wealthy industrialist, Cyrus Eaton, in Pugwash, Nova Scotia. The conferences were held mainly between Americans, English, and Russians. That year the participants conducted a "thought experiment": they practiced living and thinking in the language and relations of a world that did not yet exist—a world disarmed of its nuclear weaponry.

"One overwhelming impression of those four days remains. I lived for four days mentally in a disarmed world, with all its difficulties, and the longer I was there the better I liked it. At the end of the four days I did not at all feel happy to return to the present-day world of deterrence and counterforce, missiles and megatons. I would very seriously recommend that all military experts and political leaders who have learned to take our present world for granted should from time to time be exposed to an experience like mine. It would refresh their imaginations and enlarge their hopes."

Dyson went on to recollect the difficulty Russians have in understanding Western systems of government—the ruling party slowness, the negotiation, and the inability to enforce decisions. "The Russians must learn to live with our slowness, just as we have to learn to live with their secretiveness. Our slowness and their secretiveness are facts of history which must be understood and tolerated, but which cannot be arbitrarily overridden."

As the weapons piled up and the Vietnam War began to amass its monstrous numbers of explosives detonated and sorties flown, Dyson strained more and more ardently against the limits of official policy and the glossings of official speech. He took up George Kennan's appeal that the United States and NATO renounce their declared intention to use nuclear weapons first if necessary. "No First Use" became the slogan leveled against the nameless stooges Dyson was quoted as saying, as one breezy example did to him at a meeting in 1966, "I think it might be a good idea to throw in a nuke now and then just to keep the other side guessing." Dyson gaped. Then he and three friends banded together to write a paper, in technical idiom and for Defense Department eyes only, that demonstrated fully just how dreadful a mistake it would be to use tactical nuclear weapons in Vietnam, even from a strictly military point of view.

Dyson began to evolve his own homely doctrine for a safer world, one in which, quite properly, "peaceful countries will be well armed and well organized in self-defense" and will be helped to this condition, in his surprising view, by new, small, and devastatingly accurate weapons suited to a fixed frontiers, especially those of small independent states. Such weapons had been used to much effect in the 1973–74 Arab-Israeli War.

There is an endearing artlessness in all that Dyson the scientist did and thought in the political arena. From 1970 onward he spoke out more and more openly against the development of and reliance on nuclear weapons as the foundation of world foreign policy. He pointed out emphatically, as did many others, that that policy was driven by a kind of reverse torque: all the thought is given to weapons development and none to their use. Dyson tried his hardest to think about the practical use and usefulness of a nation's weapons.

In the seventies he found one of those simple, single concepts through which people of radically different viewpoints may nonetheless manage to see things, for a moment, the same way. Khrushchev's "peaceful coexistence" was one such; more grimly, so was "mutual assured destruction." Dyson's single concept was "live and let live."

It is banal enough, or modest and unpretentious, as Dyson claims. "Live and let live" made for flexible negotiation. It eliminated the obsession with exact parity in each quarter of the nuclear arsenal, as well as any game plans toying with the sport of limited nuclear war. With an eye on James Fallows's book *National Defense*, which describes the low morale and loss of purpose in the national army, Dyson contends that "live and let live" directs policy back to the business of defending frontiers, the type of warfare that does have a purpose and is believable in a way that plans for nuclear warfare are not, and cannot be.

Zuckerman is cutting about Dyson's concept. "Is it conceivable," he asks, "that Freeman Dyson does not know that the problem of creating an effective non-nuclear defense has been in the forefront of NATO discussion from the very start?" He concludes, with disdain, that Dyson simply lacks a sense of political reality; he does, however, agree with Dyson that the only hope for a safer future, with or without the cold war, lies in having all political leaders realize that decades of nuclear development and deployment have brought no greater safety to the world, nor has any nation been able to buy immunity.[7]

In 1990, as the migrants streamed westward through the opened doors in what had been the iron curtain, and the new Germany was embraced into the old NATO, and soldier-bureaucrats conferred with politicians on where they could put their new nuclear weaponry, the truism was still not acknowledged as a truth.

Dyson's long traverse took him from Kennedy's ACDA to the small and shabby office of his local peace movement. Zuckerman would say Dyson has become ineffectual. Dyson does have his goofiness. But he speaks for a force stronger than the rigidity of politicians; he speaks for the continuingly *progressive* force of science itself, in spite of Los Alamos and Livermore, in the history of the superpowers. "That's why, when I began as an undergraduate, I was a scientist and a socialist. Old Charles Snow was not so very wrong when he said, 'Scientists have the future in their bones.' "

The tension between science as the world's most powerful mode of thought and its political function in the cold war has electrified Freeman Dyson's career. Countering those who would let our schoolchildren believe that science is ugly, authoritarian, and destructive, Dyson's vision of science focuses on its three beautiful faces: its artfulness, its subversiveness, and its creative exchange of knowledge in the spirit of international harmony, nation speaking peace unto nation.

12

FICTIONS II

Dead Ends—End-of-the-World Movies

THE camera approaches a woman crouched on a bench, her hands half over her face, pulling at it. The camera comes so close, the face bulges out of true and the woman looks deformed, like a victim of Down's syndrome. She is out of her mind, one glaucous eye looking up at the camera without cognition, moaning endlessly, her hair dirty and matted, her skin blackened. The camera studies her objectively. It moves on.

She is sitting on a bench beside other survivors, all filthy, tattered, scorched. The woman moans and gibbers, and behind her broken muttering sounds the thin, ceaseless soughing of the wind. The camera looks up, widening and deepening the shot to take in a huddled, listless crowd of people in like case. In addition to a few armed soldiers, there is a man who is evidently a doctor, also dirty and disheveled; he makes some show of authority but is helpless to do much. He speaks directly to the camera, distraught. All he can do is separate the victims into those dying quickly and those dying slowly. Impassive, the camera listens on our behalf.

These are a few, commonplace seconds from Peter Watkins's masterpiece *The War Game,* made at his proposal for the BBC, which never showed it.

The War Game emerged from the surge in Britain of rational terror that gave birth to the early Campaign for Nuclear Disarmament (CND) and came to a peak during the 1962 Cuban Missile Crisis. Whatever else was murky about the human prospect at that time, it was clear that Soviet weapons might only be able to do limited damage in the United States, but that Khrushchev could, if he chose, obliterate Western Europe.

There has been much humbug since about the edict banning the film from the British national TV channels in 1966 after a lot of characteristically British evasiveness and hypocrisy. At the time the politically minded were suitably outraged at the usual efforts to prevent the citizenry from knowing what fatal ends their rulers might be planning for them, and a great many people crowded into the local cinemas, which were promptly screening the film, to find out what was in store.

There is no doubt that the British government, Labour at the time, *was* trying to avoid any fresh commotion over nuclear weapons that might offend its American masters and hurry Britain off the world stage even quicker. The cold war was at its catatonic moment: each side rigid with apprehension, keeping its nuclear button fully primed twenty-four hours a day. *The War Game* would only upset everybody and might even cause a tremor of uncivil disobedience in the narrow space left for such expression in the British polity.

So the film had its brief hour as a cause célèbre. It was, so the people said who had seen it and told their agreeably horrified friends who had not, a sort of glassily calm horror movie. It showed shocking things: an incinerated baby, protesters shot, the hopelessness that must follow nuclear devastation, the certainty of British defeat. It was surely intended to be a blow, however forlorn, against the fixity of cold war; it visualized for people how completely dead an end present politics had brought them to.

Shot in black and white, which cannot fail to give it now something of the flavor (and the authority) of old newsreel, the film tells the story of a nuclear attack on southeast England. Watkins used only local residents of the area and no professional actors; he was one of a group of English directors whose politics and aesthetics led them to build from the germ of an idea in collaboration with ordinary people, slowly improvising an action until it coalesced and solidified as a work of art in the process of endless and difficult rehearsal.

Watkins saw that to make nuclear war present to the post-Hiroshima generation he had to use the documentary form, whose immediacy and conventions insisted to the viewers that what they see is true and alive. For somebody with Watkins's human and moral purposes in *The War Game* making us feel that the nature of nuclear war is truly in front of us cannot just be a matter of being shocking. The horror film shocks us regularly; but the shock of the horror film is for its own sake. A person cannot turn into a fly, but if he could, it would happen like this: look, come on, *look*. Watkins was on a different tack.

At any moment after about 1957 nuclear war could have happened. By the time *Dr. Strangelove* (1964) and *Fail-Safe* (1964) had been made, the most persistent nightmare of ordinary people was that the unimaginable could happen by accident. There were two responses to this nightmare. The first, most robustly exemplified by Herman Kahn in *On Thermonuclear War* was the response of the planner-utilitarian. He reckoned up the numbers to see whether nuclear war was practicable; Kahn concluded that it was.

The second response was that of the moviemaker, who tried to imagine what nuclear war, accidental or not, would really be like.

"He who fights monsters," Nietzsche observes, "should look to it that in the process he himself does not become a monster. . . . When you gaze long into an abyss, the abyss also gazes into you."[1] The artist-documentarian cannot avoid this danger. He brings his camera to imagined horrors and must look at them calmly and objectively. Watkins had to ask his 1964 production team for certain "special effects": the accurate simulation of charred flesh on dead and living bodies, smoke-blackened faces, weeping wounds, toothlessness, and the baldness of thick scalp skin with sparse hair.

The moment of the explosion is the most thrilling; everyone looks out eagerly for that. Watkins brought off his blinding effect by overexposing the film as the bomb detonated: the bright white light fills the screen with only the pale shadow of people and pots and broken glass faintly visible in the glare. Then the wind machine comes up hard on the soundtrack and combines with the sounds of smashing glass and splintering wood and a remote scream as the householder is blown out of our vision.

Of course, there is a ghastly relish in all this, the usual delicious horror that attends all representations of violence and disaster. The artist waits for us to get over the horrified pleasure of seeing a nuclear blast without being hurt ourselves, and then forces us back to thought, and to moral sympathy.

The underlying political situation is not implausible. The United States threatens to use atomic weapons against the Chinese, who are sending troops to help the North Vietnamese army. This being before the Sino-Soviet quarrel, the Soviet Union comes to its ally's aid and occupies West Berlin. NATO forces move to eject the Soviets; the Warsaw Pact countries meet them head-on in East Germany. NATO falls back on its famously "flexible response" by firing battlefield nuclear weapons, and enemy missiles, launched to wipe out nuclear emplacements in Britain, miss their target and hit Kent.

All the cogent place-names are in the picture, even if the superpowers have to seem a little touchy for such a plot to work. Once the evacuation begins, however, the thought prompted by *The War Game* is, "Yes, it would be like this." People are piled into billets at short notice. Some householders refuse to take evacuees and are compelled to by law. British audiences saw the crowded trains and buses, wholesale exodus, gasoline rationing, and panic buying of canned goods and summoned up its folk memory and newsreels of the blitz. In any case, these events precede the big bang, to which everyone naturally looks forward with that agreeable frisson.

After the explosion, apart from a little help from what is known about Hiroshima, the artist in Watkins is on his own. There are the cinder corpses, the family shocked out of its wits, the mercy killings; the police shoot (off-screen) those who will die soon anyway. There is dullness and apathy rather than the unspeakable scenes described to John Hersey when he went to the ruined city some weeks after 6 August 1945 to write *Hiroshima*. Watkins did not include the six men walking with faces still upturned to the sky, out of which the light and heat came and melted the eyes in their heads. And he spared us the woman looking for her baby, dragging behind her like her own shadow the blackened skin flayed from her body in an instant, still attached to her rawness by one heel.

But he made us see that the two great institutions of refuge in modern society, the family and the welfare state, cannot protect their charges from such a disaster. The film told its audience quite flatly that there is no defense against these weapons. Perhaps that was why the BBC took such fright at the film—the broadcasting institutions of Britain, always squeamish about upsetting either their governors or their audience, decided against bothering either with such uncomfortable thoughts.

Much of the latter part of the film has as its only soundtrack the desolate blowing of a thin wind. The postnuclear world it depicts is helpless rather than horrible, stupefied rather than victimized; all that can be done is to slow the pace of life toward death down to a uniform shuffle. The moral of it all is the documentarian's only ethics: not, "don't do it," but, "this is what it is like." His tense is the present indicative.

Watkins's vivid imagery, however, came to dominate the popular imagination as a picture of what nuclear war would really be like. It served, that is, as an extraordinary vehicle of public education in English-speaking countries, particularly after 1980 when the peace movements, as they began to have a significant effect on the making of foreign policy, created new and even larger audiences for *The War Game*. Precisely because of official lying and timidity, Watkins's film had a long life. It slowly and genuinely changed how people imagined what might happen.

It is a pleasure to record, however, how one of the better-known if less-gifted BBC producers responded when asked by an American researcher in 1971 for her opinion of the BBC ban on *The War Game*. Her response illustrated just how deeply the old folk tales of a well-loved culture can lift up the heart in spite of the defeatists all around you.

Well, I have great sympathy with the BBC, having commissioned *The War Game* and then refusing to show it. There's no difficulty in seeing *The War Game* if you want to see it. . . . But, having lived in the South-east of England throughout the war, having seen how people behave in circumstances of war and bombing, it was an absolute slander on humanity. His observations were profoundly wrong. . . . This is not the way people behave toward each other in times of stress. . . . I think it was a *stinking* film. We don't need these emotional, left-wing intellectuals to tell us that we can destroy the world.

When I see a film like *The War Game*, I am ashamed of it, and I think the BBC was quite right to ban it. And I hope, shown in a country like this where you have not got the personal experience of seeing how ordinary people react to an extraordinary situation, that you will not believe that this is true. I'm not a left-wing intellectual, am I?[2]

II

Watkins's film was so striking because so few films at that time had tried to imagine what happened after the moment of explosion. Two useful examples, *Fail-Safe* and *Dr. Strangelove*, both issued in 1964, show with some unevenness how to get to that moment. But apart from assorted sillinesses about mutant ants and the like, only *On the Beach* (1959), the film made from Nevil Shute's worthy little novel (1958), had until then attempted to envision what it would be like when the bomb went off. It provides the contrast that highlights Watkins's achievement: the deep difference (I believe) that he made, in Britain at least, to people's willingness to continue to support the presence of nuclear weapons.

On the Beach was made by the respectably famous Stanley Kramer and had a star-studded cast—Gregory Peck, Ava Gardner, Fred Astaire, Anthony Perkins. The film is completely unmemorable. It stifles the kind of thought it is supposed to prompt. Watkins's strength in *The War Game* is exactly that he. gazes deep into the abyss and sees on our behalf what he is likely to see. John Keats once said of unpleasant subject matter in art that it must "excite a momentous depth of speculation in which to bury its repulsiveness," otherwise it *merely* repels, like the average horror movie.[3] By this standard, *The War Game* passes handsomely: it prompts us to deep speculation of exactly the kind its censors denied, to ask, would people act like this? Look like this? Could anything be worth this? Is this a near danger? *On the Beach,* on the other hand, could so little deal with repulsiveness that it excited no speculation at all.

The story takes place in the novelist's beloved Australia, which is awaiting the clouds of radioactive dust that have destroyed all life in the northern hemisphere after a nuclear war. As everyone noted at the time, the film has none of the horribly necessary details, still less those that could excite Keats's momentous depth of speculation. Nobody dies onscreen; nobody is dead onscreen (or in the novel). Gregory Peck plays the captain of a U.S. nuclear submarine who is sent north to check the levels and drift of the radioactive fallout. He calls in, inevitably, at San Francisco, and the submarine passes below a hauntingly empty Golden Gate Bridge while the periscope's view of the city shows us its silent streets. A yeoman whose family has died in the city jumps ship by the escape hatch to go ashore and die in his hometown.

The theme of the movie, insofar as it has one beyond the banal observation that nuclear war could destroy the human race, is the question of how to die well when death is certain. Answers to such a question are a matter of style rather than morality, a conclusion that Kramer himself evidently came to. So, the movie sounds no warning, nor does it have any of Watkins's documentary intention. The corpseless streets fail to suggest the decaying bodies rotting in a billion bedrooms; they are more like a picture of the world-become-the-*Marie Celeste*, a vacant globe sailing on through space to puzzle some navigators boarding from another planet in a few millennia.

Each of the characters will die folded in the frame of feelings given by their past. Well, that is likely enough for everybody. But the feelings in this case combine facile resignation with the comfortable sureties of victory in World War Two. Thus Gregory Peck's captain takes his ship to die at sea, in a mournful, contentless act of obeisance to naval duty. Ava Gardner plays a woman who has fallen for him before he can acknowledge through the pain of his loss of his family that he has fallen for her. She bids him a last farewell, then drives as fast as she can out to the headland to watch her man sail to his destiny down the waterway of sunset. The shot echoes the many such shots from war movies.

So too with the death of Fred Astaire's character, a scientific adviser much involved in the maintenance of nuclear weapons. As the cities in the north of Australia start to fall silent, he fulfills a life's ambition by driving his pet racing car to victory in the Australian Grand Prix—all around him other drivers are deliberately immolating themselves in wild crashes—screws the winning plate to his vehicle, and then smilingly gasses himself with exhaust fumes from his car in his sealed garage.

Finally, the young married couple, he a naval officer from the sub, she a new mother, feel the onset of uncontrollable irritability, the first symptom of radiation sickness, and take the government-issued suicide pills. The film ends with a slow shot of a windswept square, a banner from a Salvation Army rally flapping, "Brother, there is still time," and the dead leaves blowing. As with so many other cold war movies, *On the Beach*, passionless and soft-centered but worthy as it was, could comprehend the terrible finalities of its subject only in terms of the earlier, much less terrible military conclusion of 1945, in which the losses were all personal and had their measure.

III

The War Game took the measure of present ignorance, so thoroughly served by *On the Beach*, and told a truthful fiction. *The War Game*'s anonymous sufferers with their scorched faces are physically present in a way stars can never be. The British Campaign for Nuclear Disarmament gratefully seized upon the banned movie and showed it in countless church and village halls during its revivalist meetings between 1980 and 1984. The British government could have mitigated the satisfaction that audiences derived from viewing forbidden fruit by asking the BBC to screen the film, but the habits of prohibition and secrecy go deep in British culture.

As nuclear politics in Europe heated up in those years, however, another left-wing filmmaker, Mick Jackson, and his scriptwriter to match, Barry Hines, obtained funds to make *Threads*, a new film about the advent of nuclear war. The BBC put it out nervously in 1983, following up the first showing with an instantaneous studio discussion that included the British defense secretary Michael Heseltine and the campaign's beautiful, fluent, and quick-witted chairwoman, Joan Ruddock. (The story went that the same defense secretary was bluntly advised by his henchmen never to appear on TV opposite Mrs. Ruddock: "She's very good-looking and very intelligent; she'll cut you to pieces.") One side repeated the fatuous bromide that nuclear weapons had kept the peace for forty years, and the other repeated the dependable litany that no country could survive nuclear war and that it was more than time that Britain, an unimportant place with a very large overdraft, defended itself without nuclear weapons like almost every other nation in the world.

Such discussions are always staged by British television after politically controversial programs in order to corroborate the constitutional axiom that politics is made up of three groups: two of which are extremist and easily identified by the noise they make, one of which is balanced and moderate, and makes no noise at all. The last group believes that truth is found at the point of balance, which is why it is so quiet and well mannered. By these lights, everybody is discouraged from holding commitments of any kind, and the BBC stays out of hot water. The model derives from the old opposition, Fascism and Communism. It is a cold war premise that we on our side are the balanced ones.

Even so, studio discussion was hard put to outbalance *Threads*. The movie begins with the arrival of Soviet troops in Iran, with Iranian complicity, goes on to a murky incident in the Persian Gulf in which a U.S. submarine is sunk, and concludes with a single exchange of local nuclear weapons. Tension mounts, U.S. troops from Jimmy Carter's Rapid Deployment Force land around the Iranian oil stations, Berlin is once more at the center of the action, and the missiles begin to fly. In the ultimate war, 3,000 megatons are exploded, 200 of them on Britain. There are nine million immediate casualties, the nuclear winter supervenes, and the whole country starts to die horribly of radiation sickness. The world has rushed headlong onto its nuclear sword.

Threads is a full-length film. It follows Watkins's lead in using a documentary form for much of the first hour and in using no stars but a few unknown professionals and various of the people of Sheffield, which city loaned its streets and premises to the movie as an act of civic duty.

But *Threads* also includes an honest effort to occupy the space between the statistics of destruction and personal lives, in this case, a pair of lovers, their families, and the baby they are expecting.

The statistics of death and destruction are presented with the news of the war, appearing onscreen as bare facts accompanied by the clatter of the teleprinter. At times they are accompanied instead by the patient, carefully elocuted exposition of state videos telling the populace what to do when Armageddon comes. Such films, standard products in the NATO nations at the time, merit a display studio in any museum of the age. Like the relics of other monsters now housed in museums, they strain to the limits the powers of belief they are there to confirm.

The calm voice repeats the instruction of the British Home Office leaflet, *Protect and Survive* (1980). At times a delirious tingling sort of music intrudes; it signifies fallout. The audience is told to keep calm. Everything is precautionary, caution itself having been blown to the winds.

The unhappy families in the movie are bending to their useless tasks when the bomb bursts. When the siren wails in the center of the city, a girl clutches her hair wildly, a man runs and crashes into a woman, knocking her sprawling, a woman empties her bladder and the urine cascades about her shoes. Ninety or so miles away the shapely mushroom cloud unrolls across the sky. A boy stares upward, his hand to his mouth: "They've done it."

The blast arrives, and then the tumbling balls of fire. As in *The War Game*, Jackson gives us the flash of blinding white light by overexposing the film so that only the faint outline of the woman is seen as the window glass blows inward. Jackson knew that grand screen effects would be no good here, that television thrives on small, quickly cut details: two milk bottles melted at 3,000 degrees Fahrenheit; a cat turned to cinder; two hands at a grill first spouting blood, then turning black; flames roaring at a window, in the gutter, inside a car, across a face, pulled and tugged by the terrific wind.

That is one of Jackson's strongest images: his wind machine drives the flames horizontally, and their roaring is everywhere. The camera pulls back for a glimpse of the city skyline and the firestorm around it, then closes once more on a shop window sucked out, a man's head glowing.

After fire and gale, the dust begins to fall. The calm voice instructs us on the quantities of dust such explosions would puff into the atmosphere, poisoning it, and explains that it would slowly gather in smoky clouds through which the sun could not penetrate, below which the temperature would fall by twenty-five degrees, while the thin wind still blew, the frosts came, and the earth died twice over.

The young pregnant girl walks around the ruined city. Her parents and grandmother succumb to radiation sickness, retching and vomiting loops of bile, fouling themselves involuntarily. The grandmother dies, is laboriously dragged upstairs, and, as the booklet specifies, is laid out on the floor as best her children can, wrapped in a blanket. Her shod feet poke out of it. In time, rats come and nose busily about the corpse.

At this point the narrative has a problem with its audience. Some of what has been shown has indeed been shocking: the quick as well as the slow destruction of people's bodies, the violation of their self-explanatory pleasures (the young hero's collection of pet birds, for instance). But telling a tale of unrelieved physical deterioration to audiences sitting in comfortable living rooms soon runs up against fatigue and indifference. People have a shock limit. Once the gleeful fascination with the explosion and with the horror is done, all that needs to be represented are the noise of retching and the facts of dereliction.

Jackson and Hines keep their story going by following the young mother, her baby, and then her teenage child into a familiar science fiction trope. The world remains a wasteland under a dark sky, and the postnuclear generation turns out to be a version of the undead. Its androids kill for food and ignore the bonds of trust and love. Although *Threads* keeps its bare, bleak style well away from cliché, we have met these characters before as zombies and body-snatchers in several other movies. Maybe that's how we would be. But this is a story, and it cannot hold us beyond the point at which human sympathy runs out. This is the unsolvable riddle at the heart of all end-of-the-world movies: they can go no further down a dead-end road.

This difficulty caused the U.S. equivalent of *Threads*, shown about the same time, to peter out in the same way. *The Day After* (1983) recruited some of the stars who had noticed the danger the world was in, headed by Jason Robards. The Americans cast their cautionary tale in the conventions of a family movie. They did their best, but those conventions turned out to be so stifling that the movie never got beyond clichés: the overworked surgeon (Robards) trying to work in the underequipped hospital, his attempts to find his own home in the debris, the postnuclear hunting-and-gathering society that turns out to look very much like the old frontier.

One notable absence in all these films speaks straight from the heart of popular feeling about the nuclear danger at the center of the cold war. None of them has attempted to imagine the sequence of decisions that led to the end of the road, nor the circumstances and dispositions of those who made them. This is deliberate. We the people are those to whom the devastation happens. We learn about its advance only from glimpses of news on television. The bombers take off. Nobody says no. There are scenes of terrified apprehension and, in *Threads,* laughably small demonstrations and inadequate street speeches.

The science in these films is much better than their politics. But for thought about the then very present danger, let alone momentous speculation, we needed from them a keener sense of how it might have happened. We needed the shock of recognizing the ordinary people who might have made it happen.

13

FICTIONS III

Loyalty and Lying—Spy Stories

FRANZ Kafka is the poet and prophet of the "world without memory" created by a century of ceaseless lying by states and the organs of propaganda. Kafka foresaw the living nightmare of rule by state bureaucracy. He detailed the operations of a system that had no point except its own systematic operation: the dreary round of official questioning, the interminable waiting in blank rooms on hard benches, the dull corridors and empty flights of stairs, all merging into one endless, pointless process. After Hitler and Stalin, after *1984* and *The Trial,* these are both the appearance and the reality of state rule and its meaninglessly efficient surveillance.

Eternal vigilance being the price of freedom, watching is a costly arm of bureaucracy. To spy is to watch without being watched yourself, but it is, of course, the first concern of spies to watch other spies. Gradually, it has become the sole pastime of espionage. The subject and object of spying was spying, a perfect, vicious circle.[1]

Espionage used to be different. In the days of SOE (Special Operations Executive) and OSS (Office of Strategic Services), as the hard ending of Frank Thompson's war showed, intelligence, in all senses, was hard to come by and was needed as a matter of life or death. Whether it was Alan Turing's code-breaking efforts, the Nazis' Enigma cypher, or Freeman Dyson adding up the dead aircrews left behind in thousand-bomber raids, intelligence, once it broke through the ranks of stupidity, spoke for life and the cause that was clean.

But the structures of intelligence remained in place long after they were welcome. The dismal record of all those initials, OSS, MI5, MI6, the CIA and its secular officers in the FBI once peace broke out, would be laughable if it were ever added up in public. These agencies have had a poisonous effect on ordinary human freedom, both national and individual, and are a powerful, inert force for induration in the political feelings that contain and shape us all.

It is only during Kafka's century that espionage has become an arm of the state. Kafka foresaw the powerfully *incarcerating*, as well as amnesiac, tendency of the state. Michel Foucault, the later theorist, names ours "the carceral society" and remarks that surveillance is its latest but most dominant institution.[2] Kafka and Foucault see, in other words, that such a society has to watch and imprison itself endlessly. Totalitarianism brings with it, as the concept indicates, the steady construction of a state capable of invading and pervading in its totality every corner of public and private life. Children in Stalin's Russia and Hitler's Third Reich were encouraged to peach on the deviations of their parents. Watchers were appointed to neighborhoods and their reports taken as proof of their neighbors' treacherous nature and unwarrantable disloyalty. But *all* societies, as they evolve a state, have concurrently developed a system of surveillance.

Of course, it is true that in Britain and the United States intelligence only rarely leads to the body dumped in the waters of the harbor or the punctual immersion of those under suspicion in baths of shit. More usually, it has led to long days of interrogation in return for promises of immunity, to years of humiliation and semidisgrace, to gossip, innuendo, and the judiciously incontinent leaking of calumny and scandal. Unquestionably, the Anglo-American way is better: no bullets in the neck, no darkness at noon.

There have been far fewer corpses in Britain and the United States than in Beria's Soviet Union: to each state the surveillance system that suits it. The KGB grew straight out of the Cheka. The CIA is the enormous, inchoate creature of Washington's sort of administration: a corporation, a system of open patronage, the military partner of the FBI, an army of legitimate gangsters gunning for its enemies in a vast continent where the law is the law only if it carries an automatic weapon. Such an organization generates, certainly, a high degree of loyalty to itself and to its highly abstract image of the nation. In a society whose emblem of moral ratification is cross-examination by jury or by congressional committee, the state's bureaucracy of surveillance can permit itself to deal in truth and truthfulness only as a commodity. That is, its officers, expressing its general will, may agree at times to exchange the truth for advantages. On both sides of the Atlantic, the spy is the last word of the state itself on the supremacy of its own interests. Its reasons command what virtues it has.

II

The CIA was a product of the zeal for wartime intelligence activity and, like its British "cousins" (in the ineffable parlance of the trade), it started by recruiting bright young men from the posh universities. When the FBI under Hoover was an outfit staffed by fat cops, 25 percent of the CIA came from Harvard.[3]

The great, reckless years of the CIA lasted from the end of the Second World War to the early 1960s when the agency made a frightful mess of trying to unseat and assassinate Castro. The records tell us that assorted dizzy ideas for bumping off or putting down the great dictator were floated in the agency, including spiking his cigars with LSD, poisoning the tip of his fountain pen, preparing a dish of clams to blow up in either his face or his insides, and—best of all—emasculating him by a covert operation to depilate his beard.[4]

An organization whose pet schemes are worthy of the Pink Panther may deserve our hilarious gratitude, but hardly our respect. Up to 1961, however, its battle honors were writ large in its annals. Radio Free Europe was credited with doing much to advance the cause among its listeners behind the iron curtain; Edward Lansdale, prince of CIA counterinsurgency, succeeded brilliantly in the Philippines before he failed in Vietnam; in Iran in 1953 the handsome agent Kim Roosevelt, both dashing and cunning, levered the unreliably populist and anti-American Dr. Mohammed Mosaddeq out of power for a mere one million bucks; a year later the agency, with the help of a few bombers, displaced the undesirably left-leaning President Jacobo Arbenz Guzmán in Guatemala and put the robust Colonel Armas in his seat. Its agents almost certainly made away with the wretched Patrice Lumumba for his vaguely pan-African socialist views during the Katanga secession in the Congo in 1961. Understandably agitated by this outfit, Khrushchev formed the KGB as an opposing instrument in 1954 (a date whose lateness in the day will surprise most people).

By then, the language and bureaucratic systems of the CIA had taken on their familiarly corporate and American style. In contrast to the murky anonymity of the British heads of intelligence—whose mere existence could not be formally acknowledged in an Australian court during the famous efforts to ban publication of *Spycatcher* in 1987—the agency bosses are, on the other hand, well-known public figures that include Wild Bill Donovan, Allen Dulles, William Casey, and George Bush. The agency devised a corporate language to go with its publicity releases: "psywar" was the pseudo-scientific doctrine that complemented "massive retaliation," and for the (frequent) occasions on which agents were caught out in some embarrassing corner, they were covered, or half-covered, by "plausible deniability," a full stop in the attribution of responsibility, which nevertheless broke the circuit and the career in question—for instance, that of Oliver North in 1987.

These antics, products of the Truman Doctrine, nuclear superiority, the overweening self-confidence of the start of the Pax Americana, took their color, flair, and flatfootedness from the CIA version of American corporate enterprise, arrogance, and entire visibility. Rhodry Jeffreys-Jones says evenly, "The covert operations gift for publicity was particularly felicitous in a democracy, a type of society in which so much depends on persuasion."[5] However one might wince at the tone of this, it was certainly true that the CIA made its successes central to the cold war of the 1950s by an ingenuous combination of murderousness, spendthrift derring-do, and what used to be called when I was a soldier "swinging the light," the boastful telling of tall but credible tales.

The CIA belongs much more to the history of publicity than the history of foreign policy, which is why it turns up here as the shaping spirit of the spy story. The spy story, in turn, is *the* split symbol of cold war: the imaginary living out of war values in the decadent laboratory of clandestine action.

The balance between decadence and chivalry is a nice one. It tilted steeply toward decadence in the United States after John Kennedy came to power, when, on 16 April 1961, 1,500 hapless Cubans waded into the mud of the Bay of Pigs alongside the well-made new roads that lined the landing ground and of which the agency knew not a thing.

In the sixties, the CIA sank deeper and deeper into arbitrary and self-protective intrigue. As the public learned of its ineptitude and duplicity, especially as a consequence of the loss of faith in the Vietnam war, the agency could no longer offer itself as the brave icon it once was of American mastery of the "great game" played on a world stage against the Last Enemy. For many, it became an object of post-Vietnam disgust, a cause of revulsion from the foul-smelling underside of power.

After lying about it steadily for a year, the CIA's involvement in the 1973 Chilean coup and the murder of Salvador Allende came out in 1974. Thereafter, the agency, however active it remained, pretty well vanished as a force in cold war propaganda. The same year, Director William Colby sacked his infamously half-mad head of counterintelligence, James Jesus Angleton, and the CIA's two decades of self-importance were over. When it had mattered, the agency never knew what was happening: it failed to predict Castro's assumption of power in 1959, the Tet offensive in Saigon in 1968, and the Yom Kippur War in 1973.

Angleton is an emblem of the CIA's fall. Tall, gaunt, a crashing Yale snob, virulently right-wing, he won his decorations feeding dollars to the Italian right in 1947.[6] He held simple Fascist views of the kind espoused by Ezra Pound, whom he admired; he believed the Sino-Soviet split was a plot to mislead the United States. When a wretched defector named Yuri Nosenko came to the United States with a box of secret tricks in 1964, Angleton's instinctive reaction was to assume he was a plant. The poor man was put under "white noise," disoriented by darkness and turned-back clocks, locked away in a tiny cement vault built just for him in Virginia, and interrogated steadily and naked, for *three years*. When after all that time Nosenko stuck to his story, the CIA let him go. This was Angleton at work. He was off his head. He belongs, however, solidly in a section of this book that treats of fiction and falsehoods and the deep truths of each, rather than in the real history, its events and crises.

The truth is that the spy networks of both sides did almost nothing for real history. Utterly absorbed in one another, their only information-gathering activity that seems to have inflected foreign policy has been their routine reconnaissance of surveillance systems, which to be sure, make a difference: they bring back photographs of Soviet missiles in Cuba, or they get shot down, like the unfortunate Gary Powers, and cause an international incident. But the potent significance of "intelligence" and "security" lies not in what they are but in what they seem. In a war fought for nearly five decades with propaganda as well as bullets, the story is the thing.

In Arnold Wesker's admirable play of 1957, *Chips with Everything*, a group of British boys goes through the compulsory two years of military service that all of us enjoyed from the end of the war through to 1960 or so. A senior officer explains to the raw recruits how much better armed and more militarily ready the enemy is than they are. "Remember, a Meteor is more important than a Library." One of the eighteen-year-olds then pipes up, "Sir, if the enemy is so much better armed than we are, what's he waiting for?"[7] The spy services are kept in phony action exactly so that such a question cannot be asked without revealing the questioner as one who simply does not understand how delicately matters stand, how the balance of terror hangs by a whisker, how "confidence," like the confidence of stock exchanges, cannot bear too much reality lest it wane and shrink, and the enemy will at last know it is time to pounce.

For Americans to feel secure, it hardly matters whether the many tales of CIA adventures are true or false, so long as they are plentiful. The national capacity for fantasy is no greater in the United States than it is elsewhere, so far as I know, but it has its distinctive form. The psychotic nationalism of one large segment of American fantasy life is fed by the alleged deeds of spies, fed even more, it may be, as the cold war action shrivels down from Vietnam to the cowardly launch of a few dozen rockets over Tripoli or the invasion of Panama. The mad zest for gun-toting is stoked by adventures like the coup in Santiago, the bloodthirsty dreams of Miami Cubans, and the novels of Tom Clancy. By the same token, Britain's prurient, death-wishing nostalgia for power turns the facts and fictions of the spy trade to a peculiarly British kind of treachery. Ancient, contorted, and rank-smelling Britain, lost in the endless corridors of its class and imperial deviousness, devised a state whose first value is its secretiveness. The secret of the extraordinary visibility of the spy in British political and popular culture is a secret.

III

Power in Britain is symbolized as a sequence of points of access: the door, then the next door, and the next, until you come to the door through which members of your class are not allowed to go. The same with the telephone. You ring up power and are warded off or allowed through by the subordinate—the secretary, the clerk, the aide.

In the United States, by contrast, access, even to grand viziers of state, is impressively easy. Power is based on either money—lots of it, with no intervening symbolism at all—or coercion—the cop, the gun, the bomb. American foreign policy, for instance, has few instruments of negotiation apart from a checkbook and B-52s. British foreign policy has customarily been a play of surfaces and secrets to which it admitted the members of its empire on a strictly regulated basis: the Indians, for example, got further in than anyone else and as a result knew how to make the most successful exit. Britain only reached for the machine gun when dealing with people too unimportant or too unsubtle to respond to the hierarchy of doors: Zulus, Boers, people like that.

Secretiveness is an expression of power. It is a commonplace that the notorious spies who defected to the Soviet Union at the first heights of the cold war—Kim Philby and company—were all schoolboys at those fiercely class-protected and monastic institutions so falsely described as "public." Indeed, calling these private schools "public" neatly reflects the openly assented-to hypocrisy of the institutions themselves. They exact allegiance to ideals of public service while remaining foundations dedicated to the retention of private power. Within them, especially in the 1930s, the period when Guy Burgess was a boy, the formative years of all ruling-class men were spent in an atmosphere of strictly regulated secrecy.

You moved up the school ranks in a sequence of admissions to sacred spaces: the country towns in which such schools are placed were marked by strict, invisible boundaries. Drink, cigarettes, the working class, girls, disapproved-of evening pastimes (movies), the wrong food (fries), girls, were out of bounds. So was the domain of those senior to you. To move anywhere, you knocked and asked permission. Rules written and unwritten were absolute; no reasons were given for them. The order of all preindustrial societies, Durkheim tells us, is implicit, mechanistic, nonrational. Order *is*. So too with English public schools. Rumors abounded, especially of forbidden activities—drunkenness, sex, adventures after hours. Only those with power knew what was going on, in degrees marked by their seniority. Boys in office—the school prefects—knew more than their juniors; staff—the masters—knew a little more; the head knew everything and was a sacred figure, terrifying by virtue of his omniscience and powers of execution.

Plenty of English writers have reminisced about those days, of course, and it's to the point that they mostly hated their school experience and counted themselves on the *bien-pensant* left. George Orwell, Cyril Connolly, T. C. Worsley, as well as the famous poets of the left in the thirties, W. H. Auden, Stephen Spender, and their circle, all stated or implied that the only way to respond to the class brutalities and fatuities of English public-school rituals—once you had seen through them and, what was more, carried the stigma of having done so like a necessary badge—was to defy the institution politically and join the left, even the Communist party.[8]

The very structure of these schools created a small, intellectual, dissenting coterie. Some of its members became the great British Marxist historians after 1945, but the best-known names chose the vocation of traitor.

They were on the left because they could violate their schools from that position. Bred, in the extraordinary way of English public schools, to a state of generalized and yearning high-mindedness—as though, in Jim Hunter's words, forever rehearsing tenor arias to an ideal and unattainable goddess[9]—and alienated by the obvious repulsiveness of English snobbery and venality, some boys found their glowing ideals fulfilled by the similarly powerful silhouette of 1930s communism. Given that, to any intelligent eye, Fascism was the dark threat of the time—and after Spain, the actual monster at the gate—then this mixed chorus of dissidents, from Louis MacNeice to Frank Thompson to James Klugman to Donald Maclean, was unanswerably in the right about British ruling-class complicity with Fascism, pusillanimity in Spain and at Munich, and utter callousness toward its own working people, their poverty, slums, hunger, and lack of jobs.

So when war came, and even better, when Hitler opened the Barbarossa offensive against Soviet Russia and did away with the painful anomaly of the Molotov-Ribbentrop pact, all men of this class and politics could rejoice. The secrecy and otherness that they had flaunted at school could now be pitted against a public enemy. An imagination educated on high, unfocused ideals and scenes of individual heroism could find, in this historical moment, a mode of action that satisfied politics and fantasy, chivalry and the dialectic.

The purest version of solitary heroism on behalf of the socialist future is the parachutist from SOE. But in his case, a hero's death and the passage of time have absolved him of all ambiguity. The British spies, however, were dedicated, competent, and, when necessary, brave. Anthony Blunt's courage in a small boat loaded with explosives coming home from Dunkirk has since been canceled from the annals. They now pose in statuesque contradiction, for the figure of loyalty.

IV

Spies provided international episodes for all the half-century of cold war because they bewitched their audience with the spell of the double life.

They were so satisfyingly contradictory. For a start, they were the only political heroes in a culture in which all politicians had become villains. As the spy holds silently to his passionate convictions, he has to live the lie of seeming to detest those convictions by holding to their opposite. The earnest Communist proves his loyalty to his secret cause by being the best all-American or blue-blooded Englishman. Thus, Kim Philby and Bill Haydon.

The spy's life is private, invisible, as unwatched as possible. Simultaneously, however, the spy carries the burden of the historical significance of the nation, the values and meanings for which he denies himself private satisfactions. Even better, the double life of the spies, in the melodramatic form in which it is narrated, whether in fiction or in reality, demands of them that their most important lies be told so that they can be trusted in the most intimate circumstances. It is therefore no surprise to find that so many spy stories give off such a smell of sex.

Sex remains the location of both our deepest commitments and our most spontaneous betrayals. The spy makes such betrayals and commitments not only in the name of love and desire but also in the name of freedom and nation. His betrayals have the rich, savory taste of sex-and-politics, hence their appeal for the conscientious classes.

The crude way of tasting this ambivalence is found in the standard thriller. Novels such as *Gorky Park* or *Red Storm Rising* allow the reader to relish the usual sex and violence while retaining his (and it's mostly his) crassly unqualified self-righteousness. The subtler thrillers modulate upward for subtler tastes. In historical truth and in historical fiction, double agents had of necessity to weave elaborately false versions of their lives, lies that had to be lived to be believable. Those lies corresponded to the lies covering their colleagues who remained agents in the service of only one country, and whose lives were therefore divided only once. No one was more solicitous of these colleagues as he betrayed them to his enemies than Kim Philby.

Or so it was said. But nobody knows how much damage Philby and his semblables truly did. The cold war went on, unaffected. In such a world, truth, fiction, and facts had all been made convertible by all the denials, assertions, and counterassertions made with such sanctimony.

Somebody has been lying. Or it may be that *everybody* has been lying. The dismal truth about the popular appeal of the spy story is that its readers don't much care either way. The stories are one complex measure of how deeply modern states can press propaganda into the everyday imaginations of their citizens. Because those states and their surveillance systems on this side of the iron curtain have accumulated a huge sediment of almost comically right-wing demonology, the spy stories they have spawned have had a cultural influence far exceeding their political weight.

Ideology has been crucial to the maintenance of state power on both sides of the cold war. For the West, ideology sustains vitality only insofar as it is energized by nationalism. Nationalism is kept on its toes by espionage: the spy proves how spry we have to be in defense of our country. The spy is the apotheosis of the enemy, and the enemy is socialism. What is more, the spy is everywhere, and seductively irresistible.

Anthony Blunt was a particularly happy embodiment of this thesis. He was extremely cultivated and a genuinely fine scholar—his study of Poussin is a classic. He was homosexual, wealthy, solitary, and a confidant—as Keeper of Her Pictures—of the greatest star in the British patriotic firmament, the *Queen.* Nobody could have fitted the fable better.

An envious admiration for duplicity and a relish for censoriousness: these are the familiar, unadmirable traits so vigorously exercised by the spy story. Lying and secrets are immanent in the facts of power. As Dostoyevski's Grand Inquisitor reminds Christ after his adventitious second coming, men simply do not want and cannot bear either too much freedom or too much truth. The power of Holy Mother Church reposes in her miracles and mystery, which taken together confer authority. This has been the message of the political right since the French Revolution.

V

The surveillance arm of the modern nation-state includes the propaganda necessities of that state. Effective surveillance is largely a matter of the management of consent. Obviously, modern states manage consent by managing ignorance: they tell people what it is best for them to know. "People who know" know more, far more, than us. They let us know what it suits them to release and what most happily sorts with what they want us to expect.

Voiced today, such ideas are sure to attract brisk accusations of conspiracy theory. But the history of U.S. politics since Tonkin Bay in 1965 by way of Watergate in the early 1970s to the scandals of Reagan's second term all testify straightforwardly enough to the actuality of government conspiracy. A contradiction inscribed deep in American culture, and in the technology that expresses it, is the one between the state's requirement of secrecy and the vast amount of information inevitably produced by the freedom of access to so many diverse electronic sources—television, radio, computers, public records, telephones, cassettes, satellites, photocopies, listening devices, and all the rest. As a structural necessity, for instance, U.S. Senate hearings are perennially pulled between state secrets and the policy of open information.

As I noted earlier, the security services are uncompromisingly and stupidly right-wing. Kim Philby made speedy headway while expressing straightforwardly pro-Fascist views as his form of cover, because such opinions ensured that he would be readily accepted as a recruit—and accepted, moreover, in the middle of an anti-Fascist war—by British superiors who themselves held virulently right-wing opinions. Furthermore, most of these superiors were terribly dim: when intelligence became absolutely central in the Second World War, the intelligence services in Britain were opened up for the first time to intelligence itself, in its happiest sense. It is painfully clear that when those competent newcomers making their contribution to the war effort left again after 1945, the only intelligent people who stayed were the moles. The story we are told, on the other hand, is that since 1945 only our supremely clever watchfulness has warded off the depredations of these equally clever, devilish traitors.

In Britain the public agents of this activity are the spymaster-journalists. Working in a state whose origins and institutions permit a masterful control of propaganda, these writers police the outreaches of ignorance management. Their eponym is Chapman Pincher.

Pincher's books and Frederick Forsyth's spy works, taken together, bring out all that is most rotten, most arrogantly self-deceiving, most bigoted, about cold war culture in Britain. Pincher's regular disclosures in the Beaverbrook tabloid, the *Daily Express,* and his well-known, garrulously written books, like *Inside Story* (1978) and *Their Trade Is Treachery* (1981), command wide reading: these two books were in their seventh and fifth printings, respectively, in the United States by 1990. They work by indirection and allusion to create a sense of startling revelation, familiar from so much nominally investigative journalism, in defense of our best freedoms. The allusiveness is studied:

> Though Sir Harold Wilson now seems deeply perturbed by the extremist penetration of the Labour party, he was inclined while in office to dismiss warning about it by the security authorities. I understand that he was told how an MI5 agent had heard a Russian intelligence officer claim from memory that thirty-one Labour MPs were "full Party members completely on our side and who will do anything to help us." Wilson is said to have dismissed the information as "reds under the beds" propaganda. . . .
>
> MI5 has evidence from defectors to the effect that some crypto-Communist MPs are under such close day-to-day control that they are used to ask parliamentary questions calculated to damage the interests either of Britain or her allies. Soviet bloc intelligence officers are detailed to make a close study of Parliament and to make use of its privileges, wherever possible. They will frame an embarrassing question based on intelligence material and induce one crypto-Communist MP to ask it. A second MP will then ask the even more embarrassing supplementary question.
>
> Some of the MPs are even named—by their code-name—in KGB radio traffic. The late Konni Zilliacus was one of them, and Driberg was another. I greatly regret that, at this stage, I am unable to name others who are still alive and shelter behind the libel laws, as it is difficult to induce any intelligence source to appear as a witness, as MI5 itself knows only too well.[10]

How could anyone—*anyone*—take this rubbish seriously? Every page reads like this. Nobody could possibly remember the details or the relative significance of each figure in this torrent of muffled gossip, all got up in the snobbish baroque of embassies, London clubs, houses in Belgravia, lords, knights, and "highly placed sources."

It is easy to understand how the intelligence services give it all so much time. They were and are terrified of being cut out of the American circuitries.

After the discovery that Communists had penetrated British intelligence during the war, the CIA—especially under James Jesus Angleton—threatened in a fearful huff never to speak to the British again. That would have been as terminally humiliating to British power as the country not having its own three or four nuclear submarines. In a frenzy of abject self-justification, the British government elites of both parties promised Angleton and his successors that they would do anything to keep themselves in favor. This gratifying subjugation was completed in 1985 when legal trade unions were outlawed for anyone working at the British government's communication headquarters at Cheltenham.

These maneuvers have been fueled by the constantly bubbling current of anticommunism that has been kept on the boil in the sensibilities of the British and American nations throughout the cold war epoch. Anyone trying to maintain a sensible indifference to the ludicrous tattle reported by the likes of Chapman Pincher in Britain and Robert Lamphere in the United States inhales a whiff of the bubbling brew, and the government soon has no trouble with its latest measures to enhance security. An unnecessary, barely controllable, and very expensive espionage bureaucracy has been transformed in popular consciousness into handfuls of brave, underpaid men and women hiding in foreign capitals, constantly threatened by Communist treachery, which they overcome in spite of efforts by spineless, gullible, and uncomprehending politicians to deal feebly with the ruthless Soviets.

This is the Angletonian vision. When the prize defector of all time, Anatoly Golitsyn, came across in 1961, he claimed that the KGB was absolutely *everywhere*.[11] The Labour party was riddled with agents. Yugoslavia was still a Soviet pawn, Tito notwithstanding. China was still in Moscow's pocket. *All* other defectors were KGB plants. As for MI5, well. . . . Angleton, who was in a position to send great electric surges of *grande peur* and anticommunism through the political system, lapped it all up. Golitsyn's gripping story overshadowed the reports of paranoia, incompetence, and petty rivalry being made by whistle-blowers like Philip Agee, and it flattered the appalling bigotry and megalomania of J. Edgar Hoover.

The judicious management of news about such men, together with a steamy, inaccurate literature in which truth cannot be distinguished from fiction, all served well the national myth in which Russians ("Ivans") were indeed hard, brutal, cold, and ugly; in which the enemy constantly threatened to uncover the secrets of our defense; in which the way of life of the law-abiding and civilized consumer could only be maintained by special agents, inevitably less civilized, roaming the mean streets of Bombay, Berlin, Budapest, Baghdad, with shoulder holsters and their strong, cynical loyalty to the nation. The theodicy of the literature teaches that God is a capitalist and a liberal, that evil is communism, and that the end of the world may come in the form of nuclear war but that this would be worth it if our freedom remained intact after the bang.

VI

Spy stories are exercises in a debased nationalism. Nationalism has been the besetting danger of the world for nearly two centuries, but in the cold war it was a rallying point against socialist *inter*nationalism with its imagined hordes of a vaguely Mongolian or Tartar outline who would level all individualities in a common gray mass. Perhaps the feelings that give shape to the spy stories could more accurately be termed patriotic rather than nationalist. A patriot is motivated by single, simple sentiment—love of country. Love of country is a passion rather than a belief, although it is certainly the source of strong motives. Love of country by and large works against any very detached or critical views of that country. When criticisms have in honesty to be faced, patriotism becomes violently selective, repudiating some realities, embracing others as the ones worth fighting, even dying, for.

The books on the spy shelf are mostly patriotic. When Dr. Johnson uttered his famous dictum about patriotism being the last refuge of the scoundrel, he spoke from the Enlightenment's view of disinterested reason; he knew that releasing the deep atavistic passions characteristic of patriotism could flood reason and judgment. Patriotism is charged up by nationalism; once the vital passions of patriotism are released, great images become household idols and national flags are flying everywhere. Patriotism is never busier than during periods of national decline.

The rhetorical figures of British decline during the four and a half decades of cold war have been largely imperial. As its power has leaked away, Britain has artfully combined condescension and toadying toward its imperial successor across the Atlantic and waged cold war on the terms that would best keep alive the celebrated "special relationship." It has rewritten the history of freedoms prized from the clutch of its reactionary ruling classes as the story of its own exemplary progress toward democracy. Drawing remorselessly on the happy instance of the country's good luck and considerable courage in avoiding defeat by the Nazis, it represents itself to its American masters as the key link between a United States of Europe and the United States of America.

In this sustained and intricate policy, British spies stand as sentries of British independence and as soldier-heroes of the cold war. They are talismans of the country's standing, both as a world power and as an anticommunist stalwart. Thus, when Pincher writes from his Elizabethan farmhouse in *Inside Story*, "I have spent a large slice of my life investigating and trying to expose the machinations of the extreme Left against the interests of the country I love," he has the scoundrel's way with him of conserving patriotism as the domain of the right, and designating the left as necessarily divided in its loyalties and probably treacherous.[12]

What is at stake is an ideological struggle over national identity, which in Britain is inseparable from the struggle over class identity. The force field of the cold war takes all socialist sympathies that *start* from premises of nations divided in themselves and promises of the great freedoms of internationalism and twists these sympathies into expressions of treason. It cannot be doubted that the four infamous spies from England did much to transform this myth into history. But it also cannot be doubted that British preoccupation with them has lost all political balance and become a national madness.

VII

The most haunting and pervasive meditation on these ambivalences, on the icons and ethics of decline, and on the meaning of political loyalty, is surely to be found in John Le Carré's trilogy, *Tinker, Tailor, Soldier, Spy* (1975), *Smiley's People* (1978), and *A Perfect Spy* (1986). These books have a quite extraordinarily large public: twenty-five printings of *Tinker, Tailor* between 1974 and 1988; thirteen of *Smiley's People* since 1980 in the United States alone; comparable figures for *A Perfect Spy* quickly followed. All three books have been widely translated, as well as televised in the slow, loving, leisurely manner that now marks internationally best-selling TV serials as posh products ("Masterpiece Theatre," as PBS puts it so happily).

I call it a trilogy because Le Carré's scheme is plainly a strongly ruling-class, intelligently skeptical threnody on the figures and ground of England's descent from the world stage. In *Tinker, Tailor* the mole in the inner circle of the security services wreaks genuine damage: at the coup de theatre, the Soviet army in Czechoslovakia is on red alert, two networks of agents are, in the jargon, rolled up and shot, one of the British heroes, a gentle giant of absolute probity and impeccable trade-craft, is shot in the back, tortured, and re-trieved—all at the instigation of the traitor at the center of Cambridge Circus.

These events have urgent political moment, but characteristically, the only violence in Le Carré's plangent elegy is recollected in the tranquillity of a chat in that modern confessional, the front seat of a car, between unheroic, enig-matic, pudgy George Smiley and the gently gigantic victim, Jim Prideaux, Oxford Blue, war hero, brave spy, and faithful friend, now teaching in a tiddly, down-at-heels private school on the Quantocks.

By the time we come to *Smiley's People*, however, the action on the world stage has dwindled down to a single duel. At the end of the first novel, after Smiley has patiently tracked down, by means of long immersion in the files and quite without a cloak or a dagger, the traitor and found him to be the idol of the service, painter, scholar, lover, consummate bureaucrat, English gentleman, then—in Le Carré's words—"one part of him broke into open revolt against the other."

> Haydon had betrayed. As a lover, a colleague, a friend; as a patriot; as a member of that inestimable body that Ann loosely called the Set: in every capacity, Haydon had overtly pursued one aim and secretly achieved its opposite. Smiley knew very well that even now he did not grasp the scope of that appalling duplicity; yet there was a part of him that rose already in Haydon's defence. Was not Bill also betrayed? Connie's lament rang in his ear: "Poor loves. Trained to Empire, trained to rule the waves. . . . You're the last, George, you and Bill." He saw with painful clarity an ambitious man born to the big canvas, brought up to rule, divide and conquer, whose visions and vanities all were fixed, like Percy's, upon the world's game; for whom the reality was a poor island with scarcely a voice that would carry across the water. Thus Smiley felt not only disgust, but, despite all that the moment meant to him, a surge of resentment against the institutions he was supposed to be protecting: "The social contract cuts both ways, you know," said Lacon. The Minister's lolling mendacity, Lacon's tight-lipped moral complacency, the bludgeoning greed of Percy Alleline: such men invalidated any contract—why should anyone be loyal to them?[13]

(Isn't the answer to that question covered by "professional loyalty," and isn't *that* a vicious oxymoron?)

Such morally simple-minded writing perhaps catches something of the essence of those class traitors, who were brought up, as they used to say, as "future leaders." The profounder question—which Le Carré muffles throughout the trilogy with his imagery of the wastelands of Europe, the sheer tiredness of its secret watchers, the wise, removed skepticism of his hero—is, what are the consequences of duplicity and treachery, of being formed by secretiveness and lying, for the actions of a later generation? What plausible political response is there, as the light fades on the cold war–like scenery, to so baleful a fatuity as this, purporting to be a memo from Kim Philby, in Frederick Forsyth's *The Fourth Protocol?*

> In summation, the entire Labour party of Britain now belongs to the Hard Left, either directly, through Soft Left surrogates, through intimidated Centrists or at the holding of a fast emergency meeting of the appropriate committee; and yet neither the rank and file of the Party membership, nor of the unions, nor the media, nor the broad masses of the old Labour voters seem aware of it.[14]

Le Carré, to give him his due, never speaks such confident rodomontade. In *Smiley's People* the insistently low-key plot leaves Smiley facing his lifelong opponent in the KGB, Karla, in neither an ideological confrontation nor on one of the extremities of the great powers at which their secret patrols have been meeting and maneuvering dangerously for a foothold. Humiliated by Karla over Bill Haydon, barely holding his own in other engagements, Smiley this time squeezes his rival into defection by revealing his discovery of the existence of Karla's beautiful, schizoid daughter illegally hidden in a Swiss asylum where her care is paid for out of embezzled rubles. Smiley forces his enemy into treason by threatening to expose the consequences of "excessive love" and its cognate guilt. His anticlimactic triumph avenges the murder—with the only bullet in the book—of an elderly colleague who had got onto Karla's secret.

Le Carré sees the moral point of all this; after all, ruthless treachery in the name of virtue is his theme. But he cannot connect it to the larger history that he knows is lurking in the hinterland of his novels. He is only firm and convincing on the details and facts of moral uncertainty and dividedness, of the lying and deception that are the normal weapons of states and that become second nature to their intelligence officers. The perfect spy is so perfect that by the end he does not know whom he is spying on and, under-standing his own redundancy, does the rational thing and commits suicide. To Le Carré, his end is poignant; to us, it is just good sense.

Le Carré's appeal, and the source of his power over his enormous follow-ing, is that he gives to the dreadful tedium of minding the files and sitting out the meetings the possibility of enchantment. Le Carré's office intrigues, his footsteps down dark, drenched London streets, his faithless conversations in cars, evoke the lived routines of our domestic and bureaucratic lives, but he gives them edge, pungency, taste, most of all, the taste of secrecy and threat-ened loyalty. Decorated with the apache slang Le Carré invents with such facility—the lamplighters, housekeepers, legmen, tradecraft, burning and sweating—his offices suddenly loom large with portent. If all his books did was make the deadliness of most office life more interesting, good luck to them.

In fact, they give people (me, too) a lot of pleasure. Why go on about them? Because they keep in active circulation the ludicrous charade that Britain remains at the table of great powers and, what is more, merits being there. More generally, the Le Carré trilogy, even though it takes the inefficacy of this illusion to its conclusion and leaves the perfect spy with nothing left to do but commit suicide, still endorses the adolescent world picture of the English public-school boy. (Le Carré is quite open about this: he freely admires the stammering, brave, gentlemanly ineptitude of his hero in *The Honorable School-boy* (1977), a minor work in the Smiley canon.) The circus of spies is depicted as no more than a bunch of school prefects, one of whom has disgraced the honor of the old school. Smiley is Mr. Chips, the faithful deputy who never becomes headmaster—faithful to pointless scholarship and respected for it, too, by his uncultivated pupils—the man whose enigmatic persistence and wise worldliness bring us through. Smiley's is, for Le Carré, the ideal mask of command, the wisest of wise men, beached by history.

Le Carré never grasps the loyalty problem. Writing a sort of spy novel in a very different voice, Raymond Williams has a character, once perhaps a double agent for the Soviets in the intricacies of computer software surveillance, explain to the illegitimate son of another spy, now hunting for his true past, that

> " . . . the political equations were exceptionally complex. Traitor, without doubt, is a definable quantity. There are genuine acts of betrayal of groups to which one belongs. But you have only to look at the shifts of alliance and hostility, both the international shifts and within them the complex alliances and hostilities of classes, to know how dynamic this definable quantity becomes. There are traitors within a class to a nation, and within a nation to a class. People who live in times when these loyalties are stable are more fortunate than we were."
>
> "Not only in times. In places," Gwyn said.[15]

The fascination of spy stories hinges on the delightfulness of duplicity and the reassuring but always fraudulent belief that loyalty is a simple, once-for-all choice. In his pair of two-hander plays, *Single Spies* (1988), Alan Bennett pays those Cambridge spies firmly on both sides. He devised a play around the real-life (as they say) meeting in Moscow between Guy Burgess and the Australian actress Coral Browne, then in the city playing Juliet. (In Bennett's play—speaking of treachery—she is playing Gertrude.) She listens to his reminiscences of England by the hour, together with his endless playing of an ancient 78 of "Who Stole My Heart Away." On leaving, she crisply concedes his charm but not his case. "You pissed in our soup and we drank it" (she is deservedly celebrated for her vivid candor). But being also a kindly woman, she takes pity on the lonely Burgess and agrees to go to his various tailors, shoemakers, and outfitters in Savile Row and Jermyn Street back in London to order his new clothes.

The tailor takes the order with well-bred impassiveness. The pajama maker turns priggish and gets what is coming to him from Coral Browne.

> ASSISTANT: I'm afraid, madam, that the gentleman in question no longer has an account with us. His account was closed.

CORAL: I know. He wishes to open it again.

ASSISTANT: I'm afraid that's not possible.

CORAL: Why?

ASSISTANT: Well . . . we supply pyjamas to the Royal Family.

CORAL: So?

ASSISTANT: The gentleman is a traitor, madam.

CORAL: So? Must traitors sleep in the buff?

ASSISTANT: I'm sorry. We have to draw the line somewhere.

CORAL: So why here? Say someone commits adultery in your precious
     nightwear. I imagine it has occurred. What happens when he comes in
     to order his next pair of jim-jams? Is it sorry, no can do?

ASSISTANT: I'm very sorry.

CORAL *(Her Australian accent gets now more pronounced as she gets crosser):* You
     keep saying you're sorry, dear. You were quite happy to serve this client
     when he was one of the most notorious buggers in London and a
     drunkard into the bargain. Only then he was in the Foreign Office.
     "Red piping on the sleeve, Mr. Burgess, but of course." "A discreet
     monogram on the pocket, Mr. Burgess." Certainly. And perhaps if
     you'd be gracious enough to lower your trousers Mr. Burgess, we could
     be privileged enough to thrust our tongues between the cheeks of your
     arse. But not any more. Oh no. Because the gentleman in question has
     shown himself to have some principles, principles which aren't yours
     and, as a matter of interest, aren't mine. But that's it, as far as you're
     concerned. No more jamas for him. I tell you, it's pricks like you that
     make me understand why he went. Thank Christ I'm not English.

ASSISTANT: As a matter of fact, madam, our firm isn't English either.

CORAL: Oh? What is it?

ASSISTANT: Hungarian. *(He exits right)*

CORAL: Oh, I said, and thinking of the tanks going into Budapest a year
     or two before, wished I hadn't made such a fuss.[16]

One dearly wishes that that good Australian manner could bring things to a
close. But it can't, neither in Britain nor in the United States.

VIII

In the United States the intelligence services must be kept under some kind
of public watch. This requirement leads to the courtroom and congressional
hearings we know so well, but it's something. It may even increase public trust
in the usefulness of those services, for better and for worse.

No one doubts that modern nation-states will necessarily have secret services, nor even that they need them. But as the *Spycatcher* scandal of 1987 made so publicly clear, the obtuse officers of these services, especially during Britain's decline from power, had a demented view of the world, believing that communism was forever poised at every door to steal our hearts away along with our secrets without anybody even noticing. We would wake up to find Soviet tanks in the streets of London as they were in Prague and Budapest, and with a government of Labour fellow travelers installed in one smooth movement by the Politburo.

Making personal freedom and individual rights the supreme values is bound to shake the roots of loyalty and honor, institutional virtues right through. As I have suggested, the attraction of spy stories in Britain and the United States is similar to the attraction of adultery to the extent that each of these activities turns on real allegiance and strong feeling. In the world of perfect consumer freedom, there are no allegiances, feeling is replaced by style, life itself by lifestyle. More recent spies can more easily be given their historical dismissal. It is said that James Walker, who sold U.S. naval secrets to the Soviets for eighteen years and was put away in 1984, did the most serious damage to American security since Klaus Fuchs. He did it, simply, for the money, and the money went, naturally, on his consumption of a lifestyle. If he did so much damage, however, what, as Wesker asked thirty years ago, did the Soviets still wait for?

As Bennett tersely puts it in his introduction to his plays, "Of course Blunt and Burgess and co. had the advantage of us in that they still had illusions. They had somewhere to turn. The trouble with treachery nowadays is that if one does want to betray one's country there is no one satisfactory to betray it to. If there were, more people would be doing it."[17]

Well, that is where we are. While the collective imagination is caught up by the divided loyalties and broken honor of Kim Philby and Bill Haydon, the enormous structure of lies and false crises and sexual tattle is kept up in the name of pointless mutual surveillance. It is then used by the brazen liars and the buccaneering traitors who manage the content of our public conversations and the human values they should be renewing. It is so used not to keep at bay real enemies, but to silence and mislead our peoples and, not incidentally, to betray the best of our past.

Hinge of an Epoch

PART III.

Fringe of an Epoch

14

Tet, Prague, Chicago, 1968

T HE order of things is only settled later; at the time, when history is being made, what it looks like is just one damn thing after another. All through the cold war, as in any other kind of war, people kept telling each other that it could not go on forever. But as the decades turned, the cold war frame of mind and feeling became set in a cast-iron psychology and an inflexible economics.

Maybe this is how eras happen. Dominions, principalities, superpowers assign themselves enemies and organize their money and their military accordingly. For the United States, the Soviet Union was the ready-made enemy toward whom it could point the extraordinary beam of its truculent self-confidence and its amazing powers of destruction. For the Soviet Union, the corresponding motion of its political culture locked its systems rigid with apprehension.

If we look for explanations of how these systems came to be, how these subtle, sluggish mechanisms drove the ships of state, we find no conclusive answers. The cold war was a perverted monument to human creativity, a creatively ruinous combination of passions, interests, money, and technology; it was the nightmare of the past weighing down the brain of the living becoming manifest in the murderous rigidities of the period. Sometime very early in 1946, *everything* conspired with everything else to wage cold war and consolidate that frame of mind and feeling that held so much of the world and its power for good and evil immobilized on a single political topography.

And then it all began to slip out of the control of the two puissant nations whose terrific bulk sat athwart the world and held their willing and unwilling subordinates motionless. Parts of the world began to stir again with their own life. Although the superpowers were quick on the scene with advice, guns, and checkbooks, and even quicker to claim new allegiances from one side of the revolution or the other, a different history began to break through the tundra.

At each eventuality, in Prague or Phnom Penh, Santiago or Cairo, the superpowers sought to claim dissent for their own or choke it off with bullets and money. Sometimes—as in Prague and Santiago—they succeeded for a long season. Yet sometimes it seemed as if even native peoples who had long acquiesced in their subordination, who could be taken for granted as loyal conscripts in the armies of the cold war, had decided to disobey the officers and call the whole thing off. A mass defection suddenly began in the extraordinary year of 1968.

Within a few months American soldiers and citizens surged together in strange and incredible alliances against the fighting in Vietnam. The entire population of Czechoslovakia was coerced into acquiescence by the tanks of its very own ally. The children of the Parisian bourgeoisie became intoxicated on the heavy wines of Marx and Mao and Proudhon, and the specter of the revolutions of 1848 stalked the West.

Let us cheerfully say of all this that a new generation, which had grown up under the shadow of the bomb and shivered in the winds of cold war, was declaring itself fed up with its parents. A different frame of feeling was coming through: there seemed to be less and less point in being killed thousands of miles from home for a war that could not be won; or in doing as you were told by boys in Soviet uniforms who were no older and no better off than yourself; or in agreeing to the mad charade of nuclear deterrence when you lived in Harlem or the Watts ghetto; or in trying to study in the overcrowded squalor of Nanterre.

The parents didn't like it, of course, and put the youngsters down, as parents will. All the same, something shifted in 1968; thereafter, the cold war held on its way for another two decades in spite of so much courageous resistance to it, not because it ran smoothly with the flow of the times.

Before 1968 was over, the currents of change leading up to it had been, more or less violently, deflected and broken up. Easily the most important current was the one that terminated with the Communist offensive in South Vietnam at the feast of Tet, an event whose long aftermath would irrevocably change the American nation.

II

Early in 1954, by a brilliant coup, Gen. Vo Nguyen Giap surrounded most of the troops of the French Expeditionary Force at Dien Bien Phu, a small town near the Laotian border about 150 miles due west of Hanoi. Most of the 70,000 French soldiers surrendered in defeat; *la gloire* and the *tricoleur* went down in the mud.

If Ho Chi Minh and the new Democratic Republic of Vietnam had pressed all out for control of the whole country as soon as the French surrendered, they would surely have won it. But the Vietminh, Ho's ruling coalition, lacked experience in great-power dealing, and they were persuaded to a Geneva conference attended by China and the Soviet Union on the Communist side, and France joined by Britain and the United States, on the other. The comrades shoved the Vietnamese into a partition treaty and a demilitarized zone, scared them with threats of American intervention if they did not do as they were told, and promised them money. A 1956 election on national unification was agreed to, at which point the Americans got huffy about an accord being reached in spite of their forecast that the conference would fail, and they went off home without signing anything. The Americans had come up with their domino theory about Communist expansionism only four years before; they had also already transformed the algebra of "containment" into the dynamics of "rollback"; Vietnam offered the next best arena for trying out the theory after the equivocal battle honors won in Korea, Berlin, and Guatemala. Their British and French allies had not yet suffered the aberration of Suez, and Britain, the trustier of the two, was winning its anticommunist war in Malaya. John Foster Dulles marshaled his clients into the Southeast Asia Treaty Organization (SEATO) as soon as the ceremonies at Geneva were over and agreed with the French to install Ngo Dinh Diem as the leader of the new, entirely unofficial state of South Vietnam. The structure and the dramatis personae were now in place for the titanic war between the dark armies of communism and the forces of the light shining forth from Washington.

As we have all learned, however, the characters in this simple drama did not stick to the script. After the 1954 treaty, the Democratic Republic of North Vietnam bent itself quite steadily to the task of industrialization, which it implemented without the doctrinaire heaviness of Poland or Hungary, and without breaking faith with the age-old pattern of hamlet-sized agricultural cooperatives. In nationalist South Vietnam, Diem and his gaunt, conspiratorial brother Nhu, with his beautiful, scandalous wife, appealed for and received an immense torrent of dollars, which they spent on the lavish refitting

Figure 14.1   Kennedy briefing the press on Indochina (The Hulton Picture Company)

of the army under U.S. advisers, a large bureaucracy to staff a nonexistent government, new secret police, detention camps, and anticommunist propaganda. The U.S. money also brought in a flood of American imports—cars, scooters, refrigerators, more cars, toiletries, prefabricated restaurants—all the bewitching machinery of a consumer culture with which to manufacture an utterly dependent economy.

The distribution of the money was factious and familial. Elderly Americans must now turn puce with embarrassment when they recall what they once said in praise of Diem, before they waved in the assassins. But Diem had learned from his colonial experience that running a colonial office was simply a matter of referring all budget problems to the home office and keeping the new jails nice and full of political prisoners. The money poured in, and much of it funded the graft and corruption that Diem himself created in the country labeled by the Americans "Republic of Vietnam." The small cabinet rarely met and in any case was composed of all his relatives; each took charge of a dollar allocation and spent it on his (or her—Madame Nhu and Madame Le Xhan, Diem's wife, were celebrated spendthrifts) department and aggrandizement. The money never reached the peasants, but naturally their children left the villages to follow the magnetic pull of Saigon's wealth.

Diem looked the part: portly, grave, expressionless as a Balinese king. In a sharkskin suit or his much-medalled uniform, he paced about Saigon and Washington reciting Confucian precepts and intoning banalities about moral leadership. His brother Nhu dreamed up a mad, Moonie sort of religion called "personalism," which traded in vacuous pieties about philosopher-kings and the sanctity of the individual culled from God knew what freshman philosophy texts. This stuff was solemnly taught to policemen and schoolteachers and went down well in *Time* and with the CIA.

To the Americans, Diem's regime appeared to be what they expected of their Asian protectorates: an inscrutable caliph, a venal cabinet, a strong police, a supine peasantry, and an army they had trained themselves. The man they appointed to advise Diem on their own terms, Edward Lansdale, was just coming off his successful rout of the Communists in the Philippines. He had driven the party guerrillas back into the hills, where they remained until the revolution and paved the way for Ferdinand Marcos, a strong leader, who was equally zealous about the suppression of communism and the maintenance of the American bases. Lansdale stood squarely in the American dream of an Asia whose economics, democratic forms, eager consumerism, and virulent anticommunism would presage the end of ideology and the permanent installation of liberal capitalism. Men like Diem and Korea's Syngman Rhee were to be the architects of that future. Later they would be joined by such libertarians as Messrs. Marcos, Pinochet, and Manuel Noriega.

Diem was an early player in a now familiar narrative. He was chosen out of a modest field to front the American policy of either containment or rollback, whichever fit the ambition of the moment. He was acclaimed in the press and talked up by the State Department. He was swamped with U.S. money and goods, indiscriminately poured in without much relation to either need or investment. His army, equipped in a similar manner, was encouraged to try out its lethal new toys on Communists.

Such figures have been regularly coaxed through the sequences of American statehood: declaration of independence, constitution, presidential elections, two-party government. The Communists dart in and out, killing and getting killed. The Americans count the bodies, the locals' and their own. At a certain arbitrary point, they conclude either that they have won or that the figurehead has failed in his office, has traduced American ideals, and must be ditched. In the latter case, the wretch is shortly found shot by his fellow officers.

Diem, an unappetizing and pitiable character caught up in this pattern, was one of those who ended up dead in a white suit and a bloody little alley. But by that day in 1963 the United States was irrevocably transfixed by its own actions and purposes. Diem was dead; the anticommunist crusade no longer needed him or his successor. Yet as the flow of weapons and pointless imports increased, the southern part of Vietnam still could not be called a state, and the enemy came and went in the dark, arming itself from the prodigious excess of American weapons, gradually bringing its own victory nearer.

There are two little tales that illustrate the combination of brutality and ineffectuality that caused Diem's downfall in particular, and the American defeat in general. Neil Sheehan, whose splendid book about the war does so much to restore the American civic virtues of straightforwardness and truthfulness, tells the first story.[1]

Diem had been encouraged by the Americans to establish elite regiments whose dash and flair would, by their visibility, raise the morale of the whole inept army. These so-called Ranger troops wore tiger-head badges on their shoulders and like all elite military units, ran to snappy little idiosyncracies in uniform—Colt revolvers in cowboy holsters, silk scarves, high-heeled boots, and riding breeches.

In 1962 Col. John Paul Vann was the U.S. adviser in a sector close to Saigon. He was in Vietnam for most of the decade until 1972, when he was killed in a helicopter crash. For most of his time there he was, in Sheehan's judgment, one of the most intelligent critics in the military of the military's own tactics.

When Vann first watched Captain Thuong of the Rangers at work, he was angered beyond horror at the horrible *wastefulness* of Thuong's methods. Thuong wasted the good that treating National Liberation Front (the Vietminh's successor in the South) prisoners well would have done the propaganda effort, what was called "the other war," to win, in the cliché, the hearts and minds of the people.

One day in July Vann arrived to check on Thuong; lurid tales about his conduct toward prisoners had been circulating. Thuong was also an exceptionally brave and capable soldier, and he had skillfully taken seven prisoners that day.

He was a burly, ugly man, with a snub, wide-nostriled nose and very thick lips below his dark glasses. He had learned his parachuting from the French, and much else besides. He carried, as well as his revolver and a thick leather thong that doubled as a whip and garotte, a wide-bladed, wood-handled Bowie knife that he flashed before his silent prisoners.

Suddenly, as he paced before them telling them to tell him all they knew, he halted. Thuong stepped close beside one young man, slighter and smaller than the others in his black Vietnamese pajamas, wound the man's hair into his hand, pulled his head back so that the throat stretched bare and muscled upward, and pressing firmly across it with the shining blade, sliced it open and across so that the great vessels gaped cleanly for a second, before the blood surged from every cut tube. Thuong dropped the man where he fell, gargling and choking as he tried to pull his head down and hold together the sliced arteries and raw flesh. The murdered man took perhaps half a minute to die, the blood pumping copiously out onto his black shirt and the ground beneath him.

The shock was absolute. The prisoners trembled where they stood. Vann was made angry by shock. He screamed at Thuong and the other adviser present. "Tell him to cut that shit out." Thuong's knife flashed again as he cut a second throat, and Vann, beside himself with anger, screamed at him again to stop it. Unimpressed, the Vietnamese captain, now blood-spattered himself, turned insolently away from Vann and cut open a third man's throat before wiping his knife on his breeches and walking away. Vann was left gasping and shaking.

When the Front executed its enemies, which was often, it used the method common to all Communist killers: one or two bullets at the base of the skull behind the ear. They then sought to explain and justify the murder to the villagers. They were never whimsical. Murder was wasteful, and Ho's famous *Twelve Recommendations* to all his recruits forbade it. It lost friends and influenced people against the cause.

The second incident that brought home to Vann, and Sheehan, just how badly things were bound to go in Vietnam was a skirmish that took place on 2 January 1963 at a hamlet called Bac, about fifty miles southwest of Saigon.

The hamlet lies on the rich agricultural mud of the Mekong Delta not far from the main canal, on the edge of the Plain of Reeds. The flat fields with their sodden crop of rice are irregularly divided by dikes bonded together by low trees—bamboo, fruit, banana—and jungly undergrowth anything between four and ten feet thick. There were known to be Front guerrillas working from the hamlets, and the Army of the Republic of Vietnam had been, after much procrastination and phony foraging, cajoled into launching an action against them.

The Front guerrillas were carefully dug in, knowing an attack was on its way and ready to withdraw, after their usual tactic, when it became necessary. Vann and his Vietnamese colleague had ten Huey helicopters with which to land a battalion of men in several shuttles. Four of them were knocked out by small-arms fire in the first engagement. The South Vietnamese soldiers and their officers took cautious cover and declined to move. U.S. Sgt. Arnold Bowers, one of the advisers present, rescued a wounded aviator from his helicopter, disentangled a dead one from another, and running dangerously across the line of the Front's enfilade fire, dived into the muddy protection of the dike beside a South Vietnamese lieutenant. The lieutenant and his commander stopped Bowers calling down an air strike in case they got into a worse corner than they were in already.

Finally Bowers obtained napalm rocket fire onto the Front's position. The napalm burned its huge orange bloom, and the heat hit the dike where Bowers was crouched. The South Vietnamese saw the flames, cheered for victory, and stood up to watch. Two fell back dead. The rest dived back into the mud. The Front had not moved.

Colonel Vann, circling suicidally in a light aircraft only 1,000 feet above the fighting, called up a company of light M-113 tanks (armored personnel carriers). Their commander, a careful man but an honest soldier, refused to attack across the oozy fields. The Front knocked out another Huey.

The Front had about 350 men armed with light infantry weapons in action along the dikes beside the village. They were faced by a battalion, the company of armor, the remaining helicopters with their rocket and strafing power, and any fighter-bombers within call. They were edgy after living through the fury of aerial bombardment. Their extraordinary discipline held. They stopped firing; the Americans and the South Vietnamese officers assumed that the Front had withdrawn. When at last the armored tanks labored over the brush-filled ditches and approached the Front's position, the guerrillas aimed at the driver's head just poking out the top, and at the gunner of the heavy Browning behind. The South Vietnamese soldiers were demoralized; they were very poor shots with those heavy weapons. Their flamethrower failed.

The armor edged hesitantly forward until it was within twenty yards of the dike. The guerrillas drove them back by jumping out from their cover and chucking grenades onto the tops of the carriers. The combination of deafening bang, the Front's bravery, and flying chunks of metal daunted the dauntable South Vietnamese at once. They reversed at top speed, slinging a flail of mud over the infantrymen hiding close to their tracks. It was 2:30 P.M.

When South Vietnamese Gen. Huynh Van Cao brought up an airborne battalion much too late, the Front shot them as they jumped from their planes. When the Americans tried to shell the Front's lines of retreat, Cao, desperate to stave off reprisals by the enemy, fired four shells per hour.

The Front lost eighteen men and retreated in good order. The South Vietnamese lost eighty and spent amazing quantities of plant and ammunition. They lost some more when they shelled their own men in Bac the next morning. It was a small battle, but it was a rout.

The first stage of the war in Vietnam ended with the defeat of the French. The second ended with the casual deposing and assassination of Diem. The hamlet of Bac stood for the previous eight years of venality and cowardice on the part of the South Vietnamese elite, both civilian and military, and it stood for the cautious self-preservation and lack of ideological commitment on the part of their mercenaries. It also, of course, stood for the discipline and courage of the ideologically staunch cadres of the National Liberation Front. (The Americans nicknamed them Cong, abbreviating the Vietnamese word for Communist party, but the Front insisted that it was a coalition.) That day in Bac, the Front guerrillas drove off a hugely more powerful force, wounding it much worse than they were injured themselves. That night they withdrew through their villages, in good order, taking their dead and their empty cartridge cases with them. Early in 1963, the Americans were beginning to lose.

III

After the fall of Diem, the subtle strategists of the North realized that they could win, but they also realized that the Americans were certain to raise the stakes. In 1963 there were slightly over 16,000 American military and their official camp followers in Vietnam. The Front consulted its masters in Hanoi and enlisted direct military help; the North went to China and with Chinese help began to widen and to metal the Ho Chi Minh Trail, which led from the People's Republic of China across the northern tip of Laos, eastward to Hanoi, and down the narrow stalk of Vietnam toward the South. At that date, the guesses go, only one in twenty Front soldiers came from the North; in the South the Front had never been more energetic or penetrative. Its amazing system of underground tunnels warrening the entire landscape of the South was packed with rice, salt fish, cooking oil, ammunition, morphine and bandages, guns, and the assembly kits of the land mines that wrote the signature of the Front across the aerial photographs of the day.[2]

Meanwhile, back at the ranch, both Maxwell Taylor and Robert McNamara had begun to lose their breezy optimism that the war could be won by present methods, and the National Security Council began to believe the only alternative to defeat was to bomb the North. Being a liberal world power, however, the Americans needed some justification for this action. By late 1964 Maxwell Taylor was ambassador to Saigon; his commitment to the domino theory as blithe as ever, he called up his heaviest cliché to clinch the momentousness of the moment. "If we leave Vietnam with our tail between our legs, the consequences of this defeat in the rest of Asia, Africa, and Latin America would be disastrous."[3]

Lyndon Johnson's cabinet was clear that it would find its pretext for raising the stakes; it came in August 1964. Before that, a bunch of ineffable cowboys from the CIA had run provocations in the North, and the navy had sent patrolling spy ships to within four miles of the shore stations of the Red River Delta in the Tonkin Gulf. At the same time, the Bundy brothers, McGeorge and William, dreamed up a resolution to put before Congress that would ratify anything the administration might care to do the moment it got an excuse. They earmarked the bombers and the bombs, ordered that the troops be massed for transport, and began to construct a grand lying project for the deceit of the American public. Finally, they sent a Canadian envoy to tell the Hanoi government to lay off the South or face the bombs. Hanoi sensibly offered to join a neutralist coalition in Saigon.

More cowboy raids followed in the Red River Delta. The USS *Maddox* set off with electronic equipment capable of triggering air-raid alerts all along the coast, and the North Vietnamese patrol boats duly obliged by firing on the *Maddox*. Congress passed the Tonkin Gulf resolution like lambs, and all was ready for a new and much more pyrotechnic phase of the cold war, at its hottest moment.

The problem of a plausible leadership in the South hardly mattered any more. General Khanh came and went and came again as president before dismissing himself and being replaced by Nguyen Van Thieu. The generals under Thieu ran everything from the opium trade to the U.S. consumer durables trade; Madame Thieu kept the president's hands clean by looking after the family reserves herself. When the Americans' zeal for elections finally carried the day, Thieu and Ky fell weeping on each other's necks after agreeing on who got which job. But the voting went wrong. Thieu could not even rig a ballot. He and Ky got in, but a lot of votes went to people— Buddhists and the like—who were not supposed to get any, and mysterious fistfuls of pro-Thieu voting papers turned up in thousands at the end of the count. Miffed, Thieu put his opponents in jail.

But Thieu was a sideshow, a front, or just plain incredible. More important, around the beginning of 1965 the indefatigable conceptualization teams hired by the State Department—the people who gave the world the "strategic hamlet" program and were to follow it up with "forced-draft urbanization"— (or driving peasants into concentration camps) came up with a key criterion to test national nerve and military resolution. It was "credibility."

The word has become an acid cliché, dissolving in its bitter flavor older and sounder values such as trust and truthfulness. It was part of that radical demoralizing of the language of politics that would put the Front at such a linguistic advantage. The pious Dean Rusk advised the president that "the integrity of the U.S. commitment is the principal pillar of peace throughout the world. . . . If that commitment becomes unreliable, the communist world would draw conclusions that would lead to our ruin and almost certainly to a catastrophic war."[4]

The American war effort became incredible all right, and American "integrity" turned out to mean integral payloads on bombers and bomber sweeps integrated by banks of sensors, cameras, and infrared detectors that in 1969 guided 130,000 tons of high explosive per month to complexly differentiated targets. But these efforts would take some preparation; in 1965 it was enough to raise the number of U.S. soldiers in Vietnam to 175,000 during the summer months. At the same time, the bombing of the North began. By the end of 1966 there were 500,000 American soldiers, sailors, and airmen in South Vietnam.

All Johnson wanted was to silence the enemy so that he could go back to his home improvements. He wanted, after all, to be a great social reformer, just as Richard Nixon, in his sly, insidious way, wanted to be a foreign policy pioneer. But both shared the overwhelming confidence of their subordinates that the world was their creation and that they held the levers of power that would ensure that it went along with American priorities.

This settled and mutual assurance among those who made the decisions about Vietnam in Washington and the senior people in Saigon who told their juniors to kill and be killed is what is now so stunning. How *could* they? one asks; managerialism was the answer.

Managerialism was the ideology of the new class of power elites in modern industrial societies. The doctrines of managerialism dictate the technical imperatives that politicians take.

Managerialism was much aided by the free hand that the geopolitical climate permitted before 1968. Given its hinge position in the cold war, the Vietnamese campaign was little regarded by either China or the Soviet Union. The People's Republic sent copious supplies of ordnance and vehicles south and gave unhindered passage to the northerners to drive down its mountain roads from their supply dumps safe across the border. But Mao himself sacked Liu Shao-Chi from the central committee of the politburo, of which Liu had been a member since the revolution, because he threatened the Americans that China would join the war on the North Vietnamese side in order to defeat the imperialists. Mao then launched the crazy experiment of the Cultural Revolution in 1966 just as "Rolling Thunder," the U.S. bombing of the North, reached its paroxysm and Sino-Soviet relations were finally sundered.

So, too, the Soviets, always suspicious of China, made their arrangements for the North at Geneva in 1954 and were plumping for peaceful coexistence just when hostilities got their hottest. Apart from the routine knockabout from Tass, they did hardly more than gaze at the Vietnamese comrades and their redoubtable leader with timorous apprehension. Le Duan of the Hanoi politburo ferried back and forth with counsel of sorts, while the Americans veered between a view of Vietnam as now the stooge of China, now of the Soviet Union. But the major war that the United States waged against the Vietnamese nation was always far less important in Moscow than the nuclear weapons that war brought to their gate.[5]

Military managerialism, blithely disregarding the facts of geopolitics, lowers over the epoch as its brooding spirit and paramount frame of mind. The Americans were its certain and insouciant executives. William C. Westmoreland, general in charge of Military Assistance Command Vietnam (MACV) for the crucial years of the war, was a graduate of West Point *and* Harvard Business School. His way of thinking speaks for itself:

> I see an army built into and around an integrated area control system that exploits the advanced technology of communications, sensors, fire detection, and the required automatic data processing—a system that is sensitive to the dynamics of the ever-changing battlefield—a system that materially assists the tactical commander in making sound and timely decisions.[6]

His rapture over efficiency, irrespective of its human consequences, has (literally) demoralized intelligence. It makes him co-celebrant of the rites of slaughter in Vietnam with his Secretary of Defense Robert McNamara:

> That is what management is all about. Its medium is human capacity, and its most fundamental task is to deal with change. It is the gate through which social, political, economic, technological change, indeed change in every dimension, is rationally spread through society.
> . . . The real threat to democracy comes not from overmanagement, but from undermanagement. To undermanage reality is not to keep it free. It is simply to let some force other than reason shape reality. . . . If it is not reason that rules man, then man falls short of his potential.[7]

The extrusion of morality from the systems of politics, the identification of politics with technical management, the managerialist executives freely hiring from humanist universities, all this shocked the academies, causing the fiercest altercation known to the history of American universities. But the managerialists marched on.

Not the least repellent of the moral consequences of this freebooting attitude of the power elite was its not unusual contempt for the public that employed it and only very slowly learned to distrust it. But automatic recourse to lying—lying hard and all the time—about the conduct of the war became policy not at first because desperate men wished to conceal their failure—that came later—but because the managers believed the public would not understand the issues.

"The problem-solvers," Hannah Arendt writes, "did not *judge;* they calculated."[8] Among the many lessons of the war is the vulgar moral that sheer numbers, especially those enumerating the systems of production, are in

Figure 14.2   Lyndon Johnson displaying his operation scar to journalists, 1967. (Drawing by David Levine. Reprinted with permission from *The New York Review of Books*. Copyright © 1966 Nyrev, Inc.)

themselves meaningless. The colossal quantity of American firepower bore no relation to its success; the B-52s, the tanks, and the artillery unloaded *fifteen million tons* of explosive onto the country between 1964 and 1972, over twice as much as the entire U.S. armed services spent on the whole surface of the globe in the Second World War. Early in 1967 the U.S. Army alone was producing two-thirds of a ton of printed reports *per day*. With the most productive *and* wasteful political economy the world has ever known, American managerialists came to believe that sheer production, and production for the purposes of destruction, would impose their will on global circumstances.

The same credulity extended to American losses. By the time U.S. forces left Vietnam, they had lost 3,689 aircraft (fixed-wing, in the jargon) and 4,857 helicopters at nearly $250,000 apiece on the 1972 weapons market. As to the corpses, like so much else in military intelligence, there are wide discrepancies in what was coined in the Vietnam War itself, in a vicious neologism, as the body count. It was guessed that one million of the enemy had been killed up to 1972, but no one will ever know how many of the routinely disbelieved patrol reports were included or discounted, nor whether the dead, those bits of black pajamas and red meat, were enemy, friend, mischievous child, or warrior-child. Estimates of the civilian dead in the South before 1972 run to 1.35 million. Only on the black marble of the Washington Vietnam War Memorial is the number of the American dead precisely recorded: 58,132.

The three years of fighting before the Tet offensive in January 1968 rooted the gigantic overproduction of destruction deeper and deeper in Vietnamese soil. It drove the Vietnamese off that rich soil into camps and onto the streets of Saigon. In the camps they lived off what was left to be handed out after Thieu's generals had kept and distributed the graft. On the streets the girls became harlots, their mothers became serving maids, and the men became peddlers, hucksters of American beer and ice cream, Japanese cameras and cars, Taiwanese clothes, and Colombian and Bolivian cannabis and cocaine.

The management of firepower according to the approved criteria drove the whole gigantic machine at a mad pace. If there were not quite enough bombs to fill the bellies of the fighter-bombers, they still went out to drop what they had, because efficiency and productivity demanded a given number of sorties. So, too, with counting and recording recaptured and "sanitized" hamlets; the numbers finally turned out to be so ridiculously (and fatally) inaccurate that the policy was dropped. But still the administration would have its will, as Samuel Huntington so notoriously said:

> The depopulation of the countryside struck directly at the strength and political appeal of the Viet Cong. . . . The Maoist-inspired rural revolution is undercut by the American-sponsored urban revolution. . . . the United

HERBLOCK'S CARTOON

"Admiral, You Have Been Summoned Here On Charges That You Have Failed To Spend Beyond The Call Of Duty"

Figure 14.3   Cartoon on Nixon's defense expenditures, 1972—from Herblock's State of the Union (Simon & Schuster, 1972).

States in Vietnam may well have stumbled upon the answer to "wars of national liberation." The effective response lies neither in the quest for a conventional military victory, nor in esoteric doctrines and gimmicks of counter-insurgency warfare. It is instead forced-draft urbanization and modernization which rapidly brings the country in question out of the phase in which a rural revolutionary movement can hope to generate sufficient strength to come to power.[9]

The same mad logic so outweighed the capacity of unfinished docks and warehouses to handle the pace of imports that half the ocean-going merchant fleet of the world was waiting outside Saigon Harbor for many months. It also set the rigid patterns of U.S. military tactics.

In this, as in everything else, the Americans were as unlike the Vietnamese as could be. The Front, however, knew their Vietnamese men and, either by instinct or hard experience, also understood something of American methods. When the Americans flattened the face of the country, the Front dug and lived and fought below its surface. When the dreadful bombing of the North left some craters 100 feet deep, they used the newly exposed gravel to mend the Ho Chi Minh Trail.

Beyond the innumerable instances of their resourcefulness, the dead and captured soldiers of the Front's People's Army of Vietnam bore witness to a close, dense, and very striking military culture. The Front helped bereaved families with money and food. It mobilized women, largely behind the lines and in food production, and trained the peasantry in wartime technology and economics. There was no corruption. In the South the children of the peasants were drawn irresistibly to the garish, magnetic street life of American-made music, junk food, plentiful dollars, and the G.I.s' cheerful, casual friendliness. In the North and in the Front-controlled villages, the children were taught the practical maxims of local, Confucian knowledgeability and home-cured Marxism.

As things went badly for the Americans between 1965 and Christmas 1967, only McGeorge Bundy and Robert McNamara in Washington finally said so, and only John Paul Vann in Vietnam. Rolling Thunder, which would last, with intermissions, nearly three years, proceeded. Westmoreland asked for more and more men, and although he never got what he wanted from Johnson, he was sent—according to the protocols of negotiation—a fair share of what he wanted. The famous "credibility gap" opened between what U.S. military intelligence said was happening and what soldiers and peasants lived through in the field.

But the Democratic Republic and the Front were led by Marxists, who made a theoretically necessary link between the rural and urban populations.

Mao taught them that the guerrilla swims in the water of the rural peasantry; Lenin taught that power lay ultimately in the cities, the headquarters of energy, communication, and money. As usual their theory was sensitive to reality. The U.S. dollar was floating upward on a fast thermal of inflation; the urban bourgeoisie distrusted and despised Thieu. In spite of the appalling bombardment in the North, the politburo of the northern Communist party and the leaders of the Front felt strong enough to try for urban insurrection. As 1968, the Vietnamese year of the monkey, approached, they planned a huge urban uprising to coincide with the Vietnamese new year holiday at the end of January—the Tet offensive.

IV

By the end of 1967 McNamara's departure was imminent and his skepticism well known. Everyone else in Washington and in the command offices of South Vietnam gazed entranced at their printouts. General Westmoreland, shrewdly replacing those advisers who told him what he did not want to hear, deployed his considerable managerial skills with flip charts, block graphs, and his impressively craggy profile to show that monthly body counts were up to 7,000 VC, that half of the Front and North Vietnamese units were not ready for combat because of illness, hunger, and shortage of ammunition, and that the Communists were deserting to the South in droves. The war of attrition, a menacing phrase borrowed from the First World War was, he said, working out well for the Americans.

Westmoreland had said all this to Congress earlier the same year. He had his own political touch, he was an imposing national celebrity, and Johnson needed him. In full uniform, he brought the House to its feet, saluted first the Speaker, and then, turning smartly around, saluted the packed assembly before marching quietly out, leaving the chamber to pounding applause. Impressed by the power of kitsch, Westmoreland flew back to Washington in November 1967 and told the National Press Club,

> We are making progress. We know you want an honorable and early transition to the fourth and last phase [of the war].
> So do your sons and so do I.
> It lies within our grasp—the enemy's hopes are bankrupt. With your support we will give you a success that will impact not only on South Vietnam, but on every emerging nation in the world.[10]

Ten weeks later, the Front struck in large numbers, and with complete surprise, in every major town in the South but one.

It was a fearful shock to the Americans. Many intelligence reports had told them that something big was up, but not this. The North Vietnamese army had concentrated troops heavily around Khe Sanh in the highlands near the 1954 armistice line. Westmoreland, eager to fight on the terms he understood and had learned in Korea and Europe, poured a huge bombardment on them through the thick mist and drenching rain. Almost half the American infantry and armor in Vietnam was massed below Khe Sanh.

In the South the Tet offensive began at 3:00 A.M. on 30 January. Although Gen. Fred Weyand, one of Westmoreland's deputies, expected trouble in Saigon, no one was ready for the scale of the offensive. A North Vietnamese division swept into the ancient imperial capital of Hue, struck the Marines' flag, and ran the gold stars of the old Vietminh up the flagpole. In Saigon the local population connived as actively as was necessary to allow the Front to bring eleven battalions into the city and to pass heavy machine guns, light carriers, and mortars through the dense shantytowns surrounding the city. In 1962 less than one and a half million people lived in Saigon; by Tet 1968 the city had swollen to nearly four million people, driven off the land by terror, by free-fire zones, by forced-draft urbanization, by arrogance and cruelty.

Given the jammed, chartless maze of temporary shanties built out of the detritus of the war and junk dumped by the American troops—flattened Coke and Bud cans, torn-off car doors, duckboards, tarpaulins, tents—the Front had even thicker camouflage in Saigon than in the rain forest. It struck almost simultaneously in thirty-six provincial capitals and sixty-four market towns. In Saigon a Peugeot truck and a rundown taxi stopped on the boulevard by the U.S. embassy and nineteen young commando sappers jumped nimbly out and blew a large hole in the embassy wall. The ambassador, Ellsworth Bunker, in his grand villa just down the road, no less nimbly hopped into a Marine 10-ton carrier in his dressing gown and trundled safely away. The commando unit died to the last man in the embassy gardens.

General Weyand had kept back enough combat units to hold the fort. The Front and the Hanoi commanders had hoped for a general uprising in the cities against the Thieu regime. They knew that the Americans would hold the command and communication centers. But they hoped to destroy the Republic of Vietnam as a valid state, turn all the urban masses their way, and thereby be in a position to negotiate an American withdrawal because the Americans would have nobody left to support.

The American war machine went beserk. "The Cong started it" was their spokesmen's only excuse, and no doubt it did. The Front did not get its urban uprising; in its official evaluation it recognized what a sickening blow it had dealt the Americans, but it also acknowledged that its own clandestine organization in the cities, its agents, informers, cadres, and storemen, had largely vanished. They were killed or captured as the storm of the American response pulled them into the fight. The Front lost far more of its most experienced men than it could afford. But its discipline and flair held, and would hold for another two years.

Philip Jones Griffiths, a Welsh photographer working in Vietnam at that time, tells a story about a Vietcong soldier whom he photographed being given a drink of water by three G.I.s. The Vietcong had been wounded in the stomach three days before, the wall of his abdomen being slashed open by shrapnel. His intestines fell out but were undamaged, a little brown dust settling on their blue membrane. The soldier held them to his belly, grabbed a shallow enamel bowl from a peasant woman, and tied it over his pendant burden.

> As he was being carried to the headquarters company for interrogation, he indicated he was thirsty. "OK, him VC, him drink dirty water," said the Vietnamese interpreter, pointing to the brown paddy-field. With real anger a G.I. told him to keep quiet, then mumbled, 'Any soldier who can fight for three days with his insides out can drink from my canteen any time."[11]

The heaviest fighting of the Tet offensive was in Saigon and Hue. The Vietcong fought street by street on the west side of the capital for two weeks, while overhead the American choppers and fighter-bombers rocketed and blazed away with bullets and fire with no regard for what they hit. They burned 20,000 houses to the ground and killed 6,000 civilians. And all the time the American public watched on television the boys they had sent to save Southeast Asia for democracy doing so by burning it and cutting its citizens in half with .50-caliber Brownings.

The battle at Hue and Khe Sanh would make those names resound in the national imagination like Guam and Bataan from the Pacific war. While Westmoreland faced the North at Khe Sanh, with all his divisions drawn up for a Korean-style fight to the finish, the North Vietnamese army went quietly around his flank and occupied Hue.

Hue was then a small provincial capital with about 140,000 inhabitants and a resonant name in Vietnamese history. It was the old imperial capital of the Thai Hoa emperors and had been deep with ancestral lore for the better part of a millennium. The palace and the noble buildings at the center, massively constructed and as lovely as Calvino's fabled city of Diomira, were, it is said, deliberately risked by the northerners in order to prove that the Americans would not only destroy the country's children, they would also destroy its ancestors, its desires and memories, so that one day, if the Americans were allowed to stay, there would simply be no country left, only the dead pools of water in the craters, the skeletons of houses, the stripped trees, the homeless.

The Americans obliged. They razed the city to rubble. Its ancient twenty-foot walls, six feet thick, were leveled, along with everything else. The Marines, retaking the city street by street, lost 200 men; 7,000 Vietnamese died, in about equal numbers of soldiers and civilians. What was left of the city was then comprehensively looted, and the Vietnamese buried their dead in the craters.

The American G.I.s had shot their bolt. It would be a hell of a year: the Front and the North Vietnamese regular soldiers would come back to Saigon, more carefully but fiercely, in May and August. From then on, however, it was only a matter of time before the U.S. half-million quit the country. The bombing of the North went on. Westmoreland asked for another 200,000 men. At a key meeting at the White House on 26 March 1968, Dean Acheson, still a crazy cold warrior after all those years, still arranging to kill people from the chief executive's suite, summed up the mood of those present and declared they could not hold up the regime in the South any longer.

On 31 March Lyndon Johnson told the nation that bombing would be limited in the North to the access roads. He also announced that he would not run for president in the November election. The force of ideas and the force of circumstance had combined to shake profoundly the worldwide belief in the strength of American weapons and the weight of the American will. But in Vietnam there were millions of tons of bombs and many cadavers to go yet.

V

Something happened; you can't say it didn't.

Look again at a historical map of 1968, and as you see the little flashes of gunfire wink out from the area at the tip of old Indochina, move your eyes up; it seems then that the flashes of light stretch out in a line from Saigon and Hue and touch many other cities in that year, thousands of miles away. The lines of light cut across the oceans and touch cities and thrones and powers: Poland, Prague, Paris; Los Angeles, Miami, Chicago; even London.

The Polish students were first out in 1968. In protest against expulsions and police liberality with their truncheons, the students closed the universities and fought back. A sextet of the intelligentsia who had been expecting a knock on the door since 1956 were fired; Kolakowski and the admirable Zygmunt Bauman left for England.

Paris was the glamorous face of European insurrection that year, Prague the grave one. As is the way with official opinion in our times, each side takes a header into a hypocritical cesspool when men and women are in the streets crying variations on old themes—"One for all and all for one," and "Liberty, equality, fraternity." If it's *their* subjects demonstrating, then they're making natural and spontaneous demands for freedom. If it's our own citizens demonstrating, however, and particularly our young citizens, then the dull voices intone over there about counterrevolutionary elements and bourgeois irridentists, or over here about "utterly irresponsible minorities" who "make any person loyal to his country weep," according to taste.

The 1967 march on the Pentagon shook the government a tiny bit; the events in Paris in May ultimately brought down the French president but never touched the state; the dissidents in Prague, however, briefly *were* the government, so that street dissent and a change of state control came together and opened sweetly into a spring of common happiness.

Once more the students were close to the heart of the matter. They had been marching and chanting and getting their heads bashed by the cops during the autumn, and they had struck a chord in a quiet, smiling, obedient sort of chap named Alexander Dubcek, who sat on the central committee of the Communist party for Slovakia. He made a careful attack on the granite-jawed leader Antonín Novotný, who had over the years made himself into Lord High Everything, including first secretary of the party, president, and commander of both the army and the People's Police.

In January Dubcek was elected first secretary; he had looked like the safe candidate to succeed Novotný when the latter's Stalinism had had its day. In spite of Klement Gottwald and cast-iron communism, Czechoslovakia had hung onto its democratic traditions and a picture of what Dubcek himself cast in an agreeable slogan as "socialism with a human face." It had an active intelligentsia and dedicated students on good terms with a serious and organized group of novelists, playwrights, and poets, including Vaclav Havel and Milan Kundera. Czechoslovakia had also sustained a tolerable level of economic production.

Dubcek was that rare bird, an honest and true socialist, and he won power. His public pieties were sincere; he gave back to socialism for a quick day or two its simplicity and hopefulness. "The party exists for the working people. . . . Democracy . . . is the genuine participation of everyone in the process of decision-making."[12] (When Brezhnev heard of this in Moscow he gaped and glowered and prowled up and down like a big bear in a loose suit.)

There were immediate reforms. Several of the old Stalinists and their police stooges were sacked. Legislation and travel were eased, opinion freed especially in the press and the cafés. In a small country accustomed to invasion, the enormous joy of open argument and jaunty political joking was reform enough for now. There was no violence.

Figure 14.4   Posters in support of Dubcek on King Wenceslas' Statue, Prague, 1968. (The Hulton Picture Company)

Dubcek, of course, listened nervously for the phone from Moscow to ring. It did. He was told to slow down, much as Nagy had been told in Budapest in 1956. (He remembered that lesson from socialist history well enough.) Dubcek was also under pressures that one man could not hold. He appointed a commission to tell the truth about the old purges and the death of Jan Masaryk. He released thousands of political prisoners and promised them solid compensation. He lined up a special party congress to which, the whole country knew, the delegates elected by the party would be overwhelmingly on Dubcek's side. The tough old subjects would be wiped out. Novotný was promptly voted out of the presidency on 18 April. He took defeat hard.

Vaclav Havel and Ludvik Vaculik wrote two of the most popular manifestos of the day and paid later for them with prison terms. Both appeared in the intellectuals' trade magazine *Literarni Listý,* which sold hugely as a result. Havel called in April for the formation of parliamentary opposition and asked for it in the name of socialism, while Vaculik called in late June for a complete cleaning-out of the bullies and toadies who had run the country for so long; he addressed his "2,000 words" to "Workers, Farmers, Scientists, Artists, and Everyone." And in a grim, gallant flourish, Vaculik ended by telling Dubcek's government that "we will support it, even if it comes to bearing arms, as long as it fulfils its mandate."

Brezhnev was astounded and told Dubcek to put the writers inside. Dubcek, perfectly aware of the extreme precariousness of his position, stuck to his principles and temporized. He tried to trim his boat by assuring Moscow that his was a socialist road, that there was no question of leaving the Warsaw Pact or of in any way threatening stability in the East.

But it was all too much for the Kremlin. They tried to make Dubcek postpone the special party congress. They conducted Warsaw Pact maneuvers in Czechoslovakia throughout July. And then, during the night of 20 August, the biggest Soviet tank force to move since 1945 rolled with that unmistakable squealing, crunching roar toward the Czechoslovak frontier.

It was preceded by an airborne landing of Soviet parachutists who had been told that they were on their way to save the Czech comrades from counterrevolution. They easily captured the airport, state offices, telephone exchanges, and, after a short battle, Prague Radio.

There had been, of course, plenty of traitors among the diehards for whom liberalization meant the loss of all their status as well as their beliefs. The unspeakable Salgovic, head of the police and Brezhnev's informer, had known of the invasion and helped plan the airborne landings; he now became chief collaborator.

The Czech people had learned the pointlessness of rerunning Budapest, but they found the forms of a new heroism. The bravest men on the central committee, led by Dubcek, refused to acknowledge the justice of the invasion. They issued a formal proclamation against the invasion warning their people to protect their lives but damning the naked "violation of international law," calling the invasion "contrary to the fundamental principles governing relations between socialist states." They were still in session at four o'clock in the morning when the Soviets burst in. They told the intruders to get out.

The Soviet officer, who could not speak Czech, went away baffled to find a senior Czech toady to do the dirty work. Outside the secretariat offices, a crowd chanted for Dubcek and sang the national anthem. About 9:00 A.M. the senior members of the government were arrested on phony authority, hand-cuffed, and removed from the country. As an encouragement to solidarity, Soviet tanks fired bursts of machine-gun tracer into the night sky at regular intervals.

But the Czechs would not lie down. The students and young people of the towns copied the motley peace groups marching on Washington by standing in front of their contemporaries guarding the tanks, mocking and pleading, "Look at me. . . . Look in my eyes. . . . Touch me. . . . Kiss me. . . . What are you doing? . . . Let's be friends. . . . Please go home." Prague Radio came back on the air from renegade radio vans moving elusively over the cobbled streets. The Czech army lent its medium-wave transmitters to what became a concert of pirate radio stations warning people of danger, telling them how to treat the Soviets—"Let us show them again and again that they are unwelcome guests here. . . . Let us give them nothing to drink, let us not show, even by the slightest gesture, any momentary pity over their situation."[13] It took the Soviets several days to track down and silence every covert radio station. One reported that people were lying down in the street in front of tanks on the move so that the Soviets, still anxious for the peaceful settlement they never found (only a sullen one), had to screech to a stop.

The Soviets needed to install a new regime—just as the Americans needed one in Saigon: they could not find it. They sent a deputation to see the ancient president, Ludvík Svoboda, sometime war hero in the Soviet Union as well as at home, and Svoboda also refused to have any truck with them, asking only to see Dubcek and the others now in prison in the Soviet Union.

In brave heart, various party branches further startled the Soviets by calling the special congress Brezhnev had been so apprehensive about for 22 August, the day after the tanks arrived in Wenceslaus Square. In a country jammed with 250,000 Soviet troops, over 1,000 out of 1,500 delegates arrived, almost all of them convinced socialists and liberalizers, all bitterly hostile to the invasion, and all acting together as an embarrassing repudiation of the lies by which the Kremlin and its creatures tried to justify it.

All the political and cultural institutions of the country stood together in dignified and courageous solidarity. That unity was the Czech people's only solace under the reassertion of Stalinist power.

By themselves, the Soviets could not patch together an adequate regime. Dubcek and the other prisoners were taken to Moscow where Svoboda joined them, still as president. The tale that Dubcek always denied was that he had been drugged by the Soviets according to their little ways. But he was ill, his gentle nature exhausted; he agreed to a wretched compromise that turned out later to mean nothing. Censorship returned in a rush; liberals were sacked, and intellectuals humiliated. Although Dubcek remained in nominal charge, the tanks and soldiers stayed. The special congress resolutions were revoked.

The country stood by its man. No sooner had the Soviet soldiers torn down Dubcek posters during the day than they went up again at night. But Dubcek was broken. He spoke stammeringly and in tears to the nation by radio on 27 August, then he stumbled out of public life.

Compared with 1956 or with Saigon, Prague was a playlet in the cold war. The Soviets only killed seventy-two Czechs. One of them was a fifteen-year-old boy painting "Go home" on a wall, and another two were a baby and his mother, who was trying to protect the boy from the trigger-happy psychopath who shot all three from his tank's machine gun.

But what baffled the Soviets and made Prague so different from Budapest was that they were resisted by both the nation *and* the party. It was *socialism* in Czechoslovakia that sought a new path to the future. The deep ambiguity of the Prague Spring ran quietly underground for twenty-one years, and its meaning, for Eastern and Western politicians still uneasy over memories of 1968, steadfastly resists propaganda and sanctimony. Its glowing image distills the spirit of stubborness, happiness, honesty, fairness-of-shares, and deciding-about-our-own-lives: the human faces of socialism.

VI

To be a student in Paris is to absorb like radioactivity the energies of a radical culture always counterposed to your own class origins. Most French men and women live comfortably enough with this faint fever hissing in their veins, but every now and again it runs to a temperature—in 1936, in 1968—and boils up into a feverish carnival of dizzy argument, barricades, and ancient, tribal cries from the first socialists, from Proudhon, Blanqui, Saint-Simon.

On 3 May a big crowd of students gathered in the great courtyard of the Sorbonne, the central college of the University of Paris, a few hundred yards from the main buildings of government and tourist sites. They were there to protest the disciplining and suspension of a small group of students who had in turn been protesting the truly gruesome conditions of study at Nanterre, a branch of the university in the suburbs. The university rector summarily called in the police to clear the crowd.

The police were led by the *compagnies républicaines de sécurite* (CRS), France's riot squads notorious for their freedom with truncheons, nasty dogs, and tear gas. There were a ludicrous number of arrests, some fighting. The students' top blew off. They called a student strike, and on Monday, 6 May, as large numbers of them marched quietly enough back from a long promenade through Paris, they were needlessly attacked again by the CRS, who were massed behind their perspex shields and their fiercesome helmets with the black shatterproof visors that removed all human features and gave them their savage license.

This time a lot of students were hurt. With fluent resourcefulness, the students formed themselves spontaneously into cooperatives. Gaily quoting the vocabulary of urban insurrection learned in their cradles they improvised barricades out of light parked cars and the paving stones of the street and used anything portable as a projectile against the bitter gas and the wrist-breaking, cheekbone-smashing truncheons.

Marching, taunting, fighting, flinging together one small barricade after another, the students kept the CRS on the run well into Tuesday, at times displaying a nonchalant bravery and recklessness quite in keeping with their great traditions. In all these engagements they were followed by an entranced throng of international news photographers and television camera teams.

The story quickly found its starring hero, Daniel Cohn-Bendit, German Jewish redhead (Danny the Red), fluent in three languages, witty, brave, vivid with gaiety and intelligence.[14] He cut through the labored cant and hauteur of his political seniors and expressed the grievance and outrage in a plain-spoken way that won the argument for the TV-watching millions. On 9 May, he declared the Boulevard St. Michel an open university. All down the street in the warm May sunshine the students sat and argued the nature of their case far beyond local grievance to the terms of life, liberty, and happiness in consumerist France.

The students played a dangerous game at the very edge of street guerrilla warfare, discovering the success of hit-and-run tactics as they went. Blinded by rage and the class hatred that quickly surfaces in all student-police encounters, the CRS retaliated with the new crowd poison devised for Vietnam, Mace. They beat on the balls a few of the students who most infuriated them until the lads passed out. It remains a mystery that none died.

But what is best remembered from the days of the Parisian May is the pointed mischief and brilliant gaiety of the students. Set aside the filth they left behind after the occupation of the Sorbonne—there were only twenty lavatories for several thousand young men and women, and a million bottles of beer. Set aside the interminable and procedureless debates. Look instead at one sample of the Parisian graffiti; turning upside down the Cartesian university and all its illiberal examination systems, one wit scrawled *"Ils pensent, donc je suis."* And in a poster, another wit parodied the recitation pedagogy of his country, as well as the latest bit of state foolery:

> *Je participe*
> *tu participes*
> *il participe*
> *nous participons*
> *vous participez*
> *ils profitent.*

Truly, imagination had taken power. From across the Atlantic, the comrades of Students for a Democratic Society replied, "Make love, not war."

De Gaulle called the students dog-shit in a broadcast. But things didn't stop with the students. They were still in occupation all over Paris and the rest of the country when the autoworkers, the aerospace workers, the miners, and the railroad workers joined them. The power workers remained at their posts but occupied the premises in the name of their comrades. The *lycées* closed. The dockworkers shut the docks. The post stopped. The country stopped. It was 24 May.

Then all this extraordinary energy, the surge of exhilarated resourcefulness with which young students and young workers discovered a few of the things they could do for themselves, could find nowhere to go. Neither students nor workers knew what to do next. Like a flood, the wide waters subsided until they ran in the usual direction, down the usual channels, to the sea. The trade unions agreed with the prime minister on an enormous pay rise of 35 percent and a reduction of the work week to forty hours. The lockjawed and wooden-headed helots of the French Communist party dithered and did nothing about revolutionary opportunity, instead denouncing everyone in sight as adventurists. De Gaulle checked on his army's loyalty in a secret trip to his commander in the field, called a general election, and won a huge victory from the anxious farmers and bourgeoisie who had watched their children on television calling into question the principles of respectability and productivity by which they lived.

De Gaulle summoned the cold war spirits up from the deep, pointed out the planless anarchy, the cheerful sex, and the simple squalor of the student left, and damned it all as Trotskyism; the French left was finished for another thirteen years. The Communist party snorted and labored in its Stalinist toils;

the other *gauchistes,* who did so much to criticize the sclerosis of state socialism to their generation, rattled the state gates and then ran away, routed.

Something stayed, all the same. Paris 1968 remains a standard to strike. It is a keystone in a critical arch spanning the Western and the Eastern states and it names the plague that poisons each. State socialism, liberal consumerism: those *can't* be the only choices?

VII

The Parisian critique caught and held for a season in the American universities. The assassination of Martin Luther King, Jr., in Memphis, Tennessee, on 6 April had connected in one gunshot the violations of human dignity in Vietnam with the violence and bigotry always simmering in gun-toting America. Besides, King had just come out vehemently against the war, in which a great disproportion of black men had been conscripted.

The essential orientation of all the protest and disorder on the U.S. campuses was the war. In 1968 it was time for the twin, giant institutions of the Democratic and Republican parties to set in motion the protracted electoral machinery by which each would choose a man to stop the war more or less quickly and cruelly.

Until 5 June the people's choice for both the Democratic nomination and the presidency would almost certainly have been Bobby Kennedy. But after beating Eugene McCarthy in the California primary and congratulating him in the Hotel Ambassador that night, Kennedy was caught by a Muslim killer as he took a short cut through a kitchen and shot once in the brain. As he fell and his cortex failed, he tried to say, "Protect the others . . . get them away. . . ." He died during the night.

Kennedy had come late and uncertainly into the presidential competition after Johnson discounted himself, but he briefly carried a great weight of hope. His brother had arrived eight years before, full of beans as president and flourishing his picture of the country's future at a time when it was temporarily quiet. Bobby came in on his brother's slipstream, having previously declared himself a devout cold warrior who took the family's approving interest in Joe McCarthy's goings-on. He was cautiously on the side of his brother's national militarism during the Cuban Missile Crisis.

It is a difficult business attributing motive to the very powerful. And yet it is hard to refuse a sympathetic response to Norman Mailer's description of their meeting—between the great writer (with his tremendous faults) and the man of power (with his).

He had grown modest as he grew older, and his wit had grown with him—he had become a funny man as the picture took care to show, wry, simple for one instant, shy and off to the side on the next, but with a sort of marvelous boy's wisdom, as if he knew the world was very bad and knew the intimate style of how it was bad, as only boys can sometimes know (for they feel it in their parents and their schoolteachers and their friends). . . . Since his brother's death, a subtle sadness had come to live in his tone of confidence, as though he were confident he would win—if he did not lose. . . . He had come into that world where people live with the recognition of tragedy, and so are often afraid of happiness, for they know that one is never in so much danger as when victorious and/or happy—that is when the devils seem to have their hour, and hawks seize something living from the gambol on the field.

The reporter met Bobby Kennedy just once It was on an afternoon in May in New York just after his victory in the Indiana primary and it had not been a famous meeting, even if it began well. The Senator came in from a conference (for the reporter was being granted an audience) and said quickly with a grin, "Mr. Mailer, you're a mean man with a word." He had answered, "On the contrary, Senator, I like to think of myself as a gracious writer."

"Oh," said Senator Kennedy, with a wave of his hand, "that too, that too."

So it had begun well enough, and the reporter had been taken with Kennedy's appearance.[15]

Afterward, remembering this, Mailer wrote of Kennedy's youth, too young for senator, let alone president. "It was incredible to think of him as President, and yet marvellous, as if only a marvellous country would finally dare to have him."[16] And then he was killed. The poor—Negroes, Puerto Ricans, elderly Jewish widows, shabby Italians, and always the voluble, disappointed, optimistic, but irretrievably poor Irish—came to mourn, waiting six hours in a long, quietly lamenting line, a mourner every three seconds edging past his exploded, rebuilt head.

The next day the train voyaged from Penn Station to Washington, its bell clanging all the way, its horn sounding sonorously. Here and there, singly and in groups beside the rails, the poor watched him go.

Bobby Kennedy had changed since the days he spent in the Oval Office when Khrushchev's flotilla was approaching. He had been changed by ambition, the sadness of power, and the facts. He turned against the war, as he had turned against American racism. Whatever was abroad in the air of 1968 filled his veins and simmered there. He had become, recklessly and truthfully, a man who was going to tell his country what was wrong with it. So he spoke, as his brother never did, for an unhypocritical truthfulness about what that country, with its fine ideals, did so unspeakably: in Saigon, in Memphis; so somebody shot him.

After he was killed, the Democratic nomination was just a big fight. As for the Republicans in Miami, Reagan blew in, Rockefeller flew in, but there was never any doubt that the job was Nixon's. The candidate discovered the silent majority in his acceptance speech and ended, "I pledge to you tonight . . . to bring an honorable end to the war in Vietnam."

The Chicago convention was the defeated side of this bland success. The Democratic delegates came to the convention at a loss after Johnson's withdrawal and Kennedy's murder. Hubert Humphrey, Johnson's vice president, machine politician, decent-hearted, if he had a heart, fatheaded, for he had a head, cherubic and utterly goofy, a politician who lived so far gone in clichés he believed them all, came to be nominated. He could not even see that, if he were, the Democrats were sure to lose. Eugene McCarthy came, certain that he would not get the nomination, knowing that he was the only man after Kennedy's death morally entitled to it and knowing also, because he was very intelligent and bitter and ironic with it, that the dark, reactionary side of America would destroy him if he ran for president. But like Kennedy (only McCarthy did it first), McCarthy had also named for what it was America's arrogance and cruelty in Vietnam and had foreseen a world in which the cold war would end and Communist would no longer be the name of the devil. He also promised that as president he would dismiss J. Edgar Hoover, the monster head of the FBI who had just been confirmed in the job by Johnson for life.

These were some of the grandees who came to Chicago. The police were already there under their boss, the barrel-shaped Mayor Richard Daley, together with a vast, inchoate, and contradictory horde of American citizens who wanted the Vietnam War to stop. They were through with Johnson— "Hey, hey, LBJ / How many kids did you burn today?"—and they knew how useless Humphrey was—"Dump the Hump / Dump the Hump." They were all sorts: upright McCarthyites in respectable suits and fitted jeans, sporting the pale blue flower of the McCarthy logo, proper, even severe; barmy yippies of the Youth International party in the now famous uniform of flared, fringed jeans, loose flowered shirts, wild Afro hair, long granny dresses in pastel prints, bare sandaled feet, heavy, loose leather belts, drifting skeins of loose hair; wild revolutionary sectarians, Minutemen and Black Panthers, with stony eyes and faces and a theory of violence; bikers and Hell's Angels in heavy studded jackets and big black boots; students, singers, lawyers, doctors, actresses, waitresses, teachers, receptionists, social workers, short-order cooks, cabdrivers—an army of respectable progressives and the welfare state alongside an army of the unemployed, the underclass, and underpaid service industry workers. Among their heroes were Allen Ginsberg, Whitman's heir and poet of the road, and Norman Mailer, reporter-genius and poet of all America's amazing motley.

They come to protest to the party that had kept the Vietnam War going since Kennedy had come to power eight years before. So they gathered in Lincoln Park and Grant Park and waited for the next, inevitable step in the drama, when the police would attack them. The police duly did so, for no particular reason except that no modern state can tolerate a crowd, only purposeful pedestrians.

Mayor Daley, or someone else, ordered the police to clear the parks and main streets. Once again in 1968 there appeared on the television screens of the world the charging lines of perspex shields, the faceless visors, the regularly pumping truncheons. Running before the police was the weirdly costumed crowd, dissolving now and then into recognizable individuals: a girl grabbed and yanked by her streaming hair, her jaw driven wide open by a scream; a boy bent below his shielding arms as a cop beat on his head and back; another boy, ashen and pointing at a policeman, his face crazed with long rivulets of blood; two people, men or women you can't tell, dragged by their clothes along the ground, cops beating on them as they are scraped along. The torrential crowds ran when they could and were blocked so that they could not. Over the parks and over the rich hotels the tear gas hung and fell. People in the crowds, whether seeking peace with candles in their hands or looking for a fight with rocks in theirs, drew the searing, emetic stuff into their lungs and gasped and wept loops of mucus from eyes and nostrils, coughing up more mucus to hang from their lips and clothes.

In spite of it all, they sang "We Shall Overcome," were sung to by Seeger, chanted in support of Ho Chi Minh and the NLF—"Hell, no, we won't go"—won't go away, won't go to Vietnam. They made righteous America quiver at the thought of socialism abroad and reassured the police that their pumping truncheons were a good and needful revenge.

So Humphrey was nominated, and Nixon won, and the hot and the cold wars went on while the delegates went home. There were plenty of people still to be killed in 1968, but by September the news was in.

Going by the headlines, the forces of oppression, tanks and policemen and politicians, won: the usual bad guys. Underneath, it may be, some better things were stored up for the future. The best part of America's polity turned against the nation's cruel adventure; the Soviet Union, Prague notwithstanding, earnestly sought American friendship. And somewhere in the veins of a generation just coming into its citizenship, the rabies of ideological warfare began to subside.

15

The Fall of Allende, SALT I, Leaving Vietnam, 1972–1973

A political axiom of the whole period was that all globally significant action was to be redefined as an action of cold war, and that all global players were to be assigned to one side or the other of the ideological force field maintained by the opposed great powers.

The difficulties always grated where the spheres of influence had their shifting and uncertain edges, as they did in the Far East, and so very pressingly, in the Middle East.

For donkeys' years, however, the Middle Eastern little troubles went to show that even during the cold war, the history of the world had other tales to tell than the old lie about capitalism versus communism. Each superpower settled upon the sands of the desert and left money and weapons with the Arabs or the Israelis who came cautiously up to greet them. The Soviet Union built a ring of supposed clients round Israel: Syria, Iraq, Egypt. Nasser spoke of Soviet magnanimity; the United States put a peacock Shah back on the peacock throne, they paid, formally and informally, for the Israeli army; but when it came to the push Sadat threw out his Russian advisers and set off for certain defeat (as he knew) at Yom Kippur in 1973. A few years later Carter and his National Security Adviser, Brzezinski, were no more than bungling and ineffectual spectators as the Shah walked off into exile, to be followed a minute later by the Soviets' terminal mistake which ended their empire when they invaded Afghanistan.

The Middle East has never been tractable terrain for fighting hot skirmishes in the cold war. It is not even much good for ideological tootings. The political allegiances are so varied, the enmities so ancient, the territorial advantages so doubtful with so much desert and the intense localization of the oil, that neither side ever knew quite what to make of it any more than their imperial predecessors had. It has therefore been the site of so many of the scariest moments of the cold war, what with the Israelis being so unpredictably ruthless, the Syrians so vengeful, Arab nationalism so deliriously self-immolating. In the Six-Day War of 1967 when the Israelis broke off the American leash and had their tanks at the gates of Damascus, Kosygin was on the hot line to Johnson threatening the use of nuclear weapons in a tempest of rage, and Johnson in turn was hot with rage at the Israelis, having had his authority flouted so publicly.

On the whole each handled their clients gingerly; both were used pretty much at will by their juniors. In the end Kissinger flew to Moscow in one of his shuttle-diplomacy sallies to stitch up the ceasefire, but all his later patch-working in the Middle East came unstuck. In 1973, both the Soviet Union and the United States discovered, not for the first time, that with all their resources and will, their flatfooted advent in any corner of the globe had unintended consequences and often went awry. In 1972, however, the Soviet Union was beginning to realize that it could not afford the gross distortions wrought upon its economy by the arms race, that concentrating its productive power so exclusively upon the instruments of mass destruction had twisted everything else out of shape. It began to discover the inefficiency and indiscipline, the drunken indifference, of its gigantic and far-flung army. Bit by bit the news came through to bearish, slow-thinking, but not unintelligent Brezhnev that someday he would have to do something to stop the mad machine.

Something of the same news struggled home from Vietnam. The troops were stoned in the field; toward the end of their tours they refused to fight. Nixon *had* to bring them home.

But the will of God's own country not to be humiliated was still strong and high. Americans would let the Vietnamese fight it out, but they would do so in good order and in their own time. Meanwhile, Nixon of all people would put the wind up the Soviets by dallying with the Chinese and would reassert the Pax Americana by concluding a triumphant arms deal that pulled off the elusive trick of calling a peace while the United States was still in the lead.

And if, during this triumphal period, countries required by cold war convention to ask permission of Washington before changing presidents or uniforms or anything else showed the slightest spasm of independence, they were put down by U.S. will and weapons.

II

Chile had, of course, mortally offended its mighty neighbor in 1970 by voting in an avowedly socialist president who led a broadly left coalition government. The CIA had confidently predicted that Salvador Allende would never win, and when he did so by a whisker, Nixon and his new star National Security Adviser Henry Kissinger agreed, without even having to say so to each other, that, in Kissinger's menacing phrase from his memoirs, they would "circumvent the bureaucracy."[1]

The prince of darkness has now stepped onto center stage. He arrived from the wings in 1968 when, having backed both Nixon and Rockefeller for president, he adroitly retained the favors of both and was assured of a job either way. In fact, he only just placed his bet on Nixon in time, having often vilified Nixon's foul-mouthed philistinism behind his back, a practice he apparently maintained at cultivated dinner parties in the posher parlors of Georgetown.

It is obvious from all the records and memoirs that Kissinger relished his reputation as a realist politician. Sardonically recognizing that, in his own aphorism, "power is the great aphrodisiac," he gambled on his combination of charm and opportunism to secure his reputation.

In 1972 he had the future of the world in and out of his hands, starting with Chile. Before Allende's victory, Kissinger had delivered himself of the following foreign policy dictum to the Chilean foreign minister, Gabriel Valdes:

> Mr. Minister, you made a strange speech. You come here speaking of Latin America, but this is not important. Nothing important can come from the South. History has never been produced in the South. The axis of history starts in Moscow, goes to Bonn, crosses over to Washington, and then goes to Tokyo. What happens in the South is of no importance. You're wasting your time.[2]

Kissinger, in holding this uninhibitedly contemptuous view of the Latins, shared with the boss the correlative view that what he wanted to happen in the South certainly would. The head of the CIA, Richard Helms, and his trusted deputy, Cord Meyer, were given ten million bucks and a clear order to get rid of Allende in any way they could. Meyer wrote, with endearing ingenuousness, "We were surprised by what we were being ordered to do, since much as we feared an Allende presidency, the idea of a military overthrow had not occurred to us."[3]

Well, I bet; or not without the president's say-so.

For although the CIA had been so wrong about the election, it had fingers deep in every Chilean pie, vineyard, drug deal, and expensive bedroom. The Senate Intelligence Committee later reported that a posse of economic journalists were brought in from all over the world to write panic into the currency. As a result, before the Chilean president was confirmed in office, domestic and corporate assets had fled the country.

For convenience, Kissinger applied the domino theory to Central and South America, as he did to Southeast Asia. Yet again, a very intelligent man, drivingly vain and ambitious with it, brings to his view of power an intensity and narrowness of vision that are indistinguishable from stupidity. Saul Bellow once called such persons "high-IQ morons." Kissinger, all unawares, was one. Like Stalin, he treasured his ruthlessness so tenderly that he could no longer tell the difference between ideology and interests.

Kissinger and Nixon authorized another $250,000 to bribe the Chilean congress to vote against the ratification of Allende. They kept the CIA working against Allende in top gear without saying a word to the wretched Edward Korry, the U.S. ambassador in Santiago who wanted to open up amicable talks with the new Chilean leader. Korry was an upright man, and anxious to avoid another farce like Cochinos Bay. He did not guess at the mad depths of his bosses' intolerance—less, it may be conjectured, a matter of ideology for Nixon and Kissinger than a question of insisting, absolutely insisting, on *having their own way.*

The CIA kept in close touch with the right-wing Chilean military officers prepared to lead a coup against Allende. The Agency was authorized to bribe the senior Chilean military with American dollars. The CIA told its station chief in Santiago that the White House had "a firm and continuing policy that Allende be overthrown by a coup,"[4] for, as Kissinger in one of his savage attributables remarked in public, "I don't see why a country should be allowed to go communist through the irresponsibility of its own people."[5]

Through 1971, plans for killing Allende became firmer. The talk of the bazaar was that "Henry wanted it." Korry said as much to the Senate committee in 1976, but nobody was listening by then; they only wished Allende could be forgotten. In any case, the White House declared economic war on Chile by ending investment guarantees, putting pressure on world lending agencies, and obstructing private investments from the United States. American aid dropped from $70 million per year to just $3 million. The World Bank did as it was told, and the CIA had all it needed—about $8 million, apparently—to get President Allende seen off. The CIA drew up the lists of those to be taken to the stadium; it stole government plans to deal with military coups; it arranged, more or less, to ensure the silence and safety of American citizens when the coup happened. And in September 1973, it did.

The days of the coup are piercingly told by Victor Jara's wife, Joan, in her

biography of the famous Chilean folksinger.[6] On 11 September he was scheduled to sing at the university in Santiago, at the opening, his wife calmly
reports, of an exhibition about civil war and Fascism. There is something
approaching a general strike, and the trade unions are calling members
together to warn them of civil danger. But normality is a powerful and
necessary sedative. Joan takes little Manuela to school. The streets are empty,
warm in the sun, quiet. The bread shops smell delicious. She gets back to find
Victor listening to Allende broadcasting on the Popular Unity wavelength.
The news is desperate, and as the radio stations gradually come under army
control, military music replaces Allende's voice during his manly farewell to
his people.

> This is the last time I shall be able to speak to you. . . . I shall not resign.
> . . . I will repay with my life the loyalty of the people. . . . I say to you: I
> am certain that the seeds we have sown in the conscience of thousands and
> thousands of Chileans cannot be completely eradicated. . . . Neither crime
> nor force are strong enough to hold back the process of social change.
> History belongs to us, because it is made by the people.[7]

Victor decided to go to the university in response to the unions' summons.
He puts their last emergency can of gas in the car. A neighbor, a pilot in the
national airline, shouts jeeringly at Victor. The neighbors living round the
little courtyard of apartments are chattering excitedly; one of the women
squats down and gives Joan Jara "the most obscene gesture in Chilean sign-
language."[8]

Senior officers who support Allende have been shot. Civilian right-wingers
are collecting beside army trucks with their personal automatic rifles slung
over their shoulders. There is excitement throbbing in the trucks; it is good
to be out and about with a gun when your soldiers are due back at your side
at any moment. Fascism is driven by fierce little throbs of hate and killing
power. It is a politics of permanent excitement.

The people on their balconies bring out trays of bourbon and sangria to
watch the presidential palace being bombed and strafed. Victor telephones his
wife, once, twice, three times; he is trapped at the university by the curfew.
Through the night there are the intermittent rounds of gunfire, the occasional
heavy blamming of automatic weapons. In the morning, alone with her
daughter, Joan Jara sees the generals on television, specific and splendid in
their uniforms, talking about "eradicating the cancer of Marxism."

There is no news of Victor. The phone is tapped. After two nights, on the
third afternoon, a voice calls and nervously says only that Victor has been
taken to the huge stadium and "doesn't think that he will be released."

"Thank you, companero, for ringing me, but what did he mean by that?"

"That is what he told me to tell you. Good luck, companera," and he hung up.[9]

On 18 September a comrade came to tell her that Victor Jara was dead, shot by the military; his body had been recognized by a party member working at the morgue. Joan was taken unofficially to identify him.

We go down a dark passageway and emerge into a large hall. My new friend puts his hand on my elbow to steady me as I look at rows and rows of naked bodies covering the floor, stacked up into heaps in the corners, most with gaping wounds, some with their hands still tied behind their backs. There are young and old—there are hundreds of bodies. Most of them look like working people. Hundreds of bodies, being sorted out, being dragged by the feet and put into one pile or another, by the people who work in the morgue—strange silent figures with masks across their faces to protect them from the smell of decay.

I stand in the centre of the room, looking and not wanting to look for Victor, and a great wave of rage assaults me. I know that incoherent noises of protest come from my mouth, but immediately Hector reacts. "Ssh. You mustn't make any sign—otherwise we shall get into trouble. Just stay quiet a moment. I'll go and ask where we should go. I don't think this is the right place."

We are directed upstairs. The morgue is so full that the bodies overflow to every part of the building, including the administrative offices. A long passage, rows of doors, and on the floor a long line of bodies, this time with clothes, and some of them looking more like students. Ten, twenty, thirty, forty, fifty . . . and there in the middle of the line I find Victor.[10]

III

In Chile in 1973 the United States roundly insisted upon its will and its way. It remains one of the grimmer lessons in political realism during the cold war. But the White House had always been able to match the Kremlin for realism, never more than with Nixon and Kissinger in residence.

Like all great courts, the White House throngs with its officers, courtiers, and supplicants, as well as hordes of rubberneckers who get drawn into the action in spite of themselves. It is a large but utterly ordinary house; it is the thrilling center of the richest and most powerful nation the world has ever known. But the duo who installed themselves after the election of 1968 were unique in the history of the building for the way they kept the effective business of the place to themselves. It was their singular contribution to the running of imperial America to construct unique arrangements for the promulgation of extraconstitutional, paramilitary forms of executive action, little trace of which remained on the record.

Nixon and Kissinger have been endlessly chronicled. The oddity of the partnership itself has been much gossiped about; the junior figure so publicly and so often derogated his master, and that master was notorious for his moodiness, touchiness, stiff inaccessibility, and solitariness. Neither had the gift of friendship, or even a very great interest in its rewards, and both were insatiable for power in its simplest, most coercive forms. Each gloried in his mastery of the endless negotiation of modern power politics and of the democratic institutions that subtend but do not conduct it. Both merely despised opponents less ruthless than themselves and always desired victory at any cost.

So Nixon won on the ticket of an honorable peace in Vietnam, and Kissinger came to his side as a kind of intellectual gangster miles away from McNamara's managerialism; a man who would fix foreign policy by avoiding both formal diplomacy and the Senate, by influencing the destinies of foreign rulers, by setting up fishy funds, by carpetbagging on a world scale.

Kissinger is famous for his theoreticism. He propounded a singular view of his nation's foreign policy as everywhere interlocking. The huge map of cold war commitments—in the Middle East, Southeast Asia, Latin America, Europe, China—was to him indivisible both politically and ideologically. The power needed to maintain these commitments was prefigured in the innocent algebra of zero-sum: what you gain, the other side loses. The keen pleasure of the game was to disguise gains as concessions, and successfully to redescribe defeat as victory.

The talks on disarmament—the strategic arms limitation talks (SALT)—had been vaguely started by Johnson and McNamara in the mid-sixties. Until then, almost the only conscious check on the mad accumulation of more and more nuclear weapons had been made by Kennedy and Khrushchev. In the atmosphere of expansive mutual relief that followed the Cuban Missile Crisis, the two leaders agreed to stop poisoning the world with particles of fallout and signed the 1963 test ban treaty.

By 1966 it had dawned on Robert McNamara—probably the sharpest of the postwar secretaries of defense in spite of his irrational belief in management as the key to human advances—that simply allowing nuclear weaponry to pile up failed to square with hopes of a safer world in which the Soviets would agree to stay out of the way. McNamara concentrated on four aspects of the problem: the rhythm of U.S. weapons innovation and Soviet response—the Soviets matched every American scientific or military development by which the United States sought domination, but only after a two- to five-year lag; the partly ideological, always genuine distinction between weapons of attack and defense—in a world stiff with suspicion, each of the superpowers dreamed of perfect invulnerability; the homegrown quarrel between advocates of the new defense system, in which a screen of missiles could be launched with the sole function of knocking out incoming missiles, and the delighted engineers who had devised a new long-range missile extension, which fitted many more projectiles onto the top of the carrier, each one of them capable of taking off on its own in a different direction; and finally, McNamara—and, indeed, Johnson—simply wanted to slow the whole head-long, insane, and bloody business down.

Out of this medley McNamara and Johnson chose two ideas for wary proposal to the Soviets: to agree that neither side would proceed with development of the protective screen known as the anti-ballistic missile (ABM) system, and to explore the broader possibilities of limiting the spread of gigantic, all-destroying, trans-world, or strategic weapons.

These were the treaty terms, genteelly discussed for three years, that Nixon and Kissinger took over. Both of them at first took little interest. Both superpowers were testing the new multiple-rocket extension glued on the front of the big missiles—MIRVs, or, in the barbarous jargon of the trade, multiple independently targeted reentry vehicles. Nixon and Kissinger wanted an arms deal only if it was linked to what they saw as the Soviet capacity to influence Hanoi and stop the Vietnam War.

But the Soviets, their economy in deep trouble, overspending heavily on weapons research, badly wanted an arms treaty. Congress in its laborious way built up in committee an enormous dossier against the ABM screen; city defense, as it came to be known, began to seem impossible.

All participants agree that once Kissinger had realized the immediate usefulness of arms talks, he acquired a quite astonishingly rapid grip on the strategic arguments. He remained (Hersh says) ignorant of the science itself, however much he pretended to know, but his speed, sureness, and flexibility with technical strategy were unrivaled. He was equally self-confident in diplomacy. Nixon wanted to change history by his foreign policy; his interest in domestic matters was very limited. He would win a sort of victorious peace in Vietnam; he would bring China back into the comity of nations as a friend of the United States; and he would bequeath his people more peace as well as more power by limiting nuclear weapons.

As the seven long sessions of arms limitation talks in Helsinki proceeded, from 17 November 1969 to May 1972, both countries heaped their nuclear weapon stocks ever higher, each going ahead at a great rate with developing the multirocketed brute of a missile. Kissinger kept his strong grip on the progress of the talks, now attaching them to the future of Berlin, now linking them to possible concessions over Vietnam. In a meeting as extraordinary for what it reveals about Kissinger's vanity as it was crippling to the serious and

permanent staff, he took a secret trip to Moscow to discuss the arms negotiations with the Soviet ambassador to Washington on his home ground. In a
weird foray into domestic politics, he tried to bully the waterfront unions in
order to clinch grain aid to the Soviet Union as part of the arms treaty. It is
a pleasure to learn that the virulently anticommunist union president, disliking
Kissinger's backdoor methods as much as he hated the Soviets, told the great
intellectual-diplomat to piss off, and Nixon finally had to exert all his presidential authority to persuade the stevedores to oblige.

The talks struggled on in Helsinki, following that rhythm made familiar
since the Korean armistice of honest and modest proposals by the reasonable
men on our side countered and blocked by our wooden-faced opponents, who
only pretend to want a safer world since their self-evidently ridiculous ideas
include removal of all missile launchers and carriers now surrounding their
country on every frontier. Part of the same story, which began so widely to
come unstuck across Europe after 1979, was that Allied governments watched
nervously for reassurance that the good old United States would keep every
hellish weapon at the ready forever, thus indicating its resolve, trustworthiness, and determination to uphold the great freedoms of NATO and West
European consumer benignity.

The way it was, each side wanted to cut its costs and give away only tiny
concessions while gaining a little more in return. Each side was convinced of
the other's advantage—in numbers of bombs, in geography, in history. In
truth, it is hard now not to feel some sympathy with the Soviet view that 3,000
miles the other side of the Atlantic was a more comfortable place from which
to argue than from one ringed by missile launchers all along the frontiers of
West Germany, Greece, Turkey, the United Kingdom, and the Mediterranean.

During 1970 Nixon decided to deploy the multiple missiles, that is, he
scattered them around the silos of the United States. The Senate voted by a
great majority to stop testing the infernal weapons, but the White House went
on its way. The Americans quashed any likelihood of agreement at the arms
talks by requiring inspection of the other country's state of readiness on the
actual missile site. It was certain the Soviets would refuse to agree to this
condition, and they did.

They came back, however, with the astonishing proposal, now pretty well
forgotten, that the United States and the Soviet Union join in informal
alliance against nuclear threats from any third nuclear power. It was a measure of their congenital fearfulness that the Soviets would put up such a
scheme. The object of their fear was China. Nixon sent them reeling. He went
to China, the first American president to do so since the 1949 revolution, and
was welcomed as the harbinger of a new era in Sino-American relations.

American obduracy at the conference table and Nixon's audacious visit to Beijing brought the Soviets to agree to a trivial deal, though it was much bruited about as a hugely important one. Kissinger came back from his secret ride to Moscow having cleared an agreement by which Brezhnev would offer back as his own idea Kissinger's sketch of a treaty to limit sharply all development of ABM systems.

ABMs were always an appetizing chimera. They promised defense against other missiles without being launched against other people. But few scientists thought they could be made to work: the collision of missile with missile would cause unpredictably high nuclear shock and devastating fallout on an indiscriminate scale. The program would renew the frenzy of each country to outdo and get ahead of the other.

When Kissinger got home he found, however, that Adm. Hyman Rickover, doyen of submarine warfare, had persuaded the navy and the navy lobbyists to pay for a new submarine—the biggest one ever made—whose vast reactor would pulse below the waters of the world and drive the great slender thing on permanent guard, its torpedo tubes loaded with the new multirocket missiles. With an adman's eye for a catchy name with a classical touch, the proud parents baptized the submarine Trident. A couple of hundred of them would be made; four would be sold to the English, to make them feel big.

It was a ludicrous, all-American decision to buy big, look swanky, and give a strong push to the Soviet Union to match this imposing mastodon as soon as possible.

So when Richard Nixon and Leonid Brezhnev met to sign the first two SALT treaties in Moscow, the treaties meant little more than that the superpowers were on relatively good terms, and that they would increase their nuclear stockpiles more or less in step. With the usual touching faith of those in power in their own elderly indispensability, they agreed that each capital could have missile screens for protection, but nowhere else.

All this made Nixon look peaceable and Kissinger like his brilliant Mercury. But more important, the grand signing that accompanied the great powers' meeting in May was happily timed to launch Nixon on his certain road to reelection.

IV

The treaty was not much to celebrate; it could not rise above the jolt to any hope of mutual trust given by the 1968 invasion of Czechoslovakia and the corresponding interference in Chile in 1973. The real change in relations was the U.S. move toward China. The foundations and superstructure of the cold war were softening and altering across a wide front. The Vietnamese would be left to stew in their own juice. Nixon, however sweaty and stubbly he was in all his public appearances, was also cocksure about his position and power. In the glacially slow pace of policy change adopted by cold warriors, SALT represented a shift that Nixon believed could ultimately make him a peace-monger in Vietnam, a missionary to China, and the victor in the cold war.[11] Johnson's 1967 meeting in New Jersey with Premier Aleksei Kosygin (Khrush-chev's successor) had been the opening; SALT was the wedge driven into the gap, China the chamber of opportunity. Nixon was now sure he had a free hand in Vietnam: the Soviet Union would hardly interfere, and the People's Republic conserved centuries of enmity toward its recent ally.

Consequently, just as the Democratic party offices in the Watergate build-ing were being broken into in June 1972, Nixon and Kissinger were keeping up the war in Vietnam by the old methods. Four years after his 1968 election promise of an honorable peace, Nixon's policy was to remove the G.I.s while they were still alive and to bomb the rest of the country into a state of concussion that would count as peace, leaving the survivors to tidy up the mess.

It might have turned out like that. The deadly duo had all the insouciance it took to blast the Vietnamese landscape to mud and its people to splinters of bone and red gobs of flesh. (Like so many of the bomb crews, they never saw the results and lacked the imagination to picture them, thereby failing the noble tenet of old humanism—no understanding without the will, no will without compassion.) When Ambassador Henry Cabot Lodge, Jr.'s former assistant and consul in Hue, Tony Lake, came to Kissinger to resign from the foreign policy team with three other colleagues after the Cambodian invasion, Kissinger drew on his stock of human wisdom. "No one has a monopoly on compassion, Tony," he said.[12]

Cambodia was briefly invaded by U.S. troops in May 1970 in order to "interdict," as the jargon had it, the Front's supply lines. The country had been suffering prolonged bombing by the B-52s for a year. A large portion of Laos around the Ho Chi Minh trail had been portioned out into the lethal bombing "boxes" and comprehensively plastered within the rectangles.

According to the inventive cant of foreign policy, whose purpose is to conceal as well as to encapsulate, this was "Vietnamization," a process related to pacification. Both efforts were intended to deny the peasantry to the Front, the idea being that the peasants would create their own Vietnamese army, while their families went to live in the dismal camps in blank places away from their ancestral homes.

By early 1972, even with his Chinese and Soviet successes assured, Nixon still needed much brighter peace prospects in Vietnam. His peace pledge had resulted in over 17,000 American deaths by that date, and his bombers had dropped more tonnage than the United States had used in both the European and Pacific theaters of the Second World War. The Phoenix program, by which systems of local police and army officers identified and "neutralized" the "Viet Cong infrastructure," had reported 20,000 agents, one-third of whom were killed in the program's first year of operation. It was plain that the mad scheme was used either to assuage local vendettas or to pick up the bounty paid to Phoenix officers who met their quotas.[13]

The Front had begun once more to emphasize that it was not going to disappear. Saigon's soldiers, in spite of cowardice, desertion, and a simple, rational reluctance to fight, were dying in larger numbers than ever before, at 20,000 per year. As Nixon successively withdrew the American ground troops and infantrymen—by September 1972 there were only 27,000 left—the enormous South Vietnamese army, one million strong by 1973, half the able-bodied men in the country, simply and absolutely depended on the protection of air-strike fighters with their terrible, accurate rockets and on the unstopping rain of bombs from the B-52s.

While the bombs fell, the negotiations continued in Paris. Some of them were secret, and some were crooked, as when Nixon suddenly disclosed that he had been dealing with the future of Vietnam in discussions with China and the Soviet Union that did not include Hanoi. Blindly confident as he and they were in great-power diplomacy, it simply didn't occur to the three powers that they couldn't run things their own way; the North Vietnamese thought otherwise.

In a bold spring offensive the northerners smashed into the army of the South near Kontum. U.S. intelligence units learned that something was up but didn't know where, and the army of the South once more scattered in front of the unwavering discipline of the northern soldiers and the Front. Thousands of southern soldiers were within reach of their families and deserted wholesale to find and protect them. Everyone had appalling tales of senior officers who quit to get back to base and protect their private, venal savings. After the panic, only colossally heavy strafing, rocketing, and bombing by the U.S. Air Force prevented the complete defeat of the army of the South—that, and the apparent hesitation of the North to believe in the opportunity it had created.

Nixon relished the fact that his famous policy of "triangulation" (another slippery slogan) put him at the peak of the triangle, gazed at by China and the Soviet Union. So, on 8 May 1972, he announced gravely, sorrowfully, that the ports of Hanoi and Haiphong would be mined and bombed to accomplish the interdiction of supplies; Soviet cargo ships had already been hit in the Red River Delta. Nixon was sure he could get away with anything international, it was only the dinks who still refused to bend to his will.

Nixon put into effect during the second half of 1972 what is now notorious as his "madman theory of history." One of his senior accomplices who went under after the Watergate hearings reported that Nixon said he wanted the North Vietnamese to believe in his basic irrationality about communism and even to fear that he might push the nuclear button, that he "might do *anything* to stop the war."[14]

Acting mad—take it how you like—Nixon ordered the most violently unbroken bombing ever known in warfare until that date. (He exceeded it in Cambodia later.) The great dikes of the North and the Red River Delta were breached unstoppably, and the towns reduced to simulacra of Hiroshima. Unknown tens of thousands died; the main population was evacuated. The domestic economy stopped. But underneath Hanoi, these astounding, cthonic people lived and worked; they built a safe underground city with hospitals and a university while their young men went on fighting.

The People's Revolutionary Government, which had been organized in the South for about three years, and its allies in the North learned early that they could trust no temporary concessions by the Americans. Short of being attacked by nuclear weapons, they knew they could carry on, and they also knew that as the American G.I.s steadily quit the South, their own recruitment picked up and their losses fell.

In the Paris talks, the North Vietnamese chief negotiator, Le Duc Tho, outfooted Kissinger by suddenly presenting a draft treaty for peace that recognized Thieu as the legitimate head of government in the South, agreed that a political settlement had to be fully negotiated later and quite separately, and included conditions for a military peace in the meantime. This was October 1972, and Nixon wanted a handsome peace immediately after his triumphant reelection in November. Although Le Duc Tho offered the Americans what they wanted Kissinger was told by Nixon to stall. He first squared Thieu and then cynically denounced the Paris talks with the usual lies about Hanoi intransigence and U.S. fairmindedness, blah, blah.

Bitterly angry, the North Vietnamese published the transcript of the double-crossing. Unmoved, the great voting public returned Nixon with a humiliatingly large majority over Sen. George McGovern.

The administration promptly inaugurated its second term by stepping up its bombing and going back on its agreements. Nixon was indeed becoming madder and withdrawing to an ever greater distance from ordinary human exchange. He wanted more bombing, and he got it. Hanoi and Haiphong were finally attacked in a spasm of crude terrorism. No one thought the attack would do any good, and the B-52s, bombing straight through Christmas in a way insolently calculated to bring out the demonstrators in the liberal capitals of the world, paid no attention whatever to what they hit. They did fearful and indiscriminate damage, but they were also resisted by exceptional displays of air combat by the air force from the Democratic Republic. A lot of B-52s were brought down, and the U.S. pilots were taken prisoner.

It was a very brave show by the North, resisting a gratuitously vile and pointless act of petulance by a president now indeed beyond reasoning with, served by a gifted man whose quick intelligence was quite vitiated by vanity, arrogance, and the sheer thrill of coercive power.

Negotiations began again after the bombing was stopped in response to a public outcry that at last did credit to Americans for holding certain truths to be self-evident. There was to be a ceasefire at the end of January. All American troops would be gone sixty days later. "The United States will not continue its military involvement nor intervene in the internal affairs of South Vietnam."[15]

The Americans finished "Operation Enhance" in a rush, pouring weapons, supplies, prefabs, tanks, aircraft, ammunition, spares, and food out of the mammoth cornucopia of their plenty. Untrained, afraid, exhausted by the war, bereft of their land, their markets, their homes, the South Vietnamese stared at the gleaming piles of stuff and waited for the end. Even the revolutionaries thought they might get some help from the United States, and Nixon strung them along for a bit, mentioning $3 billion in one note sent to Hanoi, the World Bank in another. In the event, of course, Vietnam never received a penny, and the United States prevented its obedient allies from giving the wasted and blasted land any aid from that day forward.

16

EVENTS IX

SALT 2 and the Invasion of Afghanistan, 1979

THE almost twelve years from the start of the Tet offensive at the end of January 1968 to the invasion of Afghanistan by Soviet tanks at the end of December 1979 saw the great powers slowly move past the zenith of their domination. It was not a sudden thing, like the very end of the epoch would be, but we can see now that the firmament of the nations turned widely into new configurations. We can see Europe chafing for the first time against its dependence on the United States; in America the states themselves being at deep odds with one another as power and money migrated south and west; and the monolith of the Soviet Union begins to show tiny fractures as the domestic economy starts to fibrillate and as its Islamic and Russian Orthodox churches stirred tremendously.

The cold war captains felt their power slip a little and stretched themselves to reassert it. Not for the first time, the reassertion of power involved complicity with the enemy. The old nineteenth-century term "détente" was dusted off to describe the mutual dispositions of superpowers in a world proving gradually unresponsive to their assertions of power.

One manifestation of this was Kissinger's term "linkage." His negotiations with the Soviets went forward in a complex network "linking" concessions in nuclear weaponry to restraints on Hanoi, joining deals in Moscow to balancing deals in Beijing, connecting ceasefire lines between Israel and Egypt with neutralization in Angola or Somalia.

There was a necessity and a madness behind his maneuvers. The delicacy of the cobwebs of nuclear response make all those sitting in the center, however much they love it, madly sensitive to every trembling on the line. But they are of that world so completely that they do not feel the changes that come over peoples and nations unconnected to the cobweb. It never occurred to U.S. officials that Hanoi would shake off the Soviet Union and that it also felt a long historical enmity for China. That Iran would rouse itself into a religious paroxysm and throw out the Shah quite without benefit of Soviet advice or recourse to cold war precepts also fell outside the rules of the game plan. Neither did it occur to Soviet chiefs of staff that Afghanistan would prove, as it had to so many past masters, entirely intransigent to Soviet surface-to-air missiles and would send Soviet troops, stoned, on their way.

Nixon and Brezhnev were too profoundly and inflexibly formed by the political institutions that embodied and enacted their worldview to adjust themselves. Détente was a rune to be cast over the problem of the arms race; it would clear a little space in which to breathe more comfortably. After time and justice finished off Nixon, after defeat in Vietnam, after an international treaty that at least openly acknowledged nuclear danger, after dinner in Moscow and tea in China, the United States and the Soviet Union backed off just before the new decade opened and once again lowered the temperature on cold war.

II

During the Vietnam War the opposition to it had largely congested the boulevards of the eastern and northern states. But power, money, and age had shifted south and west. In a vulgar diagram of social forces in the United States, the liberals could be placed in the top half, the conservatives in the bottom.[1] Frostbelt versus sunbelt, they say, and in the sunbelt the old anti-communism and raucous nationalism fermented and boiled over again during the 1970s. Ronald Reagan of California was to be its charioteer in four years' time, but in 1976 the Georgian Jimmy Carter won the presidency; in the queasy national mood, Carter struck a balance between those who saw the Vietnam War as having been a wicked waste and those who saw it as a Democrat-induced disgrace and looked forward to the resurrection of a GOP that would be, once again, proudly all-American, anticommunist, fully armed, and rich. Their time was coming.

At this not very great distance from his short stay in the White House, there is much to like about Jimmy Carter; but whichever way the historical tide was running between 1977 and 1980, it was not going in his direction. Carter introduced the admirable notion of "deep cuts" in funding for the largest weapons of the nuclear stockyards and Americans believed that he meant to leave their country defenseless. He made a modest, reasonable fuss about the worldwide abuse of human rights, the key political concept by which the Constitution trumps all others, and the Soviets thought he was interfering in their private affairs. When he tried to reduce the arms trade and its inevitable tendency to inflame trouble in the countries that bought the weapons, he offended the manufacturers at home and the happy killers abroad. Most tactless of all, he brought to public notice what dissenting academics had been saying for years, that casting the defense of American ideals, themselves real and admirable, into the posture of virulent anticommunism had committed the United States to some very unpleasant allies who had a perfect disregard for those ideals.

A bunch of gangsters in sharp military uniforms—Pinochet, Park, Mobutu, Marcos, Somoza, the Shah—counted themselves friends of the United States. Carter believed he could break with the psychotic secrecy, lying, and mistrust of the Nixon years and stand up for candor and cleanliness in his foreign affairs. He did not sense that the peak of U.S. power had been passed and that the world would change of its own accord. Rather, he had been trained in that most generous-hearted of American impulses: to bring its fruits to the poor of the world and to teach its idealism to the world's cynics.

His own staff thought otherwise. He appointed a creature of Kissinger's as national security adviser, Zbigniew Brzezinski, a self-righteous, punitive, and narrow-minded Pole who was deeply marked by his country's historic hatred for the Russians, and who launched a new national mission to destroy the Soviet Union, this time economically.

Carter's decent impulses and respectable intelligence were no match for the experienced advisers who surrounded him nor the irresistible torque of the "iron triangle" surrounding *them*—the heavy pressure on the administration from the Pentagon, Congress, and the weapons industry.

He had responded to a postwar strain in popular feeling of longing for fewer weapons and a quieter life. So in 1977, true to his word, Carter canceled the latest bomber then in development, the B-1. But the helots of the trade persuaded him to invest in a new line of smaller nuclear weapons that could do a fiendish job without exterminating life on the planet. Instead of "massive retaliation," which had at least made the two superpowers so rigid with fear they could not move, the nuclear strategists came up with "flexible response." Carter dutifully announced the production of a new line in guided nuclear missiles, the cruise missile, which was nice and cheap and, according to the brochures, smart enough to fly across Europe for several hundred miles only 400 feet above the ground to hit a chosen target only a few feet high.[2] He signed Presidential Directive number 18 setting up special forces to extinguish Third World "brushfires" (with or without the host country's invitation). He pushed up defense spending at the insistence of Brzezinski, who, heeding his first master's voice, plotted to keep so firm a hold on the President's attention that finally he drove Secretary of State Cyrus Vance, a worthy peacemonger, to resignation. The plan was to link together a deadly triad of nuclear defense: cruise missiles, the Trident submarine misbegotten by Admiral Rickover, and the new MX missile, a colossal monster that would be parked in a subway system in faraway and salty Utah.

In spite of himself, Carter initiated a new phase of the cold war at a time when it would otherwise have been losing all natural momentum. His own peace policies and the nation's long-standing principle of giving aid to anti-communist efforts worked against each other to make matters worse. He upset American Jewry by holding hands with Gromyko as they both coaxed Sadat and Begin toward peace between Egypt and Israel. He backed Somalia against Ethiopia in its invasion of the lone and level sands of the Ogaden, before the Somalis were kicked out by Mengistu's Soviet assistants. Meanwhile, the CIA was paying Jonas Savimbi to keep the Angolan war on the boil. Brezhnev became exasperated and set out a new ring of big weapons in Europe, weapons that the Soviets had been cooking up for some time and had no intention of wasting, whatever happened with the strategic arms limitation talks. He, in his turn, was ignoring the meat queues in Moscow and the people standing in line in Talinn, Estonia, to buy bones. He did nothing about one-fifth of his country's food production rotting before it reached the shops. He left his mammoth army to drink and to sell its domestic appliances on the black as well as the official market. All he kept to was the old cold war proverb—never give the enemy an inch if you can take him for six.

So Brezhnev missed a chance to save his republics, and Carter came to 1979 having comprehensively put the wind up the weapons constituency with his intention to cut deeply into the piles of bombs. He still hung onto a diminishing version of this hope as the bargaining at the talks veered backward and forward over which side would ditch what. Back in 1977, at formal meetings between Vance and Brezhnev, the U.S. negotiators had reached for a bigger victory than SALT 1. Carter, naturally enough, wanted to outdo earlier agreements and bring off a treaty with a flourish; Brzezinski wanted to screw the enemy to the floor.

So they cheerfully proposed that the Soviets agree to not only much tighter constraint on the big intercontinental missiles with their gremlin loads of independent weapons capable of taking off on their own, but also to drastic reductions—400–600 frontline units, 400–500 intercontinental missiles—while the United States merrily proceeded with as many cruise missiles, up to a range of 1,500 miles, as it liked. The new proposal opened figures that had been agreed to by Gerald Ford in Vladivostok only three years before; there was nothing in it for the Soviet Union except the U.S. concession of abandoning the vulnerable and fatuous MX missile. The Soviets turned the proposal down flat.

Surprisingly, the American mission was surprised. Carter fell into the old threatening obliquities: "Obviously, if we feel at the conclusion of next month's discussions that the Soviets are not acting in good faith with us and that an agreement is unlikely, then I would be forced to consider a much more deep commitment to the development and deployment of additional weapons."[3] Brzezinski rode a solemn rocking horse around the American press and said that the MX system would soon be "extremely, extremely threatening" to the Soviet Union. So it was another two years before the second treaty was signed and rendered immediately null by Congress and Afghanistan.

During those years, the arcane bidding and counterbidding over levels of different weapons went on and on in the dignified, circumlocutory dance of diplomacy. Gromyko came to New York, and Vance went to Moscow. Each side measured ceilings and subceilings; Gromyko suggested an absolute ban on all new intercontinental missiles, but the Americans wanted to hurry along with their nuclear subway system, so both sides settled for just *one* new missile. Agreement was reached on the number of killer warheads on each giant missile—a mere fourteen, each seven times the size of Fat Man.[4] Cruise missiles were written in and written out. The treaty almost foundered, and Vance and Gromyko salvaged it in Geneva at the very end of 1978.

Among the tangled details of the treaty, the important truth was implicitly acknowledged that some kind of equivalence would now balance the mountains of destructiveness stockpiled in each country, even if the new cruise missiles made for a permanent condition of hair-trigger readiness. The SALT 1 treaty implicitly recognized what had become an unalterable parity. Even unratified the SALT 2 treaty confirmed this mutual judgment.

But it was not a good year to tell that to the Senate. The sunbelt cold warriors were on the yellow brick road, and they saw *any* agreement as throwing away a U.S. advantage the country had held since 1945. The American Conservative Union broadcast films against the treaty on thirty major television stations. Although Carter obtained his summit meeting with Brezhnev in the summer of 1979 and the two men signed their modest piece of paper, he had little chance of persuading Congress to ratify it. In a ludicrous incident, American military intelligence discovered a new military combat brigade in Cuba; before it came to light that this was only a training brigade everyone had known about for years, the nation was briefly hysterical over the imminence of invasion and conquest by 2,600 hearty Cuban Communists.

The second biggest story of the cold war in the year of its frosty renewal is big precisely because it came from a different history, one quite independent of the cold war tale. In the Iranian revolution the United States and the Soviet Union both glimpsed a world in which neither their interests nor their ideologies had any place at all.

Given that the CIA had fixed the coup that enthroned the Shah and had sold him quantities of U.S. weapons for sack loads of petrodollars, no one could have foreseen that so dutiful a client would have ever spun so comprehensively out of control. But after the Shah instructed his troops to fire on strikers and rioters, killing hundreds of them in Tabriz, Esfahan, and finally Tehran itself, the worm turned upon him massively.[5]

Neither SAVAK, the Shah's secret police, nor huge wage rises could turn it back. Brzezinski promised the Shah that the United States would never desert him and put the Sixth Fleet on alert; but finally the Shah boarded his own 727 in tears, with a little urn of Iranian soil, and flew into oblivion. The non-Muslim world learned a new ecclesiastical rank when Ayatollah Khomeini stormed furiously in to maintain unwaveringly that the United States was the real enemy, the Shah was its creature, and the country must on pain of death return at once to the teachings of Ibn Roschd and a twelfth-century obedience to the divine word of the *Quran*.

The implacable new leader fitted none of the conventions of world diplomacy. He brushed off Brezhnev's men when they tried to poke their noses in and claim revolutionary fervor for their own, and he kept the capital in a frenzy of xenophobia. When a group of his firebrands broke into the U.S. embassy on 4 November 1979 and took its fifty-three occupants hostage, Khomeini gave them his harsh benediction.

It took Jimmy Carter all of his last year in office to prize them out. The young bloods, with no thought for the cold war, had thereby lost him the election and ensured that Ronald Reagan's inauguration as president—during the ceremony the hostages were at last returned home—would be accompanied by another marked drop in the global political temperature.

III

The Kremlin won nothing from Iran, but it might have been tempted to gloat at U.S. discomfiture if the next country along had not timed its own fateful insurrection for 1979 as well.

Afghanistan is next to Iran, and both are Soviet neighbors. Both are Muslim also, although the Afghans are mostly Sunni and the Iranians mostly Shiites and, in 1979, obedient to Imam Khomeini. But the upheavals that hit

Figure 16.1   A captured Soviet tank, Afghanistan, 1980. (AP/Wide World Photos)

both countries were wholly unalike except insofar as both Carter and Brezhnev gained only humiliation from the very different part each played in his
loyal ally's revolution. Brezhnev died in office, and Carter lost his job; it was
left to Reagan to jeer at the mess the Soviets were in, and to First Secretary
Gorbachev to admit it, and get out.

The Soviet invasion of Afghanistan was jam for the resurgent cold warriors,
rich, comfortable, and at home in Arizona, San Diego, and West Palm Beach.
It looked such a straightforward case of Soviet aggression and expansive-
ness—the first outside the borders of the Warsaw Pact since 1945. After a
national leader was murdered during the invasion and a communist stool
pigeon took over, the Afghan people, with few weapons and no hope, immediately took up the fight against the great bear. It was so conveniently the
Soviets' "Vietnam" and to keep it so the worthy Ronald colluded with Con-

gress to provide checks to the Afghan rebels amounting to over $600 million by 1985.[6]

Vietnam changed the American people's trust in their government's foreign policy, possibly forever, and by the same token, Afghanistan, along the murky and obscure channels of unofficial communication in the Soviet Union, ultimately brought about the rise of Gorbachev and the end of empire.

By December 1979 Brezhnev was old and ill and pretty well inert. His heir-presumptive, Yuri Andropov, had high grades for toughness and correctness in applying the leader's doctrine that no client could deviate from state socialism of its own volition. Andropov had acquired his leadership credentials as ambassador in Budapest in 1956. But the mortally ill Brezhnev committed him to the disaster of helping the stricken ally, and Afghanistan became the ugly avatar of the end of cold war and the end of the union of Soviet socialist republics as well.

As the United States had done in Vietnam, the Soviet Union foundered on ignorance and arrogance. Its generals reckoned, as generals will, that they would victoriously leave the country in a few months. They stayed eight years and left in disgrace. The Soviets went in, as they had in 1956 and 1968, to keep a socialist ally up to the cold war mark. This time they also felt they needed to protect a flank open to the new dementia of Islam and to the uncertainties of the American-backed military in Pakistan just beyond Afghanistan. As Frances Fitzgerald remarks of the American staff officers in Vietnam, the Soviet generals could not have passed even a freshman class in the history of the country they occupied and fought across.

Only 12 percent of Afghanistan's land mass of 160 million acres can yield a crop, and half of that cultivable land is empty, unfertilized, and unirrigated. Roughly half the population of fifteen million is Pushtun, and the other half is divided more or less equally between six tribes that exact tribal allegiance first, in the old feudal way—Tajiks, Hazaras, Uzbeks, Turkomans, Hiratis, and Nooristanis—and are loyal to Afghanistan, if at all, second. The rural people of Afghanistan are terribly poor. In 1978 the average annual income per head was $157. There is one doctor for every 16,000 people, and no hospitals in the countryside. Half the babies born in Afghanistan die before they are five. Three-quarters of the population is illiterate.

They are devoutly and peaceably Muslim, but also belligerently independent. They live in a stony, windswept, gray, and dusty land that teaches its children extreme hardiness, faithfulness to one another, uncomplaining endurance, a natural egalitarianism, gallantry in war, and the ethics of vendetta. Raja Anwar, in his caustic, respectful book, speaks with disgust of the way daughters are sold as wives, murdered by their fathers for adultery, silenced, hooded. He speaks equally of the ceremony of war between villages in which the debt of past deaths must be, but never can be settled. He pays proper respect to the roots of Afghan conservatism and to the system of village councils, the *jirga*, which are the court, the parliament, the foyer of news and storytelling for the rural poor. These were the landless groups, in thrall to their landlords but proud, that the Communists might have bent to their purpose using the sort of sympathy and practical tuition Ho Chi Minh brought to the Vietnamese villages. But the party was stupid and in a hurry, and it failed.

The Afghans are no nineteenth-century European nationalists. Then, as now, they were seven tribes, and then, as now, one in five Afghans was a nomad, hunting, grazing, gathering, avenging, and living by rhythms that vanished from Western Europe a thousand years ago. Their first progressive ruler was Amir Amanullah Khan, who in the early 1920s brought his wife out of purdah, renounced his droit du seigneur, befriended the new Soviet government, and tried to lead his people into the modern world.

He failed, of course, and fell, and the country returned to its traditional ways. When the Second World War was over, three men found the last ashes of the old Khan's progressivism glowing and blew them back into a kind of life: Hafizullah Amin, Noor Taraki, and Babrak Karmal. In 1965 they founded the People's Democratic party.

Over the next ten years they jostled and elbowed each other in the usual way of revolutionary parties, especially ones with no popular support or even a coherent theory of what to do. They were on terms with the prime minister, Sardar Daoud, a man of royal connections and genteelly reforming tendencies. Karmal split with Amin, and Amin recruited Taraki as an honest man of the people (which he was) and titular leader of the party. The *Khalq* (people's) group took the ascendant.

In 1973 Daoud assumed power when the King indiscreetly pushed off to Italy on holiday. Daoud became president, his rule was amiably conceded by the king, he brought *his* wife out of purdah, and he shook hands with Brezhnev.

Daoud was a diplomat. He welcomed Kissinger to Kabul, and in 1977 he waved away the Pushtunistan problem and made friends with the socialist Bhutto, briefly in charge in Pakistan. And then, perhaps because he had no domestic policy at all for coping with famine and plague, and darkness, Daoud sat down to a cabinet meeting in April 1978 and found his palace surrounded by tanks.

He was shot, or shot himself; nobody knows. There was a bit of fighting because the planning for the coup by Amin and Karmal, still at odds with each other, was messy. But the armed forces, nearly all of whose officers had been trained in the Soviet Union, had been squared, and Amin became executive head of the new, avowedly Marxist-Leninist government, with Taraki as its ostensible chairman.

The Americans swore that the Soviets flew the fighters and fired the rockets that won the day, but in fact the Afghan party ran its own revolution.[7] What it never did was, as the slogan goes, educate, agitate, propagandize. It never taught the people, least of all those out in the gray, barren countryside, and the people remained indifferent to the party's fatuous claims.

Within the party, a classical power struggle was fought to the death between Amin, who had the power, and Karmal, who had the nerve. The innocent Taraki was caught between them. He was a fine revolutionary emblem, the nomad and pauper who had educated himself until he could be sent off on empty errands to meet Gromyko and Castro. But when it came to it, his disciple Amin framed and executed him, according to the classic protocols of Stalin.

Land reforms came to nothing. The party had made no provision for a state bank. Not only did it fail to educate, it failed to use the tribal councils, to win over the institutions it had to have if its political slogans were to do anything but rouse the tiny audience in the Marxist court of Kabul.

The rulers got nervous. Amin signed a treaty with the Soviet Union agreeing that Afghanistan could call on Soviet military aid in time of need. The Soviets were thinking of the United States, but Amin was thinking first of Pakistan, now under the appalling General Zia as well as the additional burden of American infiltration of guerrillas across the old frontier on the Durand Line. Secondly, he thought of Karmal.

Nervousness was well founded. The revolutionary government's ineptitude and tactlessness caused major riots among peasants all over the country and all through 1979. The Iranian revolution had fired religious zeal even among the Sunni; several hundred rioters and bystanders were killed in the town of Herat. The Hazaras turned out against the Pushtuns, quickening ancient hatreds. The central committee quarrels became shriller, and Amin decided to assume power at just the moment that Karmal decided to eliminate him.

Amin escaped assassination when the hit men came for him by rolling downstairs to safety; he ended up, very briefly in power. In November 1979 he requested that 10,000 Soviet troops be put on standby in order to make his power a little firmer.

They duly arrived two days after Christmas, but not at all as he expected them. They came to depose Amin, with extraordinary clumsiness, and replace him with Karmal.

It is hard to see why. Karmal, of course, had been toadying up to a Kremlin secretly divided by the problem of who would succeed Brezhnev. Karmal had also built a network of allies in the army. But the Soviets came crashing in when they could easily have waited for a pretext and avoided all the sanctimonious criticism to which they were rightly subject.

With an unexpected regard for their victim's health, they put knockout drops in Amin's curry, intending to whisk him away. He felt queasy, ate little, and so came around inconveniently early. In any case, he thought the Soviet troops were there to help *him*. The Soviets were flummoxed by this contretemps. Karmal brought up his tanks willy-nilly and opened fire, and Amin was found dead with a bullet in his brain. Brezhnev promptly ordered his troops to maintain order, and the deed was, incompetently, done. The license for spewing out cold war propaganda could be zealously renewed all over the United States.

Karmal was as bad as his word. He lied profusely, as Amin had done, about how he came to power. (His tale about Amin being in the pay of the CIA is pure *Animal Farm*.) He lied about freeing 15,000 political prisoners from jails that would only hold 5,000 (and in any event, he filled them up again pretty soon). He had no answer for the conscript soldiers whom he drove to mutiny by keeping them under arms beyond their demobilizations dates, nor for the hundreds of thousands of peasants who repudiated his rule and declared civil war on the Kabul government.

The Kremlin had expected to hold a client state in place by replacing an incapable stooge with a competent one. That they were completely wrong about the qualities of each is not the only familiar detail in a farcical, lethal plot.

They were, of course, pulled deeper and deeper into Afghan history. The Turkomans, the Nooristanis, and the other tribes came out in a rural peasants' revolt against both the Kabul government and the Soviet intruders. With the usual superpower belief in the technology of destruction where the world's starving are concerned, the Soviet military sought to put the peasants back in place using the conventional encouragements of napalm, rocket strikes, and village atrocities. Doris Lessing reported from the spot a grisly tale of rebel soldiers tied back to back by Soviet troops, drenched in gasoline, and left ablaze.[8]

But they could not and did not win, tanks, napalm, and all. Brezhnev died, Andropov followed him into the next grave plot, and little wizened Konstantin Chernenko hadn't the experience or the wisdom to withdraw. Thus his generation lost its authority, his army lost what confidence it had, and the way was open for the new class led by Gorbachev to make or break the union. By holding grimly to Afghanistan according to the old rulebook of cold war, Chernenko made his contribution to ending it.

The Americans helped by working themselves into a fine frenzy. They declared that the Soviets were after "warm-water ports" in a landlocked country, and President Reagan decided on the run that the Gulf was an American frontier that needed Presidential Directive number 52 and a capitalized Rapid Deployment Force to protect it. The United States put new air bases in Pakistan and copious weapons in the hands of the dauntless guerrillas, including the surface-to-air missiles that turned sporadic skirmishes into a ten-year war. Fifteen thousand Soviet soldiers died in Afghanistan even more pointlessly than the G.I.s died in Vietnam. Tens of thousands more learned to while away the war with the only reliable crop on the Durand Line, heroin. And the Afghans, pulled hither and thither by the different tribal leaders from Turkoman, Nooristan, and Pushtunistan, pretended to themselves and the world that they were fighting a holy war, a *jihad*, when in fact the bodies of the world's poor were once again being butchered to teach a cold war moral. So they died at will, and walked in millions to Pakistan, and died, and died.

17

BIOGRAPHY V

The Spy—Philip Agee

PHILIP Agee was back in the United States for the first time in sixteen years. During that time the Justice Department had promised to arrest him if he showed up and had hounded him at every border station in any country in which the United States had one of its many proprietary interests. Agee had somehow dodged across such stations, but as he drove down to the little Canadian frontier hut for this crossing in 1987, the old excitement indistinguishable from terror came and transfixed him as usual. He had hunted up and down with no luck for a dirt road that went across to the United States, one with no officials in cavalry hats ready to lean friendlily into the lowered window. Losing patience, he decided to face things—whatever they were to be—straight away. His car trundled over the heavy wooden sleepers of the bridge, the man looked casually in and asked, "What nationality are you," Agee replied "American," and it was done.

Surveillance of a nation's interests abroad is an area of state bureaucracy in which moral conduct has been formally separated from national safety, in which the state has exonerated itself for not honoring the ethics that hold together civil and public life. And yet from its inception the American state has spoken nobly of its obligation to human freedom and happiness and has grounded its justification of itself in its defense not only of a geographical realm but of a human and moral one as well.

So when a young man who enlisted to be a spy, to do his country's shady business of watching, betraying, suborning, and lying, suddenly woke up and decided to tell everyone how the system works and who the men are who do the work, there was a terrible to-do. His employers in the CIA first denied everything; then tried to discredit him by the usual method of calling him a liar, coward, drunk, and commie; then simply tried to get their own back for nearly twenty vengeful years. Nonetheless, Agee got the story out and would doubtless have been murdered by tougher state employers over the water. Having got it out, he assumed his honorable place in the line of history's whistle-blowers.

Agee was born in 1935 into a well-off Florida home. With a settled parental business to inherit, the pleasant benefits of country club life beckoning, and a degree in philosophy at Notre Dame (his possession of which has puzzled him for years), he got off to a flying start in life. "I didn't want to go into my father's business, so I took my ideals as a glowing-eyed young man into the secret service.

"The Romanticism of the CIA was very strong at the beginning [it will be remembered that the CIA was one product among many of Truman's 1947 National Security Act]. . . . There was a delectable flavor of power, knowing the secrets of other countries, manipulating events there. The people who did the work then were either liberals or Manicheans—anticommunist either way—but, of course, there must always have been plenty of apolitical people in for the job, and lots who lost belief in the work but had to keep going to pay the mortgage."[1]

Agee joined "the Company" in 1957 with cover as an officer cadet for the air force. He was bright, good-looking, well bred, fit, and eager. The CIA could not have had a better recruit. He was trained in all the spooky stuff at Camp Peary, Virginia: the electronics of surveillance, the legendary brutalities of unarmed combat and karate, and how to frame Soviet officers, foment local trouble and strikes, subvert youth and student organizations, and handle handguns. He also learned to use the paraphernalia and jargon that litters the gamey cultural undergrowth of spy movies and paperbacks. The jargon was a bit less cockney than John Le Carré's, but much of it is the same: tradecraft, false flags, dead drops, and casenames.

Late in 1960 he was posted to Ecuador, which was in its customary state of mayhem, and told to watch the loopy dance of its local politics, to keep a particularly sharp eye on the leftists, the students, the trade unionists, and other unsound characters, and above all never to lose sight of the Cubans, eponym of the devil and all his works. There's a keen pleasure in watching those unaware of your watching, especially if you know secrets about them. When later in Montevideo and Mexico City Agee was able to plant baby transmitters in the beds to which Soviet officials would shortly take their lovers, then watching and listening, picked up a comic charge of prurience as a bonus.

There's lots to do; especially planting a phony report on a suitable quarry that, when found, revealed imminent revolution planned by Cubans and locals. Agee wound the fraudulent paper into an empty toothpaste tube, giving it an impeccably spearminted security fragrance. After a hitch or two, the news of its discovery breaks in all the papers. Rapture in the Company office.

These larks keep the boys happy. Mostly what CIA agents do is watch the passage through the country of absolutely everybody. The mountains of paper and taped records beggar belief. Agee was posted to Montevideo in 1964, was promoted, recruited a Cuban, shoveled money into Chile to buy support away from the Allende left, and began to think the heretical thought that if his side put as much effort into Uruguayan land reform as it put into actions against the left. . . . Then Lyndon Johnson invaded the Dominican Republic, and this smart young CIA captain wrote in the office file next to Johnson's insolent justification the ancient epithet, "bullshit."

The high jinks with radio-transmitted sex in Soviet bedrooms continued, and Agee began to show undependable squeamishness about torture. In 1966 he put in for a transfer to Mexico City in preparation for the Olympics, to be held there two years later. He also drafted a letter of resignation to Richard Helms, telling the CIA chief just what was wicked about the Company. Agee had an ardent heart; he had been much impressed by the political lessons given him by a beautiful Mexican. "My letter was going to go straight to the chief. It was all insubordination but I meant it. It was going to tell him what he knew and what he didn't know, that the Agency was a waste, a huge waste of money and effort and of the *idealism* of men like me—I was fired up by Kennedy's idea of the Peace Corps for a little while until I realized it was just another piece of hypocrisy on the part of the United States to pretend it was doing good when all it did was interfere in, run the lives of poor people and poor countries who would get on much better without us. And how long can you take part—as an individual or as an institution—in corruption and cruelty without just becoming a weapon of state cruelty yourself?" He did not send the letter.

Agee enjoyed Mexico City, where he could see at work some much more efficient and comprehensive logging of Soviet activity from a lot closer up. But he finally made his break after seeing what the U.S. neighbor and sound ally was prepared to do to the harmless students joining the hopeful crowds of 1968 to protest against a wretchedly poor country spending all those millions on the Games while the babies, the beggars, the addicts, and the old starved in the streets among the cockroaches.

Agee's life as a man would become the photographic negative of his life as an agent. He dedicated himself to a rite of purification: he would publish a book that faithfully described the agency's methods and, moreover, named as many names in its service in Latin America as he could think of. He was thirty-three and had gratuity, bags of energy, and not a little courage. He had also experienced a kind of conversion, of a sort that switched the current of the blood of many Americans at that time: switching their worldview from one side to the other, so that they saw all that their country did in another pattern. It was, however, rare to go through such a conversion while in the service of the Company.

"The agency, 'the Company,' it simply serves the interests of a ruling class, it's a class weapon. It works everywhere on behalf of capital. The big houses of capital want to know what's going on in the countries where they have their interests. They can instruct the State Department in what they want to know, and it sends the agents out. They link up with the business associates of U.S. houses wherever they are sent. Since the United States has always regarded the southern half of the continent as legitimately its own, the agency is simply everywhere there. And the enemy is always the same, it's communism, or socialism. So it's the poor. Socialism only means that the poor get something for themselves, a decent life, a safe hut to live in. And capital and the ruling interests must stop that happening. Once you understand all this, there can be no question of staying on."

Agee saw what he saw with a directness, honesty, and ingenuousness that drove him to bear witness in a loud and voluble voice. His bravery had plenty of recklessness in it, and his recklessness was not without swank. But what guts and stamina he had.

As soon as the agency got wind of his plans, easily picked up from an indiscreet and boastful letter he wrote about them to *Marcha,* a Uruguayan leftist weekly, it began to marshal on a lavish scale all its resources of denial, belittlement, and elimination. It is plain that what motivated Agee was revulsion at the job, touched with a slight priggishness the certainty he felt was for him the guarantee of his truthfulness. It is just as plain that what motivated the agency in its efforts to stop him from publishing was, after all its frothing about security and death threats to its officers, mere revenge.

The agency must have spent a fortune from its giant reserves simply on tracking Agee's whereabouts and trying to find out what he was going to say. He was peripatetically trying to work from official records and press cuttings in both Paris and London during 1971, straining to turn his suitcases full of memoranda, official statements, plausible denials, textbooks, and out-of-the-side-of-the-mouth titbits into a publishable book. After several misfires, Penguin in London gave him a contract and a solid advance. Long before then, however, his own money ran out. He had had to live off vague pickings from the Marxist Parisian who was going to publish the book but let him down. Then he scrounged from a beautiful and buxom blonde named Lesley who helped to keep the struggling author alive by regularly leaving 500-franc notes from her "heiress fortune" under his plate at the café.

Lesley turned out, inevitably, to be a tender CIA plant. Agee discovered a full-scale bugging kit attached neatly to the lid of the ancient typewriter she had loaned him. By this time he was on the run: being watched by the cormorant surveillants into his hotel, checking out and trying to dodge them, holing up in the odd flat with a girl he picked up in a café. When the story of his venture made it into the newspapers, the CIA began its gossip about his alleged drunkenness, his wenching, his maudlin and self-pitying confessions to KGB friends. It discovered by trawling neglected passenger lists that Agee had spent six contented months in Cuba after leaving the agency, and that, too, went to the papers as confirmation of Agee's true color.

Agee finished the book in a delightful little hamlet on the estuary of the Fal in England. The agency warned him that it would keep Penguin from publishing the book, but the publishers stayed staunch. *Inside the Company: CIA Diary* was published in England in 1975, and then the hoo-ha really began.

Agee fairly piled it on for the agency's discomfiture and the newspapers' eager scribbling. He announced that he would be preparing an index of *all* the Company's men he could identify worldwide, and he asked for collaborators. The first continent in the index would be Europe, and the book would be called *Dirty Work: The CIA in Western Europe* (1978). It would have a sister periodical, the *Covert Action Information Bulletin*.

"I wanted people to *know*. If they knew as much as I could tell them about what the agency did, it could be made largely ineffectual. Especially if I could recruit others to tell what *they* knew. Then it just couldn't work properly."

So he became an agent in negation. Where the good agent is invisible and wholly clandestine, the reverse agent is prominent and wholly public. The force of his work derived from public acknowledgment and the power that contemporary culture gives to celebrity. The louder the fuss made by the CIA, the more people bought *Inside the Company*. (There have been nine printings of the American edition in ten years and innumerable translations, including one, as one would hope, into Russian.) The London press and their New York colleagues flew down to Cornwall to Agee's little cottage to find out what would happen next, the U.S. government tested its prosecution procedures, and more and more people devoured his solid compendium of all the CIA did to other countries.

It was part of a turning of the tide, after which, in humdrum and domestic ways, an increasing and motley army of people was moved to fight against both armies on the plains of the cold war, and against the mendacious men who commanded them.

In 1976 and 1977, as the Americans considered how to land Agee in jail and get the CIA its revenge, the gates of frontiers began to slam shut in front of him and behind him at the same moment. The Dutch gave him brief sanctuary before the screws were put on them, one assumes from Washington, and his permission to stay was revoked after he published a breezily tactless article in a British samizdat. The Germans bustled him out with a show of legality. The French took no trouble over legality; they stuck him in a mucky clink without benefit of lawyers and ejected him onto a farm track in Belgium before dawn the next day.

The British were unwarrantably prompt in toadying to their senior partner. Having left Agee in peace to write his book, they gave him time to buy a house in Cambridge for himself, the sons of his first marriage, and his sometime guerrilla partner, a torture victim of a passing excess by CIA-trained cops in Brazil. Then the British Home Secretary, the amiable poltroon Merlyn Rees, after making a gobbling kind of defense to the House of Commons, signed Agee's deportation order.

There was a British sort of commotion and one appeal after another until the law lords did as they were told by Rees and his puppet masters; Agee was chucked out in defense of the maxim that liberal states shall do as they please with their liberties, especially if the United States tells them so.

Agee carried on the good work dauntlessly. The U.S. government gave up trying to indict him when it discovered that in doing so he would gain rights of access to the documents that would prove their own legal transgressions against him. He ploughed across the Third World and liberty capitals like Stockholm, speaking up for rights and describing his own life not without some well-earned and garrulous egocentricity as an instance of the blightedness of U.S. foreign policy both before and after his apostasy. "The United States has always been a class preference. Look at the Declaration of Independence. It's written by the men of property. Nothing there about blacks or slaves or women. Or children. . . . So I've joined forces with as many people worldwide as I can to fight for human rights and against power. It's a common struggle . . . the cold war has been a way for national elites to keep their power, to mystify people about a fictitious enemy. It's been mystification."

Agee has followed that excellently American vocation—calling his country to account for its stewardship of its own principles. In so doing, he sometimes exasperated even his stoutest defenders, as when he phoned through in 1980 to the U.S. embassy in Tehran and tried to deal by himself with the Iranians over the hostages. But he kept doggedly to his uncomfortable fight for freedom and for a more open and innocent world.

Eventually, Philip Agee married a beautiful ballet dancer and won rights of residence in Spain and Germany.

The Justice Department, rather than publishing CIA secrets and skulduggery in the open, eventually backed off. They kept Agee's passport, but they called off the dogs. You could say that Agee won the day.

18

BIOGRAPHY VI

The Journalist—Neil Sheehan

THE axiom that in any war truth is the first casualty held as never before during the cold war. Cold war propaganda proved as lethal as more formal declarations of hostilities. When the two superpowers declared they were not doing what in fact they *were* doing, they killed the more energetically so that the truth could not get out.

In the Soviet Union, China, and the satellite countries, truth was only a cynical grace note in newspaper titles. *Pravda* was just the bullhorn of the Kremlin. What the government said and what the press said were interchangeable, even if U.S. Kremlinologists had to read the paper to interpret the nuances in the official statements.[1]

In the United States to a great extent—and in Britain completely—the press fell into continuous and supine obedience to the government. In Britain the national press has toadied to political power ever since the barons took over around the turn of the century. Only the *Daily Herald* broke with the devout litany of anticommunism; in the last decade of the cold war, the British yellow press, led by Rupert Murdoch's *Sun*, became disgustingly rabid in its commination of anybody sympathetic to the left.

In the United States, as Noam Chomsky tells us, anticommunism was supported as much in what the press did not report as in what it did.[2] Until the Vietnam War started to go wrong, the big nationals, especially the elite opinion-formers, the *Washington Post* and the *New York Times,* fell in pretty dutifully behind the offices of state.

But the American press never spoke in the predominantly sycophantic and obedient cold war voices of the British. In addition to scrupulous and independent papers like the *Boston Globe* and the *Christian Science Monitor,* the United States also had the intellectual institution of the grand columnist: the Walter Lippmanns and Joseph Alsops to whom Saul Bellow did honor in his cold war novel, *The Dean's December* (1982), with the figure of Dewey Spangler. Such men had their stuffiness, no doubt, but they spoke serious prose about the country's interests and ideals without too much regard for the sensibilities of the great. The president read what they had to say, Lippmann on Berlin, Harrison Salisbury on Korea, Alsop on Cuba, and treated their often critical opinions respectfully.

The English journalists who have spoken up the most courageously for the civic virtues have been loners like James Cameron and Rene Cutforth—men reporting from the wharves at Inchon in 1950, or from the belly of an aircraft coming in to Tempelhof in 1948. By contrast, Lippmann, Salisbury, and Alsop were usually speaking out next door to the offices of state; they were heeded.

Neil Sheehan is a carrier of that tradition. When for the first time in American history the great columnists turned entirely against the conduct of the state, the doing was almost all his. Brought up by Harvard to honor the nation's ideals and the university's code of exact and truth-telling scholarship, he found the traditional way of criticizing what the state told him quite insufficient to the historical moment, and to its crux in Vietnam. He had to find a quite different method of journalism to handle not only the state's licentious mendacity but also its willful self-deceit. Sheehan's best-seller about the Vietnam War, *A Bright Shining Lie* (1988), is the product of that new form of journalism designed not only to report on both the cold and the Vietnam War but also to turn the constitution of the state against the state itself. To do this, in turn, Sheehan himself had to be the kind of man he was—brave, patient, humble even, in the interests of steadily contemplating his corner of the human condition.

Sheehan had arrived in Vietnam in 1962 at the age of twenty-five to report the war during the most confident years of U.S. involvement, while the country was taking over the role of the French. He worked for United Press International in Saigon until 1964, did eight months in Jakarta for the *New York Times* as the civil war that was to lead to the massacre of the Indonesian Communists began there, and went back to Saigon for the *Times* for a third year from 1965 to 1966. Then he was posted back to Washington to cover the executive end of the war in the Pentagon and the White House—"investigatively," as people started to say around that time, because there suddenly seemed to be so damn much we did not know that called for investigation.

So it was to be expected that when Daniel Ellsberg, in one of the more important attacks of conscience during the cold war, came to his proper conclusion in 1971 that the secretly truthful papers kept in the Pentagon on the conduct of the war should be made public, he contacted Sheehan.

Ellsberg by then was a senior Pentagon official with a star career. He was that all-American character, the romantic puritan, a hard-working man of passionate commitment, intense feelings, and no less intense conscience. He was extremely bright, a member of Harvard's Society of Fellows, a Marine officer for three years, and one of the Rand Corporation's secret architects of nuclear war plans. After working with Edward Lansdale in Vietnam for two years, he had gone back to Rand and the Pentagon dismayed beyond words by American inability, and his own, to turn Vietnam to those virtuous means and ends that were his country's only justifications for being there. When in 1969 he read the forty-three secret volumes of what would come to be termed the Pentagon Papers (and being Ellsberg, as Sheehan observes, he read them all), he became convinced not only that the war whose affairs he had prosecuted so zealously in 1965 had gone badly wrong, but that his country was wrong to wage it.

He had smuggled the strictly confidential documents past the security desk and photocopied them, gradually accumulating a full 7,000-page collection of papers. In early March 1971 he had waited for hours one night at a Washington hotel to talk to the man closest to him from Vietnam, formerly Lt. Col. John Paul Vann, now a civilian with general officer status who was in virtual command of an entire region of South Vietnam and its army corps.

Vann, however, had other fish to fry that night and never showed up. Ellsberg went to the house of another friend in the capital, a confidant entirely different from John Vann. The friend was Neil Sheehan. After a conversation that lasted almost until dawn, Ellsberg and Sheehan made the decision that would lead the president to use all his power, mixed with his taste for personal revenge, to press for a conviction of Ellsberg for treason. Their decision would put the *New York Times* in the dock as the traitor's accomplice and would end with the historic judgment of the Supreme Court that publication of the Pentagon Papers transcended the interest of government and was vindicated by the Constitution, the highest interest in the land.[3]

Sheehan was the quiet, upright, kindly, implacable hero of the story: without beating his breast, either publicly or privately, he pressed his colleagues toward what he argued was their civic duty. He became the catalyst for a great, and at times complaisant, newspaper to live up to its professional standard. And he protected his sources, that is, he betrayed nobody. The man who had already won a civilian medal in Vietnam for "conscience and integrity in journalism" was endorsed in his action against the state by the official foundations of civil society; he was awarded a prize for "excellence in investigative reporting," and his newspaper won a Pulitzer.

Halfway through 1972 John Vann was killed at the age of forty-eight, in a helicopter accident flying above a road near Kontum in his daredevil way, defying the enemy to shoot at him. He hit a rain squall, and his pilot lost vision, and flew the machine at full speed into a copse of trees protecting a Montagnard graveyard.

Remembering Vann's funeral at the chapel at Arlington National Cemetery, staring unseeingly out of his window, Sheehan says:

"Vietnam was my first story for U.P.I. and then for the *New York Times,* the Washington end of the war was my second. I was just starting, and obviously the war was the biggest thing going on, you *had* to be there. Well, I never got away from the war. None of us did, I guess. It's far and away the biggest historical fact for my generation.

"I wanted to leave something behind besides another newspaper or magazine article. A newspaper is so transitory, the next day it's just something you wrap the flounder in. The *Times* had preserved *The Pentagon Papers* by letting Bantam publish the series as a book, but I still wanted to leave some permanent record of my own—a proper history—behind.

"I wanted to come to grips with the war and understand it. Choosing Vann as a subject was a sort of accident. My previous idea had been to write a reporter's memoir, but reporters don't lead intrinsically interesting lives except for one or two like Harrison Salisbury or Theodore White.

"The thing was, by the time of Vann's death, he'd lost his grip on reality, gone round the bend on the war. But its end was by then a foregone conclusion. U.S. air power and John Paul Vann, so to speak, held the South Vietnamese army together. When they left, that was that. And going to his funeral was like a class reunion. Everyone was there. I thought of John Kennedy, buried just outside in Arlington Cemetery, and how people said he'd never have got us in so far but who took us into the war all the same. I thought of his brother Bobby lying near him, who'd turned against the war at the end. I thought of them both as the only surviving brother came in, Edward, the only senator who'd tried to do anything for the Vietnamese people. And then Westmoreland was there. They were all there, for and against, Westmoreland still thinking they were right—this was only 1972, remember, and the war wasn't finished by a long way, even though there was no doubt how it would end. Ellsberg was there, shunned by the others, a sort of outlaw, getting ready for his trial. Lansdale, who'd won the Philippines for us, who invented counterinsurgency really; he couldn't win Vietnam. I met him at the door. We were all burying the era of boundless self-confidence, Luce's 'American Century.'

"So I wrote a memo to myself, as journalists do. Telling Vann's story is telling the story of the war. It would be the only way the country would get to grips with the whole thing. People can follow the story of one man's life, and I'd taken part in it. A journalist has to have participated. His craft is to get the truth of history in the detail of what happened. Mind you, I believe in human beings; I believe they can do things, they make a difference. Tolstoy was wrong about war. People run it, plan it, do it better or worse. Kutuzov knew the ground, and he knew the limitations of his own side. So did Wellington. At Waterloo he knew he had to hold the ground, if he could. Those were men who made a conscious difference.

"Anyway, I could bring to the writing a dimension that nonparticipants don't bring. I knew the people, and they knew me. They trusted me. That is, they knew—Bunker, for example, who followed Lodge as ambassador—that I would treat the subject seriously. And I would tell the truth. The official records of the war do not reflect what happened—in the way that certainly they did, say, for Patton in 1944. Look up the Patton papers and you get a record of reality. But the U.S. generals and the statesmen of the Vietnam War were deluded. The generals never knew what was happening (except for a man like Victor Krulak) and then, as they say, 'not knowing what to do, they did what they knew.' Westmoreland was like Douglas Haig, and when this new razzle-dazzle from the Harvard Business School came in with McNamara and Bob Komer managerializing the war, Westmoreland just fitted it into his military preconceptions, 'war-of-attrition,' 'building-the-killing-machine,' 'search-and-destroy,' 'money, men, matériel,' and all that jargon."[4]

Sheehan found his subject in John Paul Vann, the remarkable soldier, who became one of the very few Americans to study and understand the pattern of Vietcong warfare and, later, the strategy of the North Vietnamese army, to grasp in its comprehensiveness the corrupt incompetence and straightforward cowardice of the Saigon government's high command and its troops in the South, and at least for several years, to name as such the self-deluding character of American policy. Vann made his observations close up and, because he was a bouncy bantam cock of a man, insisted on describing them faithfully and very audibly to his seniors and, even less forgivably, to reporters.

The cockiness sprang from his class origins in illegitimacy, maternal drunkenness, and the grim poverty of a childhood in prewar Norfolk, Virginia. Vann's bounce and verve carried him to senior levels in the army. The same childhood was responsible for the desperate eagerness to please, rule, and seduce women of any age or appearance, a predisposition that marked Vann's file and prevented him from being promoted beyond colonel. His wretchedly unhappy wife and children in the United States were unaware of his two mistresses and one child in Saigon.

His promotion blocked by his past, Vann returned to Vietnam as a civilian in 1965 but became so important to the policy of Vietnamization that he rose to senior command as a civilian general, the first in American history. By the time he was killed, he had fallen fully into line with the American policy of bombing the enemy to pulp and blindly hoping for the best.

Sheehan spotted how fully Vann's *figura*—the line of his biography— embodied the meanings of the Vietnam War. Vann's career threw into vivid relief the little that was right and all that was wrong with the war and could do so as a gripping story intelligible to readers in the terms that enhance intelligibility in *any* story: the individual man or woman caught up in the unstoppable swirl of history, the struggle for territory or for freedom, the battling to make sense of life or death as one or the other knocks down the door.

Sheehan's genius was to see how much that one life could be made to carry. The steady center of his subject was a man. Sheehan, from his job as a journalist reporting for the country's most important newspaper the country's most important news of the day, knew all the men whose seniority and power Vann tried to bend to what he believed to be his nation's best purposes. Sheehan's long, compelling tale turns his story into history, showing from amid the excitement of the battles at which he was himself a direct observer, and from his closeness to both the senior and middle echelons of command, just how the American government and generals became like the obdurately stupid men who ordered the slaughter in Flanders between 1914 and 1918, and how they transformed the ordinary youngsters who were drafted to them as soldiers into brigands and monsters. The book is one of the classics of the cold war, a shaper of our memory of it.

Giving me his personal judgments about the war, which he does rarely in his great book, Sheehan says: "By 1962 the U.S. generals were bureaucratized and overconfident. There was an American inability even to imagine defeat. It comes out in that story I tell in the book about Vann's attempt in 1963 at briefing the Joint Chiefs of Staff in Washington on his view of how badly the war was going. Maxwell Taylor, who was the chairman then, simply struck it off the agenda. The same with his paper to Henry Cabot Lodge in Saigon. But when Capt. George Marshall caught General Pershing by the sleeve in 1917 to tell his version of the war, the general listened.

"Westmoreland was always boasting about all that he was building in Vietnam—twenty-six permanent base camps, seventy-five new airfields for C-130 transports, five more jet bases, twenty-six hospitals, inconceivable amounts of circuitry. Some of those bases were small cities. They had hot showers, cold storage, fresh meat, and Meadowgold ice cream. Westmoreland used ships as floating warehouses, he had half the biggest merchant fleet of the world tied up offshore. One day I asked him, 'General isn't there a cheaper way to warehouse ammunition?' 'Hell,' he said, 'what's money? We're fighting a war.'

"By early 1968 many of the U.S. soldiers were demoralized. Not the Marines, they never went to pieces, but the rest did. There weren't any mass mutinies, but whole units were on the edge of mutiny. There were plenty of fraggings. Junior officers had to negotiate with the soldiers. They might say, 'You're out on patrol,' and get the reply, 'No way.' There was a limit to what they could order, and if the officer didn't agree, they'd kill him. In 1965 there was no dope and all the soldiers believed in anticommunism. So did I; when I went out I was a pure cold warrior. By 1969 people were stoned in the field.

"The war-of-attrition policy and the constant turnover of everybody from the one-year tour of duty did it. By 1969 there was no NCO corps left, and they made up 'shake-'n'-bake' NCOs who, of course, received no respect. Discipline and experience went down together. It got too hard to do the job, so they *rebelled*. Everyone started to lie—everyone. The only people the U.S. soldiers respected were the NVA. They respected the enemy because he was so damn good, but they despised the South Vietnamese. Hardly surprisingly, since most of the ones they met were pimps, prostitutes, and shoeshine boys.

"Attitudes developed into a contempt for Vietnamese life. Some of the line units started to behave no better than brigands. People just couldn't believe it at home. After My Lai, people literally couldn't believe that American boys would behave as they'd learned from the movies that Germans and Japs behaved, shooting civilians at random. But it is also unfair to portray the ordinary American soldier in Vietnam as a sadist. The generals did the big killing of civilians, impersonally with bombs and shells."

Sheehan went back to Vietnam after Vann's death. He found the grove of trees where the helicopter crashed. And he began the sixteen years of research, interviews, the odd fellowship or two, writing, and rewriting in his spare time that culminated in 1988 with the publication of *A Bright Shining Lie*. (The title comes from a remark of Vann's describing the American presence in Vietnam.)

The book was a runaway success. It won a Pulitzer Prize and the National Book Award, it sat in the best-seller columns of the Sunday *New York Times* for months and then swept the board in England.

"The United States hasn't yet come to grips with Vietnam, but you can now say what couldn't be said fifteen years ago without a fist fight. Since *A Bright Shining Lie* came out, I've had an outpouring of letters from veterans.

They say to me, 'You tell it like it was, and you put me right back there. I always knew the war didn't make sense but ⁻ didn't know why. Now I know why.'"

That simple truth is still amazing. Sheehan caught and expressed a widespread American feeling that the nation had done wrong, and with that shocking openness of disposition that is the best part of the American character insisted that its wrongdoing should be acknowledged. The U.S. government still treats Vietnam abominably ("because they beat us," says Sheehan), and the plight of the refugees has yet to receive the full attention that it deserves ("and these are our people, they fought on our side"). But after all the hard and unforgiving judgments that must be made of America's conduct in the war, there is something lastingly impressive about the fact that the country should have found and listened to a chronicler like Sheehan. The style finds the man and the two become one politics.

19

Fantasy and Action—Bond, Clancy, Forsyth, Tolkien, and Kundera

THE cold war deepened the conventional separation between private and public in both the moral and the productive worlds. The political class—our national leaders—gathered the public realm entirely into their own care: "We look after all that, you go home and attend to your own affairs." Any protest that the realm of public affairs really is our business was typically met on both sides of the iron curtain with official insistence on dispersal.

To break up a crowd is to insist that everyone revert to private life; a crowd has public meaning by its nature, and the cold war state cannot tolerate it because a crowd, also by its nature, acts in a public realm that has been denied it. The state must harbor the view that all crowds formed without its permission—a permission only grudgingly granted or reserved for public festivals (the point of which is to congratulate the state for its benevolence and continuity)—are subversive. It has to take demonstrations—the very term signifies opposition to the state—personally. Its personal response, naturally more in sorrow than in anger, is to break them up, preferably with tear gas rather than clubs (make the public sick and weeping rather than bloody and angry).

Severe demarcation between the public and private realms is the form of modern life, but it carried additional force in the cold war countries, especially when the premises of the cold war itself came gradually, after 1968, into dispute and even disrepute. Our leaders then insisted by its usual instruments of surveillance and secrecy and, if necessary, by violence, that they were the only informed judges of political matters and therefore of the rightness of maintaining the cold war. This ruling class took upon itself the maintenance of nerve, the constant rousing of citizens to a madly contradictory state of passivity and productiveness. It sought to ensure the public's willingness to authorize deadly action without ever taking that action, and to ally a permanent condition of military obedience to no less obedient domestic consumption. People could keep this mad mixture on their stomachs only by taking judicious doses of fantasy-about-action, while action itself was kept at arm's length.

Fantasy-making is a fantastic business. "We fed the heart on fantasies," I quote Yeats again, "the heart's grown brutal from the fare." Fantasy can actually serve as a kind of theory. We test our effectiveness and explain our actions against what we dream of doing. If the dream is too far off, it becomes, however appealing, plain silly (the Walter Mitty syndrome). By this sensible measure, fantasy is part of our use of the available narratives of our history: we *ground* our psyches in a narrative field, thus charging our fantasy-making habits. We can then use the resulting imagery to understand our lives and to reckon up the distance between desire and possibility.

If the gap between what we want, and what we can have is too wide, fantasy becomes plain fantastic. Of course, we can never be quite sure. Both political and technological history are full of tales of fantasy coming true very suddenly. We can check private fantasies (when we've a mind to) against personal realities. The cold war and the national security state, however, provided individuals with no measure of their own significance and effectiveness. What does "we" signify? Who and what are we, the people?

Our political leaders tell us on television that whatever they have done was the right thing to do. Our only possible response is to play "let's pretend." The political spectacle is a galaxy away, in the windless space on the other side of the screen. Can we climb through and join the godlike creatures who live there? We can, of course, and do. We take the delicious medicine of identity-dissolution and enter the fantasy of perfect effectiveness and deliberate action. We get a big Chinese supper, a six-pack, a pile of videos . . .

. . . and enter the other-world of the cold war thriller, which consoles its viewers for their political helplessness. It gives them vicarious physical action which restores them to the plane of historic experience. To do this, the thriller must have an intensity capable of dissolving the corrugated self into the fantasy, otherwise something in the reader just refuses the story ("it doesn't work"). But it must also use a formula familiar to the reader. The story must balance product-reliability (mastery of the formula) and regulated innovation (creative fantasy). If the balance wavers, the reader is forced into critical reflection, his identity promptly re-forms, and he is no longer dissolved into the story. He is not fighting the cold war; he is thinking about it.[1]

II

The Bond movies embody an immanent history of cold war ideology. In the beginning, James Bond, agent no. 007 of the English secret service, was pitted against Rosie Krebs, the hard-faced almost man-woman of SMERSH, an equivalent of the KGB. Bond was played by the young Sean Connery. These early movies worked as highly efficient thrillers in which realistic personal danger and vivid bodily collision were the ground upon which the figures of espionage and secret weapons played for comic relief. The old imperialist imagery of exotic cities (Istanbul, Shanghai) and long-distance train journeys was played off against the traveler-hero's self-assurance and luggage, which fitted the perfect Anglo-Scots gent. The usual misogyny of the English private school ethos was leveled against Rosie Krebs—who closely resembled, in her granite face and mud-colored uniform, a school matron—and schoolboy sex was directed toward the beautiful spy.

The Sean Connery–James Bond movies were wildly successful at a time when the British film industry was going lame. The direct action, the calm hero, the cold war simplicities of motive, were astutely mixed with locations not yet open to tourism and automobiles still restricted by class. The action turned on gaining possession of magic secrets that would win or lose the cold war for good; Connery's Scottish not-quite-snobbishness backed by British phlegm and high tech spoke to a large number of people who liked to fancy Britain was still on the fringes of superpower.[2] When the Bond movies were exported to the United States, they went down well for the same reasons. Anglo-American snobbery had plenty of resonance in the 1960s, perhaps because it was so satisfactorily on the wane. The cold war politics in the movies were impeccable. Even so, Bond was a loner, always an irritant to M, often breaking organizational rules, a gallantly old-fashioned individualist in the era of machine espionage.

When Roger Moore took over from Connery in 1972, however, the form began to give way. The evil center began to shift from SMERSH and the Soviets to the megalomaniac master-criminal who would dominate the world. Moore acted up rather than just acted, and performed less as the English gentleman, more as the tailor's dummy. He did it rather appealingly, but the plots quickly lost all realism. The producers deliberately shed the knuckle-gnawing suspense of the early movies and, from about number four or five onward, avoided the face-tearing naturalism of the fight in *The Manchurian Candidate* or the interrogation of Alec Guinness's Mindszenty in *The Cardinal*.

The movies became elaborate and expensive jokes—at least when they were not dishy travelogues, stages for stuntmen, or little essays in technological wizardry. They have had a long run—thirty years—and have made payloads of money; but they actually stopped being cold war movies quite soon. Some of the later movies even feature collaboration with the Soviets against the cosmic plotter who has pinched a clutch of missiles. It would be nice to find that the production team and the elegantly piss-taking Mr. Moore decided to make films which, whatever their consumer and sexual fantasies, would quietly erode cold war fantasy from inside.

By *The Man with the Golden Gun* (1974), it was, in other words, impossible to feel Bond with any political intensity. The movies became a lavish sort of lark, a jolly evening out, and a measure of public incredulity. The jokey manner and the complete absence of the seriousness characteristic of earlier cold war movies mark a change in public mood. Bond's enormous audience was no longer frightened by or for its hero.

The same was not true, however, of the oeuvres of Tom Clancy in the United States and Frederick Forsyth in England, success stories of the entirely ideological second cold war. The class they wrote for in the first place clearly felt the thrills of their books with great keenness; it is a pleasure to record in this connection that Britain's sometime Queen Empress Margaret Thatcher unaffectedly spoke of having *re-read* Forsyth's *The Fourth Protocol*.

Forsyth's formulas may have got whiskers on them, all the same, he sold in the hundreds of thousands. *The Fourth Protocol*, whose solemnly allusive title refers to a series of conventions that, if broken, will lead inevitably to nuclear war and that are now, of course, all broken except for the fourth and last thread, was quickly given the full Michael Caine movie treatment.

The book tells an old, old story. In Moscow, Kim Philby is still busily helping the KGB and the first secretary to hot up a nuclear panic in Britain so that a Labour party pledged to unilateral disarmament will march to power, sweeping out the Americans and thoroughly upsetting the foundations of NATO. Of course, it is taken for granted that such a change would unsettle the *bien-pensant* classes who read Forsyth and are, prime ministers apart, at several removes from the corridors of power. But the point of such novels— like that of American movies of the 1950s such as *My Son John*, *Kiss Me Deadly*, and *The Manchurian Candidate*—is to vilify the "enemy within," the treacherously left-inclined trade unionists, disarmers, peace movement activists, and simple Labour party voters who, by 1988, refused to take seriously the idea that Ivan would be paddling up to the White Cliffs of Dover with snow on his boots and the left greeting him at the top. After the manner of the shifty new

Figure 19.1   Sean Connery as Agent 007, James Bond, of Her Majesty's Secret Service, 1971. (AP/Wide World Photos)

"docudrama" filled with "factoids," Forsyth's novels are peopled by actual figures from the British left and spiced with the studiedly actual details of names on doors and desks and weaponry, especially those of a purportedly secret kind.

Indeed, it is this sort of stuff which provide the main impact of such fiction, with its ready formulas, its assured market, its originality-plus-familiarity. Ever since Hemingway, tough guys' literature has made much of gun lore. Forsyth follows Ian Fleming in nonchalantly telling us the difference between the Colt Mark 3 magnum, the Smith and Wesson 45-06, the Beretta 92-F, and the good old-fashioned British Army Browning. Naturally, Forsyth throws in the chateau-bottled vintage clarets and very high-powered internal-combustion engines. But it is the atmosphere of debonair knowingness, of taking-murder-with-Turkish-coffee airiness, that is intended to draw the reader into the circle of those who know who is who, what is what, and what secretly will be done about it.

It's no use pointing out that such books work merely by pretending to be in the know. Pretending to be in the know is a delightful feeling and a game that everyone plays. Novels that play it with a flourish not only capture rogue imaginations but express a viciously antiprogressive, secretive, and spiteful temper that has been widespread since the end of the Second World War. The victims of that bad temper teach their children to hate such political necessities as bureaucracy, welfare, free medicine, the arts. That temper has also seeped into the climate at large where it is breathed out in the rank exhalations of the yellow press, and breathed in with the smog and fumes by the crowds in the city streets.

In the United States Tom Clancy keeps this temper on the boil. By the beginning of 1990 his novel *The Hunt for Red October*, a story publicly enjoyed by Ronald Reagan about an attempt by a Soviet nuclear submarine to defect to the West, had sold close to six million copies.

Clancy deals in the same formulas as Forsyth, and his novels are even fuller of the handgun-carrying technology allied to the cocksure and boastful masculinity featured in that large corner of U.S. culture whose household gods are machines, especially machines that kill. He speaks with an admirable American directness for the sunbelt rich who backed Reagan, for the beery rednecks in their pickups,[3] for the weekend soldiers in the British and American military reserves who love the ballistics and the six-digit radar bearings, for all those many men, and not a few women, in places like Atlantic City, Birmingham, Brussels, Johannesburg, and Brisbane who thrill to the well-established fantasy recipe of high technology, world tourism, casual murder, succulent women, branded goods, designer clothes, and ideological certainty. Such fantasy *grounds* the floating anxieties of class and sex and history in a structure of sureties that lasts for as long as the story. What happens to them then is anybody's guess, but they sure don't do anything for a more peaceful or generous world.

III

The tough eggs of the thriller market ideologize the world and its worries in the cold war terms their reader espouse. The softies have a very different narrative, as one would expect, but it too provides a fantasy of political action that the daily world denies. Its prototype author is J. R. R. Tolkien; *Lord of the Rings* (1954–55) has been running as long as James Bond.

Tolkien's irresistible appeal is that he successfully attaches the primary and all-enfolding myth of the epoch, the cold war, to a story that goes straight to the soft hearts and softer heads of a great slice of the population. These softies (I intend only affectionate deprecation—I am one such softie myself) believe in the tendernesses of individualism, the poetics of romanticism, and the rather washed-out doctrines of liberalism. Members of the soft-hearted constituency are the curators of the "meaning" industry: they guard those cultural practices that uphold individuality and domestic privacy, and they steer away from forming any picture of a larger destiny other than that vague warmth, America.

These honest worthies are English teachers in senior and junior high schools, low-grade television researchers, journalists on local papers, graduate students in the humanities, interior designers, clergy, social workers, journeyman crafts workers, and owners of health-food stores. They affirm in their systems of production the preindustrial values—face-to-face relations, working with skilled hands and eyes, baking bread, an aversion to chasing money—that signify their refusal of the modern world.

Tolkien ratifies that refusal and gives its fantasies a politics. Those fantasies, rooted in a strong and defiant nostalgia, are embodied in homely freehold citizens who combine faithfulness with independence (people like us) but are ready to pledge themselves to transcendentally higher ideals, if only they could find them. These ideals may have been found by lean, hawkeyed, sun-tanned gentleman-knights not unlike Yul Brynner, accompanied by rare, chaste, hauntingly beautiful sisters-in-arms, boyish withal, and wonderful with horses. Such fresco figures corroborate their moral choices with their fineness and straightness. The cold war is never more binding than when the pain and loss that disloyalty forces us to recognize call up a great wave of longing for unambiguous fidelity and a simpler world. (This longing is the source of the spy story's power.)

Tolkien connected the gentle identity of our somnambulist class to the pervasive forms of world politics that shaped it. The glaring opposition in Europe between communism and all other doctrines after the shock of the 1917 revolution settled deep into Tolkien's simple mind and intense feelings. Fearful of this modern world and hating industrialism, he was enabled by cold war politics to ignore the Fascism that the prevailing temperature had by the mid-1950s iced out of sight and to cast his tale in the bold and vivid apprehensions of dark versus light, smoke versus fire, ugliness versus beauty, the yeoman class against the proletariat of Mordor. And it has, of course, been much pointed out how similar the ring is to the secret of nuclear fission, and the end of Mordor to a nuclear Armageddon, which we win.

Admiring American and British readers of this fantasy have been liberal, tolerably well educated in fragmentary forms of English literature, battling at a financial level to call their lives their own, unmilitary but possessed of a keen awareness of a lost heroism, and nonpolitical but conscientious voters and watchers of television news. They will be members of the peace movement, strong in their feelings, domestic, home-loving, but independent and politically awake.

Far more than for themselves, such parents have read Tolkien through the long decades aloud to their children. They read *The Hobbit* and *Lord of the Rings* nightly in the sparkingly varnished bedrooms with the dormer windows, urgently wanting their rapt little listeners to feel the surge of that round prose go thrilling along their nerves. As father reads, his voice fills with a rich, strong, nameless longing that makes it thicken and tremble a little, he feels a great rush of love and warmth for his spellbound audience.

> There waiting, silent and still in the space before the Gate, sat Gandalf upon Shadowfax: Shadowfax who alone among the free horses of the earth endured the terror, unmoving, steadfast as a graven image in Rath Dinen.
>
> "You cannot enter here," said Gandalf, and the huge shadow halted. "Go back to the abyss prepared for you. Go back. Fall into the nothingness that awaits you and your master. Go."
>
> The Black Rider flung back his hood, and behold, he had a kingly crown; and yet upon no head visible was it set. The red fires shone between it and the mantled shoulders vast and dark. From a mouth unseen there came a deadly laughter.
>
> "Old fool," he said. "Old fool. This is my hour. Do you not know Death when you see it? Die now and curse in vain." And with that he lifted high his sword and flames ran down the blade.
>
> Gandalf did not move. And in that very moment, away behind in some courtyard of the City, a cock crowed. Shrill and clear he crowed, recking

nothing of wizardry or war, welcoming only the morning that in the sky far above the shadows of death was coming with the dawn.

And as if in answer there came from far away another note. Horns, horns, horns. In dark Mindolluin's sides they dimly echoed. Great horns of the North wildly blowing. Rohan had come at last.[4]

And the thrill of that moment courses through the listeners' skin and crepitates at their nerve ends.

What is it that stirs readers to this high, rippling, idealized, and curiously objectless flow of feeling? It is easy to analyze the *Star Wars* imagery, the Darth Vader figure, the vaudeville staginess ("Old fool"), the phony biblicality ("recking nothing of wizardry or war," "dark . . . dimly"), the Celtic nomenclature (Mindolluin), the sugared brass music, like Wagner played by Glenn Miller—easy enough: all done in a few lines like that. But such practical criticism does not answer the question. Any answer must account for a certain vast reservoir of feeling perpetually in play in American and British culture, feeling that is invoked by such prose.

Tolkien reclaims heroism for a kindly group of people rightly repelled by counterinsurgency and the shadows of nuclear war. They have withdrawn their trust in political structures but kept their faith in their country's generous ideals. They propose a partial withdrawal into an enclave culture, where they battle to keep things going on a personal scale and in a domestic setting. These factions of the polity who care for the best and most womanly side of our culture need a language that connects old chivalry and traditional courage to the feelings that keep their characters strong and upright; they turn to Tolkien.

He provides a way of imaginatively living through a range of emotions whose content and objects in actual political life are either so contradictory or so disagreeable that they can find no domestic expression. When modern politics was stuck in the wasteland of the cold war, and modernism had broken so many of its promises, the best you could imagine as an image of the action those failures had denied you was the world of Middle-Earth.

There you can return to a city-state (the Shire) without a city, exonerated from labor and with plenty to eat. There you can dissolve the problem of individuality by simply pledging allegiance to an honorable militia that is also a free association of friends (a cross between Robin Hood's merry men and Jefferson's farmer army). There you will find the two perfect fathers, Aragorn and Gandalf, the wise, true, and sacred leaders. Finally, you will find there an action that is bodiless and blood that is colorless; you will find, that is, a sort of blissful state of naïveté in which you do not have to reflect on anything, least of all your own feelings, but in which emotional*ism* carries you off to a strictly spiritual realm that, having no real world to refer to, cannot signify real life but fairly *moves* you to the bottom of your heart.

This politics is all feeling, but it excludes all hatred. Tolkien, in his sexless treble, takes the tableaux of 1914 and 1941 and transposes them to a utopian opera house in which a colder and more spiritual war may be played out as unphysical, bloodless, corpseless ceremonies, stately parades, and grave orisons. He provides a spell with which to magic out of sight the dreary routines of Washington and Moscow. The free uses of fantasy are many, and his devoted readership could have spent the last four decades in worse ways.

IV

Both the hard and soft fantasies pursued during the Anglo-American cold war were largely a matter of consolation for the vacuities of action. Fantasy on the other side of the iron curtain, on the other hand, may *be* an action.

There are a few good novels and films whose goodness is exactly that we do not know what on earth to do with them. We know pretty well how to read Tom Clancy and J. R. R. Tolkien and what to use them for. After we've swallowed "once upon a time" and "if only" they present us with cause and consequence, thought and action, in fantastic duality but in the usual novelistic way. But imagine a world of narrative committed to breaking the links between cause and consequence, and deliberately imagining a wide gap between what you think and what you do (especially if what you think is illegal and might land you in jail). Such a story, even if peculiarly puzzling, is at least memorable. As you turn it over and over looking for its meaning, multiple meanings flash out from its facets:

> The first time an angel heard the Devil's laughter, he was horrified. It was in the middle of a feast with a lot of people around, and one after the other they joined in the Devil's laughter. It was terribly contagious. The angel was all too aware the laughter was aimed against God and the wonder of His works. He knew he had to act fast, but felt weak and defenseless. And unable to fabricate anything of his own, he simply turned his enemy's tactics against him. He opened his mouth and let out a wobbly, breathy sound in the upper reaches of his vocal register (much like the sound Gabrielle and Michelle produced in the streets of the little town on the Riviera) and endowed it with the opposite meaning. Whereas the Devil's laughter pointed up the meaninglessness of things, the angel's shout rejoiced in how rationally organized, well conceived, beautiful, good, and sensible everything on earth was.
>
> There they stood, Devil and angel, face to face, mouths open, both making more or less the same sound, but each expressing himself in a

unique timbre—absolute opposites. And seeing the laughing angel, the Devil laughed all the harder, all the louder, all the more openly, because the laughing angel was infinitely laughable.[5]

I suppose Milan Kundera is pretty plainly on the side of the Devil; but one needs to know also that the angels, in his Yeatsian "magic book of the people," *The Book of Laughter and Forgetting* (1983), are, like the Communists, dedicated to "uncontested meaning on earth."[6] For Kundera, the more meanings the merrier; the malice of laughter is what guarantees the necessary latitude of life. So, too, does *weight,* good old gravity. In *The Unbearable Lightness of Being* (1984) (obviously a funny title, funny in both senses), only those experiences that repeat prior experiences have gathered enough weight to keep a person anchored in life. Without the repetitiousness of history, we become unbearably light and float away from life and love and ethics. And disappear.

As Vladimír Clementis disappeared. Gone from history. He was photographed, Kundera tells us in a salty little parable, on the balcony from which Gottwald, the new Czech president, announced the birth of the Communist nation in 1948. It was bitter cold, and Clementis solicitously loaned Gottwald his fur hat.

Four years later, when Clementis was framed for treason and executed, the famous photograph was doctored and Clementis painted out in every textbook and newspaper library. "Where Clementis once stood, there is only bare palace wall."[7] Only his hat remains, on Gottwald's head.

You can do a lot with that little tale and still never say all that might be said about it. It is a funny story in its mordant way, and it is a story about remembering as well as the eradication of memory (one is reminded of the flue in *1984* down which it was Winston Smith's duty to post inconvenient records).

Kundera's novels are indeed devoted to remembering, to laughing, and to showing how the communism of the cold war was dedicated to the enforcement of forgetfulness and the elimination of the joke.

Kundera defines kitsch as "a folding screen set up to curtain off death." Kitsch, even more broadly, is that instrument of propaganda that denies the joke. Kitsch is incapable of malice. It is a caring, killing mother. Kitsch is, of course, totalitarian. "In the realm of totalitarian kitsch, all answers are given in advance and preclude any questions."[8]

So, Kundera says, we need all the raucous past and messy love affairs we can accumulate to keep life heavy and full of remembrance. We need as many weird fables as we can find to stop these blasted angels smiling and smelting everything down to one meaning. We should beware of children, who are light in being and murderously forgetful. And we must keep fantasizing, because all the horrible presidents wanted to cancel fantasy, just as all the legions of hellish Stalinism wanted to ban the book and wipe out that strange, angular, corrugated product of the imagination, the novel.

The novel, a devious, underground, crooked, and cunning bearer of contraband, will make you free and keep you free. Kundera's artist-heroine says furiously, "My enemy is kitsch, not communism."[9]

20

FICTIONS V

Patriotism and Psychosis—The Vietnam War Movies

THE cold war period during which fantasy found an immediate vehicle for physical action in the United States was, obviously, the years of the Vietnam War. The worthy liberal heroes who mismanaged the war and misreported its progress certainly saw it as the occasion for closing the gap between fantasy about a Communist world conspiracy and action to stop it. Feeling took on physical form. The national narrative led straight into that most traditional trope, the soldier's story.

There is, naturally enough, a lag between historical events and retellings of those events in history books, novels, and films. But as this book itself witnesses, those events are themselves shaped as they take place by the complex variety of their predecessor stories, never more so than when the story is about going to war. The Korean War, despite its pointlessness, held its soldiers to their script with comparative ease because so many of them had been recently trained in a war whose significance they understood. Everybody was, in any case, prepared to find the Soviets and the Chinese convincing replacements for the Nazis.

The handful of movies that came out of the Korean War treat it respectfully in those terms. The much larger, more various clutch of movies emerging from the war in Vietnam starts in simple justification and then goes on to try to take in the war's full enormity: to balance the beauty and strangeness of

Vietnam itself against the frightfulness of the devastation; to put a value on the best and worst of America as each was played out in war's extremity; finally, to write a quiet fanfare to the common men who fought it, to boast raucously, as in *Rambo,* about their courage and patriotism, to curse, perhaps unpopularly, their versions of Lieutenant Calley and his men at My Lai.

The films about the Vietnam War are deeply narcissistic, even at their most intelligent. They register the shock to American culture of both losing the war and being in the wrong about it. But they also present faithfully but unreflexively the stiff self-righteousness that made the lessons of Vietnam so hard to learn, and that reemerged when Reagan summoned the national will to wage cold war.

There was a national psychosis in this, which the films fed and fed off. War films are box-office certainties: their strong formulas make them easily understood. They do this by setting up exclusive opposites. In conscientiously American movies, one such opposition during the cold war epoch was American/*not* American. A variant on this was true American/un-American (Communist, maybe, or coward). One of the boldest and most persistent pairings has been incorruptibility/graft (which has corresponded since Shakespeare with country/town, or nature/society). There is, simplest of all, victory/defeat and success/failure, but also home/away and violation/revenge.

Vengeance is as satisfying and pointless as the violation that provokes it. Violation—of the human body or of human rights—begins the story in all the most typically American political movies, whether good ones or bad. The lust to violate the violator is self-justifying and goes deeper than language. Vengeance is blameless. Ethics is returned to ritual.

War narratives are embedded deep in American culture, but at a time of actual war or rumors of war, they reassert themselves with exceptional strength. Communism in general and the cold war in particular challenged American honor and its good name; there had to be killings to clear that good name. Once cleared, order is restored and the tilling of the land may resume.

The restoration of order is a key motif of the narrative in which manliness is discovered and confirmed by decisively answering the challenge of action. Whether or not Ronald Reagan was moved by these imperatives (and it is impossible to believe that he was not), his most popular actions in the conduct of the cold war—the invasion of Grenada, the bombing of Libya, abusing the Soviets—all fit the story the nation so fervently tells itself about itself.

The myth of action: it is that which grew into pseudo-nationalism, and turned psychotic in its dreams. The cold war nurtured it, with its cowboy simplicities of fanaticism versus the calm and resolution of the absolute American. Its movies repeated it, until the very repetition, circling in the veins of the culture, precipitated into psychosis and made the country throb.

The Vietnam War movies, even those most nearly works of art, turn on this circle. They express and inspire intense feeling with little intelligible outcome. They presage no rational link between physical action and admirable consequence. Each of them dramatizes a striking, handsome, and memorable version of the American hero and gazes at him narcissistically. But they can give him nothing worthwhile to do.

This is the predicament of a culture that praises the active, resolute, and efficacious man in cold wartime. Action itself is too casual and devastating to be real: two or three twists of the nuclear key and half the hemisphere lies in ice-cold rubble. Psychotic pseudo-nationalism is the turn taken by the imagination in response to this condition.

II

The first Vietnam War movies fit the simplest model: popular feeling matched to popular movie. John Wayne himself felt his nationalism right through and believed, rightly, that he felt it in common with and on behalf of most of middle America. He directed *The Green Berets* at the opportune moment of 1968.

It is no doubt a very bad film, but it grossed $8 million in the domestic market alone and is still going strong in the video stores.[1] The film shows an eager, liberal-minded, David Halberstam-type journalist learning from war-weary, tough, grizzled, not-without-compassion Col. Mike Kirby how horrible the enemy is, how it *deserves* to be wiped out. The journalist befriends, inevitably, a little Vietnamese child whose father is subsequently beheaded by the Vietcong. The child herself is raped (offscreen—the film's euphemism is "abused") by five of the marauders. The worthy journalist, changing smartly out of his natty safari suit into jungle greens, comes around.

The film is really a hate-filled message (though gruff and clipped in delivery) to middle America about its disobedient children. In 1968 their children, if they were students, were marching in the peace movement; their sons who weren't students were much preoccupied by the imminence of the draft.

Sixty-year-old Wayne and his elderly cast, Jim Hutton, Aldo Ray, and several others embarrassingly halt and portly as they go about their war-waging business, tried to conjure up the perfect fit between role and character, the comradeliness that was such a happy consequence of shared and soldierly purposes. It was the clarity of the action that had made even the most routine Guts 'n Glory movie about the Second World War so convincing. The maturity of the men on the screen, the Waynes, Henry Fondas, Gregory Pecks, corresponded to something true about U.S. servicemen in 1945. Of course, those films *idealized* in the simplifying and sentimentalizing sense but also in Nietzsche's rather fuller meaning where it "does *not* consist, as is commonly held, in subtracting or discounting the petty and inconsequential. What is decisive is rather a tremendous desire to bring out the main features so that the others disappear in the process."[2] The best of the World War II movies—*Twelve O'Clock High, The Cruel Sea,* Rossellini's great trilogy—vivified the laconic courage, flat humor, and hearty friendship that were genuinely part of the war.

However, the ghosts of the old comrades and the strains of taps had long since slipped away. The peace marchers objected precisely to being told what to do by men too old and tubby to go and do it for themselves. The average age of the conscripted private was, notoriously, nineteen; Wayne's band look laughably out of place, like legionnaires dressed up and pretending they can still go through their old Marine paces.

Wayne's simple heart and patriotic politics led him to try to rebut the peace movement with a magic gesture. One minor strand in the simple texture of narratives that constitute *The Green Berets* may be pulled to show how the film wholly relies on that long-gone frame of feeling from 1946. Sergeant Provo is Wayne's ideal of the noncommissioned officer: bullet-headed, brave, almost speechless, battle-hungry. He daydreams about a memorial to his heroic death but cannot think of a suitable one. "Provo's Barracks" is no good because he is not a general, and he dismisses his various other ideas because "they just don't sing." Mortally wounded, he is inspired to whisper a last request to his colonel, and after his death Wayne arranges for a handsome sign to be displayed outside "Provo's Privy." Combining manly feeling with craggy impassivity, Wayne says in his famously creaky tones, "It sings."

Francis Ford Coppola's *Apocalypse Now,* released in 1979, was the first big movie to try to find a new frame of feeling and fit it to a circumscribing parable. As is now well known, Coppola used Joseph Conrad's very short novel *Heart of Darkness* (written in 1899) to solve his problem of matching feeling to story. Conrad's tale gave him a narrative—the journey up the river, the quest for a man whose discovery promises revelation—and it also loaned him, if not a frame of feeling, a way of pacing the awfulness of the circumstances.

In *Heart of Darkness* Conrad's narrator, Marlow, maintains a balance of geniality and no-nonsense candor probably quite close to the tone and idiom in which men of his kidney did their job in the appalling extremities of imperial outposts around the turn of the century. He goes up the Congo, passes a dreadful forced-labor camp where black slaves are worked to death, takes an old tub of a freighter upriver, bumps into some entirely hostile natives, and finds that the object of his search—Mr. Kurtz (a man, he has

Figure 20.1 John Wayne in *The Green Berets*, 1968. (AP/Wide World Photos)

been assured, of great vision and gifts) has been driven mad by his commitment to the devil-knew-what sort of indulgences and orgy.

T. S. Eliot borrowed the moment of Kurtz's death as epigraph to *The Hollow Men* only a quarter of a century later. Coppola using the outline of the tale, sends *his* hero on a ritual anabasis up a tributary of the Mekong Delta into a tip of Cambodia where Colonel Kurtz, once the jewel in the army's crown of promising officers, has gone as batty as his namesake, set up a court of perfect bestiality and solemnly self-worshiping cruelties, and appointed himself holy celebrant of horrors. Kurtz's part was played by Marlon Brando, ambiguous and cloudy star shining through the obscurity of two two-faced masterpieces of the cold war, *On the Waterfront* and *Apocalypse Now*.

Coppola had enormous difficulties with his movie. He asked for help from the U.S. Army, which asked to see the script. The army objected vehemently, in throttled prose. The script was "simply a series of some of the worst things, real and imagined, that happened or could have happened during the Vietnam war"; it wanted nothing to do with "the sick humor or satirical philosophy of the film," and it took particular offense at the film's evil presence, the Kurtz-Brando figure who has defected from the Marines and indiscriminately makes war on both the belligerents. In a weird criticism, the army said that Kurtz's conduct constituted "a parody on the sickness and brutality of war."[3]

The army wanted Coppola to be a bit nicer about them. He wanted to borrow equipment. Finally he got cooperation from the U.S.-supplied Filipino army, but they couldn't provide all that Coppola, by no means the last tycoon in the film business, wanted for his hugely conceived movie. In a grand gesture, he sent a telegram to the president protesting his "honest, mythical, prohuman and therefore pro-American" purposes.[4] Although Carter could hardly respond, and although the army was hardly in business to provide film extras and apparatus, Coppola's heroic self-regard clearly made him feel that his film was pitted against the might of the state.

In truth, the film is a majestic ruin. Coppola's imagination is fired by the two most powerful features of Hollywood: pageant and stars. The Vietnam War glowed with visual pageant: incendiaries, napalm, those beautiful girls with their cataracts of long black hair, the rain, the rich countryside, the torn bodies, and everywhere the giant, buglike helicopters, their heavy clattering din.

He also had to have a star, especially in a womanless film. In *Apocalypse Now* Colonel Kurtz is there only because Brando is; Brando is incarnated in this film as Kurtz, and only Brando survives beyond the limits of the film. We remember Kurtz as a version of that ever-present figure in our imagination, Marlon Brando. (In this movie, we cannot get over how gross—300 pounds—he's become.) Having a star like Brando made this a film about the American imagination. It is not a naturalistic story (a Conradian one) about the war. Vietnam is far less important to the film than Coppola's imagining America.

So Brando's epicene bulk looms over the movie, even though he is visible only in the last half-hour. Being omniscient, Kurtz knows that the young Captain Willard is coming to kill him, and after various mumbled sequences in which he reads aloud from *The Hollow Men*, he simply invites Willard to do his bloody business.

Willard obliges him in parallel to the graphic sacrifice of a water-ox which is sliced completely open by an executioner at a mass religious gathering outside Kurtz-Brando's palace. Willard listens with silent incomprehension to Brando's soliloquies; the keenest moment is when Brando smoothes a little water over his shaven, glabrous, and immense pate, shiningly filling the screen as the water pours down like a mountain stream into the mammoth slopes of his obese neck. The second moment that catches and holds something true about the Vietnam War is when Willard is sitting with his hands bound behind him while Kurtz is deciding which of the two of them shall die. Suddenly Kurtz-Brando appears, his face caked in a mud mask that, ambiguously, is both a savage's warpaint and a soldier's camouflage and tosses like a ball into Willard's lap the severed head of the only man in Willard's crew left alive and sane. Willard yells and squirms in revulsion, unable to move, furiously joggling his knees to try to roll the beastly thing away.

The early part of the film is accompanied by Willard's voice-over, speaking in an idiom that sets the tone of the film. Willard's inexpressive tone as he reads the Kurtz file on the journey upriver persuades us of nothing so much as his own incurious blankness. He confronts the frightfulness of the war along the river without surprise, indeed without connection. If this indifference is what Coppola intended to convey—a loss of all human interest, even in one's own survival—then *Heart of Darkness* was a misleading vehicle to choose to carry so leaden a load.

In Conrad's novella, Marlow calls at the deathbed of the dying Kurtz:

Anything approaching the change that came over his features I have never seen before, and hope never to see again. Oh, I wasn't touched. I was fascinated. It was as though a veil had been rent. I saw on that ivory face the expression of sombre pride, of ruthless power, of craven terror—of an intense and hopeless despair. Did he live his life again in every detail of desire, temptation, and surrender during that supreme moment of complete knowledge? He cried in a whisper at some image, at some vision—he cried out twice, a cry that was no more than a breath—"The horror, the horror."[5]

Kurtz repeats the same words at his moment of epiphany, but his half-pint palace, his mumbling aphorisms, his motley crowd of Montagnards trooping the water-ox to its sacrificial end, the few hanged men left to twist in the trees, these fragments of cliché can hardly shore up the ruins of the nation's imagination as it sat in judgment on its very own war.

Nonetheless, Coppola had a vision. It was a vision of his beloved country's titanic awfulness, the heavy reek of its sex and murder, its mad enthusiasms, its girls and its holiday joys, its lurid, deadly gadgets. The two most famous scenes of the film are Coppola's touches of pageant: the helicopter assault under Colonel Kilgore, and the bunny-girls dancing on a river-platform a couple of clicks from a besieged border bridge.

Kilgore is played by Robert Duval, who judges his position perfectly as secondary star to Brando. He devises for his character a slightly bow-legged cowboy sort of walk, under a cavalry hat, and takes his helicopter squadrons into battle with "The Ride of the Valkyries" blaring over the tannoy and their .50 machine guns hammering into the continent.

He never flinches at an explosion, never stoops to gunfire, tends a wounded enemy soldier, and, in a scene of wonderful insanity, orders his champion surfer onto the river bore while the brilliant incendiaries burst in sumptuous clouds of blue and orange among the palm trees Coppola's difficulty is that a country capable of producing Kilgore and Kurtz, Duval and Brando, is (as Nixon said of himself) capable of *anything*.

So, too, with the appalling scene when the bunny-girls come to dance for G.I.s in the jungle who haven't seen a woman in weeks. Arriving inevitably by helicopter, four skinny and half-naked girls with masses of curls tramp the stage in high-heeled boots:

> *Her gold crotch grinding,*
> *Her athletic tits,*
> *One clock the other*
> *Counter-clockwise twirling—*
> *It was enough to stop a man from girling.*[6]

Half raucous, half serious, the huge audience swarms onstage and in a ludicrous echo of rescue from a landing zone, the helicopter pulls out the beleaguered girls as the cheery rapists rush in.

Just up the river Willard finds the outpost, the bridge kept open so that the general can say it has been kept open. It is manned by a few stoned boys. When Willard asks where their officer is, one replies, "Aren't you?" The bridge is grotesquely decorated with loops of fairy lights, the long curves repeating the colored tracer looping beautifully into the night as the one soldier still on duty looses off his automatic.

It is a film in the magniloquent tradition of Hollywood. It is pulled between a narcissistic view of America's own gorgeousness and a fascinated revulsion at the destructive powers the country has always had at its behest and which were so extravagantly released upon Indochina. This sumptuous movie speaks in spite of itself for America's colossal self-advertisement. Willard, to our relief, gets out alive; the war, both the cold and hot versions, goes on. There's no stopping America.

III

Michael Cimino's *The Deer Hunter* (1979) stands as Ernest Hemingway to Coppola's Scott Fitzgerald. Robert De Niro plays the title character and is, as a star, a much more direct self-creation than Brando's obscure and resonant persona. In addition, the film turns upon a single trope: the reiterated theme of the single bullet that fixes individual fortune.

The film won ample awards and made a lot of money. It was immensely popular; Cimino had made a powerfully patriotic movie that did not have to conclude with military triumph. He takes the solitary hero of American frontier literature—Deerslayer, Huck Finn, Davy Crockett, Nick Adams— and brings him through fire and torture by dint of the courage and resourcefulness the hero's deep allegiance to the old, natural, and manly values has taught him. Those same values and the physical prowess they demand also lead him to rescue his friends as best he can. One is crippled in the rescue, and the other is so poisoned by the Orient that he loses the way and the will to go home. But Michael, the Deer Hunter, has stood by them and, in so doing, vindicated the loyalty and friendship his hard-muscled, soft-hearted creed demands of him. His therefore is the military triumph that Cimino endorses in his audacious coup de theatre: the survivors end the film by rendering, huskily, awkwardly, the awkward words and music of the hymn "God Bless America."

The Deer Hunter is no ruined, intensely self-conscious monument like *Apocalypse Now*. It is simply, simple-heartedly ideological. But many great films and plays have been just as unreflective (*Henry V* among them), and none the worse for it. The strongest ideological passage is in the first third of its tryptych, Michael and his best friend Nick organize the wedding of the third in their trio, Steven. All three men are steelworkers, and the movie begins with a magnificent fragment of documentary, observing steelmen at their heavy work beside the troughs of molten metal. The steelworker is one of the archetypes of industrial man, here transformed, in an apotheosis as familiar to socialist realism as to American patriotism, into a giant of independence, manliness, and working-class grace and skill, his singlet darkened with the ancient stain of labor.

The steelworks in question is somewhere outside Pittsburgh, and as all such towns truly were for the first sixty-odd years of this century, so too this idealized town is close in its bonds, intense in its fidelities, ardent in its passions, lavish in its customs and ceremonies, its food and drink. In a moving turn of the narrative, this film from the heart of the cold war makes the home of its heroes a third-generation community of Russians. The old women still speak their dialect together, and the music and festivals of the Russian Ortho-dox Church gather the wedding into a unity as glowing as the rivers of steel in the foundry.

Behind every shot of the town we glimpse the minarets and pinnacles of the Russian church and the massive black bulk of the foundry. Just up the road, however, is Michael's Walden—the blue remembered hills where he hunts deer with one bullet, pursuing the kill like a work of art.

The first section of the tryptych speaks straight to a vision of America not, as in John Wayne's 1945, uniformly American in a Norman Rockwell kind of way, but as free, and equal in the variety of its ethnic traditions. These Russians had to leave Russia to discover the freedoms at the heart of their tradition; so, too (the film implies), if they had been Sicilians, Norwegians, Hasidim.

After a grim encounter with a Marine on leave and a drop of red wine spilled on the virgin white of the bride's dress, the three men leave with the draft for Vietnam. Once again, the choppers clatter like bugs along the horizon of tropical fecundity, a Communist blows a cowering group of chil-dren and women to pieces, and the three friends are captured.

They are kept in routine degradation, half-drowned in a submerged bam-boo cage with corpses bumping against the bars. This part is not shocking: Cimino seems to have let his imagination lapse into comic-strip stuff at this moment. But the heart of this central section is the desperately thrilling sequence in which each prisoner plays Russian roulette (the name doesn't need emphasis) while the Vietnamese enemy dementedly gamble on the outcome.

Betting is certainly a crux in Far Eastern social life. Clifford Geertz[7] judges that its real meaning is always much more social-symbolic than a matter of life and death. Going by his observations as well as one's own sense of what is natural in gambling, Russian roulette would seem too abominably interesting *in itself* to be worth betting on. If Cimino were to reply to such a criticism, "Yes, but I wanted to puzzle out how one man could gamble on his life in

order to save it, and what such a man would be like," we would surely say the
Vietnam War was the wrong vehicle for such an inquiry. Death was distributed with an entirely unrandom hand in the war; it was exacted by bombs
that, if pointed at you, gave you little chance of a chance at all. And at this
juncture, the proper criticism must be made that Cimino's Vietcong are
parodies of the real enemy. They have contorted yellow faces, they are brutal
and degrading in the treatment of their prisoners, they are demented in their
mania for gambling; they are *incredible,* which is a very serious fault, both
artistically and ideologically. They may satisfy an audience that wants to feel
the beastliness of the Vietnamese enemy. But the real Vietcong had the
discipline and the popular support to win the war in a way these gibbering
torturers never could have. To puzzle out the death-or-life gamble, Cimino
arbitrarily fell back on one of the nastier cold war caricatures.

The almost unbearable suspense of the central sequence grips the audience
and makes them wriggle and sweat. Such suspense is the staple device of
American film art, and it is not a feeling capable of reconnecting an audience
with the real history out of which a narrative is idealized. When the suspense
of incipient violation breaks off, the first response is the desire to violate the
violator. This is the psychosis of revenge. Given that the violators in this
instance are screaming Asiatic subhumans, shooting them is merely a necessary purification of the world. Mad dogs etc. Balancing the rich happinesses
of the Russian Orthodox wedding against these cultureless maniacs, Cimino
plays with great force and vividness to the psychotic pseudo-nationalist frame
of feeling bequeathed to imaginative life by the cold war.

At the end, Michael goes back to Saigon after its fall to try to rescue Nick,
now a heroin addict playing Russian roulette to buy the drug. Michael appeals
once more to that affecting male love that is the strong, heartening theme of
so much American narrative: Cooper's, Melville's, Twain's. The tale of the
deer hunter might have been the tale of this love given and confirmed by
manly adventure, lost in one particular life, but still possible and good. Nick
shoots himself, and Michael leaves behind the vile Orient and goes home with
characteristic American nonchalance to see if America, at least, is still blessed
and Meryl Streep still waiting.

IV

Manly love and its context in manly action has been a natural and seemly
subject for tales of war from *The Iliad* onward. But for such subjects to be clear
and uplifting, as the American tradition prefers, the war in question must be
a just war.[8] By the time the Vietnam War movies really began to pile up in

the mid-1980s, America could not persuade itself, even in its most psychoti-
cally anticommunist and loyalist heartlands, that the war in Vietnam had
been a just war.

The Deer Hunter gets around this problem, as I suggested, first by treating the
Vietnamese enemy with loathing as folk-devils, and second by giving Michael
the limited success of his rescues—"Let's get outta here"—and a return to the
marvelous safety and happiness of home and hunting. (The last time he goes
deer hunting, he has a rush of environmentalism and shoots deliberately
wide.) *Apocalypse Now*, on the other hand, simply celebrates and vilifies in a
familiar American self-entrancement the combination of splendor and corruption to which only the richest country in history can aspire. When some
moviemakers—Oliver Stone, Stanley Kubrick, John Irvin—tried to make
Vietnam War films as close to the war as they could get, they too kept their
patriotism and their Americanness intact by holding close to masculine love
and manly action and holding at a distance the spendthrift cruelty and
technical heedlessness that were the war's true energies.

Thus Stone's *Platoon* (1986) and Kubrick's *Full Metal Jacket* (1987) both take
the combat unit as their sole field of action. In *Apocalypse Now* somebody is
running the war—the bureaucrats send Captain Willard off on his mission of
"termination with extreme prejudice." *Platoon* and *Full Metal Jacket*, however,
adhere to the strong American convention that *all* big systems are blind and
corrupt and only groups of friends, rooted in ordinary life and incorruptible
by virtue of the pieties of that life, can keep faith with honest hearts.

That convention makes possible a view of war through the eyes of the
private soldier instead of, as infallibly in England, through those of the officer
class. In private soldier movies, the grunts huddle in the rain and the green
jungle, while shattering bangs go off around them. Finally, chancily, the
choppers fly in low with their hellish clatter, and hover in swirls of dust while
the lucky men climb aboard. Elsewhere, other men stay and fight. For now,
our hero is safe.

The crop of Vietnam War movies made in the 1980s also catches and holds
the important detail about combat soldiers in the war, that they were so *young*.
The well-known pop video "19" reminded us indulgently that this was the
average age of U.S. soldiers in Vietnam, in extreme contrast to the older
soldiers of the earlier movies. The men in *The Deer Hunter* are so obviously
more mature and, being so, more admirable and impressive than the bewil-
dered boys in *Full Metal Jacket*. The waste of the lives of the fully developed,
strong adults is brought home when even such men cannot restore themselves
to the whole and healthy life of the Pennsylvania steel town.

Boys cannot lose such a full presentness; they lose a future. It is a curiosity
of these films that they do not consider the pain of loss. Loss was a dominant

effect in the films of the Second World War: the individual loss agonizingly felt but set against the greater gain of victory over Fascism. In the later Vietnam War movies it is as if the human bonds of sympathy have been suppressed—on the one hand, by the soldiers' natural lack of commitment to the war, and on the other, by the wholesale exportation of the American way of life to a battle theater where nobody stayed for longer than a year.

Kubrick engages head-on the traditional military themes of soldierly loyalties, friendships, and sympathy. The first half of his long movie follows his raw Marine recruits through basic training. Kubrick is a famously deliberate filmmaker, and he knows we know just how elite a corps the Marines are, how they are all volunteers, and how systematically the initiation, hard drill and exercise, and the replacement of ordinary domestic speech by the violent scatology and ritual vituperation of the regiment bring them unquestioningly into its form of life.

Together, the boys recite the mildly blasphemous Marine creed:

This is my rifle. There are many like it but this one is mine. My rifle is my best friend. It is my life. I must master it as I must master my life. Without me my rifle is useless. Without my rifle I am useless. I must fire my rifle through. I must shoot straighter than my enemy who is trying to kill me. I must shoot him before he shoots me. I will. Before God I swear this creed: my rifle and myself are defenders of my country. We are the masters of our enemy. We are the saviors of my life. So be it. Until there is no enemy. For Peace. Amen.[9]

They are harried ceaselessly into Marine order by their drill sergeant, played not by a professional actor but by a professional Marine drill sergeant. He blasts his charges with the obligatory passionless fury, endlessly reciting the rich liturgy of parade-ground abuse. When, most improbably, a recruit answers him back, he bellows: "Who said that? Who the fuck said that? Who's the shiny little shit twinkle-toes cocksucker down there who just signed his own death warrant?" Finally, he is shot by the squad's fat slob, whom he has goaded beyond limits, thus providing an un-Kubrick-like revenge and schadenfreude in the movie's liberal audience.

And yet, once welded into a solid fighting unit, these new Marines do not exhibit the comradeship and innocent masculinity that are genuine consequences of the training. The film neutrally observes and renders the details of individual subjection to the spirit of the corps but refuses any moral. The boys go off to war, and the brightest of them, nicknamed Joker by the sergeant (each of them is, of course, stripped of his proper name and given a Marine sobriquet), becomes a military journalist.

We see them in one short engagement in Hue during 1968, being picked off one by one by a sniper while trying to cross an open piazza flanked by shattered buildings of grim concrete. The shootings are studiedly realistic, gouts of blood spurting from the boys as the bullets hit them. The wounded die, coughing, speechless, without the director's benediction. The survivors corner and mortally wound the sniper. She is a young woman. In terrible pain, she pleads to be shot dead, out of mercy. The noncombatant journalist kills her. The film ends.

Kubrick's refusal of story and sentiment in *Full Metal Jacket* gives rise to certain level thoughts. We have watched attentively, but at some distance from the action. The young men are trained as soldiers, as they should be, with efficiency and inhumanity. They go to war, and several are killed by a woman. There is no moral, except that there is no moral. The prodigious physicality of their training, the blanking out of fellow feeling, the horrible crack-and-thump of the bullets, the (unseen) blowing off of the sniper's head, lead but to the grave. The expert rendering of war, the feelings appropriate to such highly charged words as "U.S. Marine," "basic training," "chasing Charlie," "tour of duty," and "going stateside" are left in a blank space. Our judgment on the action leads directly to a puzzled numbness.

It is one judgment on the war.

Each of the other Vietnam war movies gathers itself into one of the rich myths of American culture.

Oliver Stone's successor to *Platoon, Born on the Fourth of July* (1989), is based on the true story of a young veteran with a smashed spine and a guilty conscience subsequently fighting back home against Nixon's continuation of the war.

It is a success story. Kovic killed a comrade in Vietnam by mistake and finally expiates his mortal error by telling the boy's family. Gradually he forces the ex-members of his unit to confess the utter mess that they made of things. He leads a protesting group to the 1972 Republican convention and is belabored by the cops. He addresses the Democratic convention in 1976 on behalf of the group Vets against the War. He brings home to America the damage she has done to her great ideals. He is an antimilitary war hero.

Stone matches his tale against those strong, good feelings in his culture that men and women should stand up against hypocrisy and for their best selves. If you went to war on behalf of America—and were born on Independence Day—but still found that your great country was in the wrong, as even great countries sometimes are, you should say so, and recall your country to that love of liberty and peace that will make it the meritorious victor in its historic struggle.

There are moments, it is important to say, when kitsch is stronger and better than irony.

The debile weakness of the Vietnam War movies on the other hand is a political one, and therefore an imaginative one. Connected by profound and submarine links to the long life of the cold war, this weakness is the failure to enter into and imagine as fully as possible the state of mind of someone who sees the world in terms clean contrary to your own. It is part of the duty of an artist to make opposed states of mind intelligible—more than that, to show what it is like to inhabit them.

A significant factor in the American defeat in Vietnam and in Soviet intervention almost anywhere, but particularly in Hungary, Czechoslovakia, and Afghanistan, was the uninterest of the oppressor powers in the state of mind of the oppressed. Part of the greatness of Shakespeare's political plays is his method of crosscutting from one rebel army leader discussing his plans to another: from Northumberland and his son to the usurper king and back. No doubt each little court's being English made this easier for him to imagine; but it is a method completely absent from the best films about the Vietnam War. They are blank to the enemy's state of mind.

Such empathy is more than a matter of depicting the Vietnamese onscreen in some devoutly nonracist, nonstereotypical fashion. It would be part of a film poetics capable of representing politics, the very heart of which is the confrontation of unalike states of mind each ruthlessly determined to have its own way. American filmmakers are not even entered in this race: they take the politics out of politics in the best liberal way. They thus fail not only the test of time and art, but, incidentally, their country.

21

MAD Jokes and the Nuclear Unconscious—*Dr. Strangelove*, *Woodstock*, and *When the Wind Blows*

AMERICAN culture being the bracing thing it is, one way of handling the cold war, especially in its hot or melting down compulsions, was to joke about it. The cold war is a MAD joke. No person of taste could permit herself to take it seriously. Nor can you do much about it. It's a *cosmic* joke.

As everyone knows, the initials MAD stands for the cold war policy underwritten by both power blocs. They stand, infamously, for "mutual assured destruction." There can be no doubt that some State or Defense department funster, himself falling over the crazy comicality at the heart of his daily planning for the world's incineration, came up quite deliberately with the acronym. The balance of terror quivered on the fulcrum of that dependable partnership, mutual assured destruction.

The Cuban Missile Crisis revealed just how precarious the balance of terror was. It brought home both the mutuality and the assurance—those safe old family and insurance company words—of destruction. Those many millions of people with limited imaginations reacted to the discovery by turning to commie-hating, beating the liberty drum, or being simply indifferent. Those many millions with plenty of imagination, who had read John Hersey's *Hiroshima* and seen the newsreels, who would watch *The War Game* if they got the half-chance (which was all there would be), had little other recourse than the great tradition of American eschatological humor.

Joking can be a dangerous game and has, of course, its nasty side. But a joker, even one on the edge of extinction, is a free agent. If you can defy the imminence of extinction with a joke, you are presumably (even if desperately) your own man or woman. A laughing woman is a free woman. This is the point of much in Milan Kundera's political novels, above all in *The Joke* (1984).[1] A totalitarian state cannot permit jokes at its own expense; just try laughing at a policeman. A joke turns the world upside down and releases the joker for an imaginative and sociable moment from the going rules. The rulers rush forward and order the joker to be taken away but, laughing helplessly and pointing at them, he has done his bit of damage to the totalitarians' grip on inner and outer life.

Before the Cuban crisis, films about communism represented it as a secret menace, a scentless poison that might contaminate the purest air (the family, or first love) and, breathed in, change even the best of Americans into alien creatures. Those who were vigilant on behalf of freedom were most often characterized by laconic and parental *sternness*. They were not jokers, even grim ones like John Wayne in *Iwo Jima*. After Cuba, something changed in one current of popular temper and changed all across the war literature.[2] In the era of *Catch-22* (in both its novel and movie versions), *M.A.S.H.*, and Thomas Pynchon's amazing novel *Gravity's Rainbow*, a supple and turbulent choir of writers declared all warfare a mad joke, all military systems indistinguishably productive and destructive, and all great ideals filthy and duplicitous lies.

A bare minimum of barely trustworthy values survives this commination: friendship between men, sometimes; an exiguous personal integrity and kindness, like Yossarian's in *Catch-22;* sex (but not for long); some jokes, laughter anyway. It's a hell of a world.

A poem by the Australian-English poet Peter Porter inaugurated this change of feeling.

Your Attention Please

The Polar DEW has just warned that
A nuclear rocket strike of
At least one thousand megatons
Has been launched by the enemy
Directly at our major cities.
This announcement will take

Two and a quarter minutes to make,
You therefore have a further
Eight and a quarter minutes
To comply with the shelter
Requirements published in the Civil
Defence Code—section Atomic Attack.
A specially shortened Mass
Will be broadcast at the end
Of this announcement—
Protestant and Jewish services
Will begin simultaneously—
Select your wavelength immediately
According to instructions
In the Defence Code. Do not
Take well-loved pets (including birds)
Into your shelter—they will consume
Fresh air. Leave the old and bed-
ridden, you can do nothing for them.
Remember to press the sealing
Switch when everyone is in
The shelter. Set the radiation
Aerial, turn on the geiger barometer.
Turn off your television now.
Turn off your radio immediately
The Services end. At the same time
Secure explosion plugs in the ears
Of each member of your family. Take
Down your plasma flasks. Give your children
The pills marked one and two
In the C.D. green container, then put
Them to bed. Do not break
The inside airlock seals until
The radiation All Clear shows
(Watch for the cuckoo in your
perspex panel), or your District
Touring Doctor rings your bell.
If before this, your air becomes
Exhausted or if any of your family
Is critically injured, administer
The capsules marked 'Valley Forge'
(Red pocket in No. 1 Survival Kit)

> *For painless death. (Catholics*
> *Will have been instructed by their priests*
> *What to do in this eventuality.)*
> *This announcement is ending. Our President*
> *Has already given orders for*
> *Massive retaliation—it will be*
> *Decisive. Some of us may die.*
> *Remember, statistically*
> *It is not likely to be you.*
> *All flags are flying fully dressed*
> *On Government buildings—the sun is shining.*
> *Death is the least we have to fear.*
> *We are all in the hands of God,*
> *Whatever happens happens by His Will.*
> *Now go quickly to your shelters.*[3]

In mood and tone this poem belongs with the fictions of dead ends. It is jocular only in the sense that it is so faithful a parody of how such an announcement might go. A good parody has to be half-admiring of its subject; it has to know and understand what it parodies well enough to sound exactly like it. It cannot bring off such faithful identification out of repugnance. Porter, as the excellent writer must, speaks in the voice of his enemy quite without judging, let alone hating him.

In response, we listen both scandalized and delighted. How could he catch the intonation, the official newscaster diction and idiom, so entirely? And like a good totalitarian should, he has thought of everything, in just the correctly unexceptionable phrase ("well-loved" pets), bringing the rhythm to bear with full force on the phrase which must carry it ("massive retaliation"). The poem is so accurate that our squawks of delight peter out into the chill which freezes the blank space after the end. There is nothing to be done. That *is* how they talk. Laughter, as novelists used to say, dies on our lips. That is pretty well how it will be.

So the joke stands round the poem in its margins. The poet defies the government's newscaster to take his official duties with that portentous seriousness. Come on, man. There are prior duties. Scream a little. Run away home if you've got time. But don't, for anyone's sake, talk on and on in that officially reassuring manner as though anybody is left in charge of anything.

II

"Your Attention Please" is a terribly well-bred poem. Its figure is meiosis or understatement, the English trope. Stanley Kubrick's *Dr. Strangelove*, going for a similar joke in an American accent, is dementedly high-pitched. Could anything be wilder than risking world nuclear war over a few piddling missiles in the Caribbean? Try this.

Gen. Jack D. Ripper is convinced that the Soviets are poisoning the water systems of the United States. He knows this because he can no longer attain sexual climax and ejaculation his customary four times a day. The methods of communism, as everyone knows and fears, are to invade the pure elements—the air and the water—without our realizing it. Our minds and bodies are being unwittingly poisoned. Communism takes away our humanity—sex, family love—and turns us into robots. The state fills personal life.

Ripper was played by Sterling Hayden, who, it will be recalled, testified about his former membership of the Communist party before the House witch-hunters. Having sung his song, he got his own back in a performance of exhilarating battiness as the general who single-handedly dispatches the Strategic Air Command (SAC) on its long-awaited doomsday mission. Assuming that the sexual juices of male America are all as irradiated as his own, Ripper, in his impregnable narcissism, needs no other proof that the only just response is massive retaliation. They fired first.

The sexual comedy is black and compulsive. *Dr. Strangelove* wasn't the first vehicle for making a joke of the similarity between a phallus and a missile, but never before had the joke been worked so hard. Slim Pickens, playing the SAC pilot, rides astride the missile on its final launch; the secret (and as it turns out, unbreakable) code that finally locks the bombers onto their last bearing is Ripper's "purity of essence."

As the film's maker, Stanley Kubrick, has usually done, the military is represented not only with great panache and insight but also as completely potty. George C. Scott—he and Peter Sellers are the best actors in the movie—has a high old time as Gen. Buck Turgidson, an officer of such unremitting anticommunist conviction that he vehemently resists using the hot line to Moscow to warn the Soviets of the disaster and is contorted with uncomprehending hatred every time the president telephones through to his opposite number, Dmitri, in the hope that the Soviet defenses have blown the rogue aircraft out of the sky.

They never do. The president, an apologetic, useless sort of fellow (another craven liberal) is one of three roles played by Peter Sellers, his tours de force in the movie. His second part is the fatuous English liaison officer from the

Royal Air Force with a handlebar moustache, prattling away in Battle of Britain clichés in an effort to dissuade General Ripper from his plan. In an accurately cruel representation of British subservience to U.S. war plans, Sellers holds the belt-feed for the general's machine gun as he blasts away at the Marines who are trying to capture him and his secret base in order to abort the mission.

Seller's third role completes the fateful 'iron triangle' of Congress, Pentagon, and technology. This is the tame atomic scientist, another comic madman, a shrieking imitation of Werner von Braun with a touch, it may be, of Edward Teller. Sellers plays him as wheelchair-bound with a black-gloved artificial arm shooting out in involuntary Nazi salutes. The scientist has devised, in his Nazi snarl, a nuclear message system so secret that once it is activated, it cannot be countermanded by anybody.

Slim Pickens, commanding the last visible SAC bomber in his ten-gallon stetson, unravels the final order in its fortune-cookie codes. First Secretary Dmitri explains to the president that once the American weapons explode, Soviet retaliation will be unleashed, which he also is now powerless to prevent. In another truth-telling joke, the two complementary systems are rendered incapable of compromise.

This being Kubrick, the film is unremittingly concentrated. There are only three locations: General Ripper's secret air control base, the National Security Council's underground bunker with its illuminated wall-to-ceiling maps, and the bomber. Kubrick denies us the explosion to which, in fascinated immunity, we so eagerly look forward. The plane flies on into the last frame.

So doing, it flies out of the frame and over our heads. Abruptly, the black joke stops being funny and starts being real. The bomber joins the other bombers armed with doomsday weapons that have been patrolling the skies for well over three uninterrupted decades. Below them rave the madmen who locked these calm, crazy systems into place: the soldiers and the scientist far beyond the feeble president's control. As for us, we never even get a walk-on part.

This is the cold war theme of *Dr. Strangelove*. We had learned of our peril from the Cuban crisis, and we had also learned how little we could influence the chains of command and consequence. The swift, omnipotent systems of retaliation were (and are) immune to popular regulation. As far as nuclear weapons went—which is as far as the end of the world—a superpower's leaders could no longer ensure democracy for the people. The best they could hope for was (and is) intelligent responsibility.

Dr. Strangelove's bleak moral sorted well with the popular feeling that modern technological society had taken away human quiddity. The systematic state was determining all. As the 1960s advanced toward 1968, a small but active minority began to broadcast the message that the once-benign state that spoke for Americans was no longer to be trusted; the alternative culture was struggling to be born.[4]

III

Wherever there is a struggle over art, there is a struggle over the art of living. The local struggle over how to live differently from the preceding generation reached far beyond the suburban kitchens in which it took place toward the grand myths that framed the times.

As the number of young men sucked in by the Vietnam draft rose sharply toward its peak in 1969, their teeth were set on edge by their fathers telling them to be proper men, to go to war as the fathers had done in Europe, in the Philippines, in Korea. It was the duty of their sisters and sweethearts to kiss them goodbye, to wait for them with folded hands, to succor them when they returned, and to remember them if they did not.

This story was no longer holding personal identities in their social station. If *Woodstock* (1970) is not the most colossal anti–cold war movie ever made, it is certainly the film that most colossally celebrates the new American (and British) young. After their mythopoeic Woodstock Festival in 1969, America's youth had clearly announced to their elders that any hot fighting by conscripts in the name of cold war advantage would be done no further away than the beaches of California and New Jersey. In its three hours (of the much-cut version), *Woodstock* holds a comically self-admiring mirror up to the youthful generation. The joke, admittedly, is almost always on the older generation. For at Woodstock, in cheerful anarchy, a new, decidedly American individualism was blowing away (for a short, sweet while) the elderly, military conventions. Here was sex without marriage, without love, but with a happy giving-of-oneself to sweet feeling. Here was a loosening of identity and the boundary between men and women, so that long hair, baggy clothes, flowers, jewels, and demonstrative gestures were indistinguishably, artlessly, male or female. Here was peacefulness and kissing preferred openly and shamelessly over fighting. Here were girls giving their bodies to soldiers not to produce a new generation of soldiers but to unman them and stop their fighting. And here were dirty confusions of the bodily orifices, the mouth as glad a place for sex as penis or vagina.

Through the huge and happy crowd throbbed the music; its stars themselves were the ideal types of the crowd's new culture. The tidy distinctions of concert-going—and of day and night, of work and play—all went down in one enormous inversion of the normal world; the biggest family joke in

American history was being played on mom and dad, and on *their* moms and dads.

It was good while it lasted. After 1968, and in spite of Ronald Reagan's successful counterrevolution in the 1980s, nothing was ever the same again. The political frame of mind that contained the epoch was fractured beyond repair. Naturally, those in power contended that it was not, that nothing essential had changed in the way nuclear weapons had been keeping world peace, and that if anything *had* changed for the better, then it was owing to their own unexampled resolution and toughness before the implacable enemy. But however they prated on, the old frames of mind and structures of feeling were splitting down every cultural weld, even in bedtime stories. One picture book, published in England in 1982, serves as another happy instance that the line from political power to popular culture is not all one-way.

When he published *When the Wind Blows*, Raymond Briggs was a faintly subversive, very popular figure among the children's reading public. He had previously followed a small, rather private line in mildly scatological, entirely good-humored humor, featuring the joyously squalid Fungus the Bogeyman who thrives in dirt-as-normality and is also loving, giving, and, give or take a stench or two, inoffensive. He also celebrated Gentleman Jim, a public lavatory attendant, and his wife Hilda, who he claimed were fondly modeled on his own parents.

In *When the Wind Blows,* Jim has now retired to a country bungalow with Hilda, and immediately we rejoin the affectionate details of their lives, pictured in the sequences of small (often no more than one-by-two inches), faithful color-wash and pen-and-ink drawings in a style somewhere between English children's comics and the primitive school of the 1940s. As the publishers said when making their first acquaintance with Briggs's holograph, it is interesting how quickly one enters the story and reads on as one would read a novel or watch a TV movie. There is no condescension to the form, only a perfectly judged congruence of word and image. *When the Wind Blows* is, after all, a rare work of art.

Briggs shocks the reader by making a few, frugal breaks with convention. Pages 4–5 are a grim double-page spread depicting a dark sky and an erect nuclear missile; a few pages later, there is an even more compelling image of the dark, diagonal bulk of a submarine seen from below and moving dreadfully forward underwater.

With the national news on the radio so bad, Jim sets to building a little lean-to shelter in his house, taking doors off hinges and leaning them against an inner wall, stacking the doors with cushions and bundles, all exactly as prescribed by the British government leaflet *Protect and Survive*, which we shall meet in real life in 1980.

In the very center of the book Briggs vividly represented the bomb going off by bleeding all the color from almost the whole double page and leaving it a dazzling white except for the bottom right-hand corner, which is suffused with a faint pink. Over the next page, the pinkish white deepens to red sprinkled with shards of glass, and gradually, in the second right-hand corner, the spectrum returns and Jim's speech bubble says, "Blimey."

The repeated joke, as grim and painful as the hysterics of *Dr. Strangelove*, is placed on the gap between what is happening and Jim's attempt to follow the official social narrative that has been prepared for this moment. Each of his comic misreadings touches us with its law-abiding credulity and, often, its deadly double meaning: "The Govern-mental Authorities have been aware of this eventuality for years, so Continency plans will have been formalated long ago. . . . We won't have to worry about a thing. The Powers that Be will get to us in the end."[5]

They get to them all right. In all his loving efforts to protect and then, when protection is soon useless, to console his beloved Hilda, Jim still voices their placid trust in official diction, and in the folk memories by which he sorts and interprets the march of these chilling events. In a glowing sequence of page-filling almost-ikons, Jim summons up images of himself as a hero-firefighter in the London blitz and of a jumbled cast of the great leaders of old—Churchill, Montgomery, Roosevelt, and Stalin (the last causing him some understandable confusion). He struggles for a moment over how to respond to the arrival of the standard cold war, fur-clad, Kalashnikov carrying Soviet soldier—until Hilda suggests that they give the invader a cup of tea. As in any consummate work of art, it is not quite clear who the joke is on. A cup of tea is a disarming weapon.

From this point on, the tale becomes a piercingly real chronicle of the breakdown of their health and of the stories they have believed in. Briggs's loving humor fades to a murmur, and only his strong compassion sees us through. Jim's and Hilda's gums begin to bleed, their skin to blotch, their hair to come out in handfuls; Briggs shows them, with backs discreetly turned to the reader, vomiting helplessly.

The stories that Jim has so doggedly used to keep their lives connected to a knowable past come to a similar end. As death approaches and they lie enfeebled in their little shelter—the tiny space left that retains the meaning of home—the solace Jim finds for Hilda is in barely remembered fragments of religious liturgy and elementary school literature. The incongruity at once horribly and poignantly takes the measure of the politics that has brought them to this pass. At the end Jim brokenly quotes scraps of the Twenty-third Psalm and odd lines from "The Charge of the Light Brigade," and Hilda thanks him for their beauty. Several of the last pictures of their lean-to are wordless.

Jim and Hilda were lied to. The mild and worthy culture and politics by which they lived were abominably misused. Briggs's extraordinary tale, its deserved success splitting open the received divisions of the market, between children's book and comic-strip book—marks the point at which any belief in winnable nuclear war finally ended. It taught children and adults alike to disbelieve the wicked story that had kept them all in their places.

With a strong sense of jocularity, the British publishers sent the book to various political worthies of the cold war, who could hardly fail to be impressed and who variously praised Briggs's masterpiece with faint tuttings. Briggs had them either way. After all, his book was just a joke.

PART IV

History as Farce

PART IX

History as Farce

22

EVENTS X

The Second Cold War, the New Peace Movements, Solidarity, and Ronald Reagan, 1980–1981

THE crucial events of 1979 were the Soviet Union's blundering into Afghanistan and the American announcement that its new pet weapon, the nuclear cruise missile, would be scattered around Europe. The Soviets had amalgamated two heavy designs for nuclear-warheaded missiles driven by solid fuel—"coke-fired rockets"—and had produced a powerful new weapon. Fast, a good dodger, and fitted with separately targeted warheads known in the jargon as MIRV, it was simply called the SS-20. It could reach all over Europe.

The cruise missile was also an amalgamation of old designs but it was now pressed into service as a reply to the SS-20.[1] Its trade name was Tomahawk, a "reassuring cowboys-and-indians sort of name," Edward Thompson remarked, but it was also a "smart" missile.[2] Stored in the nose of the missile were digital maps of the territory it was crossing specified down to the details of rocks and stones and trees. Following its maps at 100–400 feet above the ground, the missile would twist and turn and, they said, find its way to its minutely designated target where it would go off with a bang and a blast seven times as powerful as the Hiroshima bomb. Of course, even a smart missile can get lost or blown off course, and if its mission is long and serious—Red Square, say—it might even miss. There were even those on the military side (it was an air force weapon) who naughtily suggested that the Tomahawk and its Tercom (terrain contour matching) might get lost if the said terrain was covered with snow, as it is apt to be in the Soviet Union, and the contours didn't match.

In spite of all, however, cruise missiles were to be' distributed around southern Britain, Holland, Belgium, West Germany, and the toe of Italy, where it was thought, mistakenly, that the Catholic majority would welcome a lot of vile explosive intended to blow up atheists.

There were other nuclear weapons afoot: a choice little twenty-to-thirty miler with the agreeably archaic name of the Lance; and the neutron bomb, which killed better with its poisonous fallout than with its blast and was accordingly nicknamed "the capitalist bomb" by the enemy because it destroyed people, not property. It was a neat joke. One can understand Jimmy Carter's relatively uninformed decision to stop its production. But the defense constituency tutted away over his inconstancy. They would have plenty more to tut over before the year was out.

Nobody among the nobs had thought about what the Soviets would make of all this. Cruise was simply there to match the bigger SS-20s, which in turn replaced their elderly parents earlier in the SS line. But the Soviets were violently upset. They pointed out vehemently that the new weapons allowed the United States to hit Moscow in a few minutes if it chose. This was no longer parity. Cruise made their world abruptly less safe.

Indeed, neither side had worked out what its actions—invading Afghanistan (much the more objectionable move) and spreading cruise missiles around Europe—would mean to the other. Carter and Secretary of State Cyrus Vance, sensitive to election year politics and the smell of gunsmoke coming over from California and up from Florida, renounced détente, blocked grain exports to the Soviet Union, halved trade, and returned to a policy of containment and arms development. Carter, in characteristically impulsive form, said to Congress on 8 January,

In my own opinion . . . the Soviet invasion of Afghanistan is the greatest threat to peace since the Second World War. It's a sharp escalation in the aggressive history of the Soviet Union. . . . We are the other super power on Earth, and it became my responsibility, representing our great Nation, to take action that would prevent the Soviets from this invasion with impunity. The Soviets had to suffer the consequences. In my judgment our own Nation's security was directly threatened. There is no doubt that the Soviets' move into Afghanistan, if done without adverse consequences, would have resulted in the temptation to move again and again until they reached warm water ports or until they acquired control over a major portion of the world's oil supplies.[3]

So the Rapid Deployment Force was duly established, with fawning applause from Britain, and Mrs. Thatcher implored her nation's athletes to register national displeasure by refusing to run at the Olympic Games in Moscow that summer. The Americans duly withdrew.

The Europeans didn't like the new music: neither their leaders nor, it seemed, anyone else who had noticed what was going on. Helmut Schmidt, chancellor and leader of the West German democratic socialists, had a caustic, unrehearsed, and victorious exchange with Carter over his falling in behind the antique drum of the cold war. Schmidt also had an election to win, and as the very well informed electorate had noticed, his country was uncomfortably placed straight in front of the Soviet weapons lined up in opposition to the much-bruited flexibility of these infernal cruise missiles. By the same token, the elegant rightist Valéry Giscard d'Estaing of France, though quite unlike Schmidt in his politics, was just as loath to let Europe become the dangerous playground of a piqued president. And he, too, had a lot of people in the streets reminding him of that possibility.

The Soviets, knowing they had put their foot heavily into the Afghanistan muck, were still startled that eight years of work at keeping the world in balance, at finding a form for Khrushchev's peaceful coexistence, should be so thoughtlessly torn up. Brezhnev was an abominably immobile old block of wood, no doubt, but like blocks of wood he was at least *steady*. (Steadiness, it seemed, would be needed when Ronald Reagan came looming up against the setting sun.) In Brezhnev's speech to the twenty-sixth party congress in 1981 he made what was really an exemplary statement of the state of Europe, if not of the world.

> We have not sought, and do not now seek, military superiority over the other side. That is not our policy. But neither will we permit the building up of such superiority over us. Attempts of that kind and talking to us from a position of strength are absolutely futile.
>
> Not to try and upset the existing balance and not to impose a new, still more costly and dangerous round of the arms race—that would be to display truly wise statesmanship. And for this it is really high time to throw the threadbare scarecrow of a "Soviet threat" out of the door of serious politics.[4]

The real measure of European anger and anxiety over Carter's strictly American response to the Soviet action in a faraway and grubby country appeared spontaneously on the boulevards of its great cities. In London, Paris, Bonn, Amsterdam, Rome, Bologna, Copenhagen, Stockholm, and Berlin, the private citizens of the Continent turned out in vast crowds to bear witness to

their rational fear that the political elites they paid and voted for no longer had
Armageddon under control. Reason and honor had become prey, they believed, to knaves or fools.

II

There had been national peace movements in Europe ever since people began
to realize which way Europe was heading in the 1920s. After the Second
World War the Campaign for Nuclear Disarmament (CND) found Britain to
be a sensible enough place to start peace work because it was the third nuclear
weapons power in the world and could no longer afford to be. In addition, the
Labour party and the trade union movement were fiercely critical of a Conservative government still so committed to the illusion of Britain as a world
power that it had perpetrated the folly of Suez. In 1957, after the double shock
of Suez and Budapest, the honest gradualists of the British left began pursuing
a little isolationism.

They retained, however, an evangelical ardor. The CND proposed that
Britain renounce nuclear weapons without any corresponding concessions
from anyone else; this was unilateralism, a moral model for the world. It was
easy at that date to see that China and probably India would soon join the
nuclear elite. (But it would have been difficult to predict the frightening
extension of ownership that had taken place by 1980.) These early idealists
were of a generation whose Britain really had mattered to the rest of the
world, and so they still believed that its good example might just conceivably
change the world.

They were much encouraged in this by their campaign president, Bertrand
Russell, last of the *grands seigneurs* of classical English liberalism, godson of John
Stuart Mill, direct descendant of the longest serving prime minister in British
history, the twentieth century's greatest philosopher of mathematics, and a
pacifist hero of 1916 when he was jailed, in a blaze of publicity, for his
deplorably unpatriotic refusal to go to war. Russell wrote to people who, as
Raymond Williams remarked, one could hardly believe had a mailbox.[5] At
the time of the Cuban Missile Crisis, and at the age of ninety, he wrote to
Khrushchev, who replied; he wrote during the Vietnam War to U Thant and
Harold Wilson, urging the cause of humanitarian reason; in a separate letter
to the North Vietnamese prime minister he expressed sympathy over the
beleaguering of Hanoi. At eighty-nine he went back to jail for his civil disobedience with a six-day sentence. He gave CND its authority and ratified its
dedication. Along with the lesser literary worthies who composed the early
CND leadership—a group of the Hampstead and Oxbridge standard-bearers

Figure 22.1   CND in Trafalgar Square, London, 1965. (The Hulton Picture Company)

without whom British public life would be much less liberal—he gave disarmers a historical weight and continuity.

After the Cuban crisis—when everyone agreed that the British disarmers
might have a point—Khrushchev and Kennedy signed their test ban treaty
and set up the famous "hot line" telephone from the Oval Office to the
Presidium snuggery. As a result, the vision of nuclear annihilation went out
of focus for a while. The weapons themselves were, of course, still accumulating, but ordinary people got on with ordinary life.

Later in the 1960s, however, a huge charge of energy was given to the
peace movement by the many coalitions against the Vietnam War; the same
energy provided the various peace campaigns with the comprehensive and
critical library of information that they had always lacked. By 1980 the peace
movements of West Germany, France, Italy, and the United States, as well as
the revived and revivalist Campaign for Nuclear Disarmament in Britain,
could bring to their activism a technical vocabulary and a formidable confidence with the figures and casuistry of deterrence theory. The crowds of
people demonstrating in the streets gave weight and representativeness to the
critical scholarship of nuclear disarmament. The books of numbers and the
numbers of marching people complemented one another.

Figure 22.2   A Greenham Common woman, leafleting outside the High Court, London, 1982. (The Hulton Picture Company)

In the autumn of 1980 the marches spread through the capitals in long, serpentine masses. From towns all over Europe the buses and trains sped overnight to the capitals, bringing contingents of ten or a dozen, one hundred or two hundred, to gather below the autumn trees on a dry October day and walk together in peaceable and slovenly dignity a couple of miles to some main square under the eye of government, high in its tall offices. The side roads are closed to traffic, and the pavements lined by police amicably shepherding the populace in its lawful objection to its lawful leaders. The road is filled for miles back from the square, the great motley of individuals turning into that single meaning, a demonstration. Above the uneven files of people and strollers the banners would be raised, carefully embroidered, simply declarative, sometimes beautiful, naming their allegiance to town and village, college and church, gradually coming together as an enormous tapestry of an imagined land whose people would always be at peace. Over them all hover and wheel the watchful helicopters.

The NATO authorities had announced that the cruise missiles would be deployed in Europe in 1983. In Britain and Italy there was no parliamentary debate on the issue; the governments simply agreed to the deployment. Then it was disclosed where the damned things would live—in Britain, in two old air force dromes in Berkshire and Cambridgeshire counties. By 1983 "Greenham Common" would be a rallying slogan all through the international women's movement. NATO also noted that, in times of international tension, the missiles would be taken out of their burrows and trundled off down leafy lanes to sites less vulnerable and less visible to the vigilant satellites above. In addition, there would be regular airings for the missiles, when such dispersals would be practiced in readiness for the great day.

It was at this time that the British government's office of information published *Protect and Survive* (1980), an advertising brochure available for fifty pence, that described how to prepare one's little home for nuclear war. It is an incomparable piece of work intended, apparently, to stiffen the fiber of cold war discourse with practical advice about what to do when history really turns nasty on you and climbs out of the television set into your living room. It contributed richly to the popular European tide of incredulity that greeted the new call to cold war arms in 1980. How can I resist quoting this small jewel of advice for the end of the world?

Limit the Fire Hazards
As you plan the fall-out room and the inner refuge you need also to limit as far as you can the dangers from heat and blast to the rest of the house. Though the heat could not ignite the bricks and stone of your home it could set alight the contents by striking through unprotected windows.

There are things you can do now to lessen these risks—

Remove anything which may ignite and burn easily (paper and cardboard, for example) from attic and upper rooms where fire is most likely.

Remove net curtains or thin materials from windows—but leave heavy curtains and blinds as these can be drawn before an attack as protection against flying glass.

Clear out old newspapers and magazines.

Coat windows inside with diluted emulsion paint of a light colour so that they will reflect away much of the heat flash, even if the blast which will follow is to shatter them.

If you have a home fire extinguisher—keep it handy.

Keep buckets of water ready on each floor.[6]

There's much to give pleasure here. I particularly like "If you have a home fire extinguisher—keep it handy." Everyone who read the thing treated it to a loud raspberry. At the same time, its fatuity not only was plain enough evidence that nothing could be done to protect your family from the weapons, it also revealed that the government *knew* that nothing could be done but still perpetuated this ludicrous lie upon the populace.

So quite a lot of honest citizens, voters, and taxpayers became angry, rather than just resigned, and went about their countries in large groups, walking up and marching down, and generally giving notice that they did not believe all that the government told them was necessary for nuclear defense when the advice on protecting themselves was to bang together two old doors and a table covered with bags of muck.

Their anger was given voice by the celebrated English historian and leftist, Edward Thompson, Frank Thompson's brother, who published by way of the Bertrand Russell Peace Foundation his majestically contemptuous reply, *Protest and Survive*. It was five pennies cheaper than the government pamphlet. Thompson suggested not only that the whole mad business of accumulating nuclear weaponry had run out of control, and that the superpowers had more than enough weapons with which to end the world, but that this overproduction, part of a drive within industrial systems that he dubbed "exterminism," itself made the ultimate war far more likely to happen than not.[7]

Thompson's terse critique changed a great many minds. He pointed out that in Europe public discussion and reasoned debate about the considerable increases in national defense expenditures had been nil. The entertainingly named High Level Group at NATO headquarters had made the decision to modernize (a word always greeted with reflex applause) its middle-range nuclear weapons.[8] It passed this decision to the Nuclear Planning Group, which gave instructions about it to NATO member governments, which in their turn accepted the decision obligingly. Then the posh press received its punctually embargoed releases and set about telling the literate populations what to think about it all.

The military told the politicians. The politicians told the press. The press told us. Thompson objected.

The result of Thompson's pamphlet—and I don't think it is an exaggeration to describe it as *his* result—was an uproar. The Campaign for Nuclear Disarmament was reborn in a trice, and Thompson himself was the leading drafter of the 1980 "Appeal for European Nuclear Disarmament," the central document of a new, popular, and enormous movement across the Continent. The first task taken on by the movement—always spontaneous, voluntary, and very broke, though regularly accused of receiving Moscow rubles in old ammunition boxes—was to prevent the installation of cruise missiles, the modernized counterweight to the Soviet SS-20s.

All successful mass movements are directed toward single ends, which must be intelligible and feasible. People can then make the necessary commitment of time and ardor to win or lose. (One such successful movement was the anti-Vietnam War movement.)[9] The case against cruise was easy to grasp as well as very strong: the missiles made the world more dangerous, not less. If you lived near them, in Oxford perhaps, or Cambridge, you were threatened along with those ancient centers of the humanities with swift and perfect extinction.

So "the People of England," to whom Thompson grandly addressed his pamphlet, stirred suddenly from years of somnambulation and almost woke up. They were treated by their own government and by certain Americans to stern homilies. When some of the Germans, Italians, Dutch, and Belgians stirred as well, it was really too much for the cold warrior fideists:

> The idea that the Soviet Union might be a very different phenomenon from a Western constitutional democracy and that in fact its ideology is such that it is unable and unwilling to concede an equal right of existence to constitutional democracies is simply too complicated for most in the peace movement to grasp. When they are presented with that idea, they do not argue against it, but instead denounce its supporters as fascists and warmongers. It means, however, that the East-West struggle, the cold war, is not only not over, but that it will go on for a very long time yet and that the West European nations, being constitutional democracies, are directly involved in that struggle. The peace movement's alienation from the democratic system . . . its refusal to accept the possibility that the Soviet Union might be totalitarian and expansionist . . . and its simplistic belief that more weapons mean more chance of war . . . add up to a monumental failure of political imagination. Whether that failure is due to inadequate education,

psychological immaturity, or a lack of experience is another matter that I shall not discuss here.[10]

Meanwhile, the elderly Brezhnev, in what seemed a more than ritual and personally meant attack, had publicly accused the United States of unreliability, of being a power "whose leadership, prompted by some whim, caprice, or emotional outburst, or by considerations of narrowly understood immediate advantage, is capable at any moment of violating its international obligations and cancelling treaties."[11] With Afghanistan at the top of the Presidium agenda, he was in a morally queasy position; but he seemed to be sincere about what he was saying.

Certainly, his remarks struck several chords in European capitals; only Mrs. Thatcher's, among the European governments, failed to register unease. And for the first time since de Gaulle led France out of the alliance in 1964, the North Atlantic Treaty Organization was at odds with the peoples it was there to defend. The U.S. leadership of Western Europe faltered in 1980, and hesitantly the continent in its western half began to resume self-propulsion. Reagan's imminent presidency would make the new Right victorious for a while. But there was nonetheless widespread skepticism in Europe about the need to keep the cold war going, and even the beginnings of a helpless hilarity at the idea that the Soviets were working themselves up into a fine careless rapture of expansionism.

III

Throughout 1980 it was very apparent that the Soviets could not even hold down their loyal allies. Poland, always the stormiest of the Warsaw Pact countries, was once again giving cause for concern.

It had had a bumpy time ever since Wladyslaw Gomulka so adroitly kept the Soviet tanks out of the country in 1956. There had been riots in 1960 in defense of the Catholic Church, which, after linking itself bravely with the state against the Kremlin in 1956, had become the natural and resilient center of opposition to atheistical oppression. After all, it had been at the center of state power for centuries until the Communists took over in the aftermath of a war in which the Church had compromised its moral standing. Now under the vigorous and polemically independent Cardinal Stefan Wyszyński, it reclaimed that authority in the face of local party incompetence, brutality, and corruption.

Throughout the 1960s Gomulka had promulgated certain economic reforms without any corresponding liberalization of social and political life. At the end of 1970 one such reform was to raise sharply the price of food; the higher prices were enacted abruptly to avoid panic-buying and were scheduled, with typical oafishness, to take effect just before Christmas. The workers at the big shipyards named after Lenin in Gdansk and after the Paris Commune of exactly a century before in Gdynia, together with the Warski in Szczecin came out on strike. There was rioting and angry looting of food shops, the police opened fire in the three industrial towns, and at least fifty people were killed.

The workers occupied their factories, rapidly improvised an effective command-and-committee structure between the factories, and called, for the first time, for new trade unions freed from the dead weight of party control. Gomulka had a stroke, and although he still fought his opponents in blind rage, he was voted out of office and replaced by the decent, conciliatory, and honestly socialist Edward Gierek.

Gierek, bravely, went straight to the seething shipyards. In a nine-hour open debate on the future of socialism in Poland of a kind peculiar to the socialist republics that would be repeated endlessly in the marathons of August 1980, he won the day with an appeal to working-class solidarity. But the workers had overthrown the first secretary and felt their power.

Polish history is weirdly recursive. Between 1970 and 1980, as Neal Ascherson's gripping history brings out, Gierek recapitulated Gomulka's term in office.[12] He started with genuine commitment to a new candor and truthfulness in society, with the party being fully answerable to its natural constituency, the workers; by 1980 the entire working class had returned to the faith of its fathers and shut out the party of its class. Like Gomulka, Gierek instituted recklessly ambitious economic reforms: he borrowed vast amounts in an effort to combine a crash program of industrial modernization with subsidized wages and food prices. His policy brought the country economic ruin, hunger, and the largest debt per head in the hemisphere. Domestic manufacturing collapsed under the weight of imports of both plant and consumer goods, while the new, tiny private enterprise farms could not deliver the food wanted by the home market.

The party could not control the economic deluge it had released. Correspondingly, it lost its grip on its two fundamental constituencies, the workers and the peasant farmers. The Church, the glowing center not only of opposition to state communism but also of all that the party daily defiled of truthfulness and trust, human sympathy and humility, steadily drew both classes unto itself until finally it contained and expressed both completely.

The Polish church would also benefit from the astonishing election in 1978 of its beloved son, Cardinal Karol Wojtyla, to the papacy. Pope John Paul II was the first non-Italian Pope for centuries, a man of extraordinary presence, erudition, gritty strength of character, and staunch theological conservatism. He was also to prove master of the international media, and a stirring orator.

He gave Polish Catholicism renewed authority; from the eminence of Rome he could speak directly for Warsaw. When he came home in June 1979, the world watched the television and joined the Poles in their loving and awestruck welcome. Millions upon millions, fully one-quarter of the nation, lined the streets and knelt to give the craggy, handsome primate in his pure white robes their fealty with tears of joy. The Pope met Gierek in the Belvedere Palace, where Gomulka had outsmarted Khrushchev in 1956, and gently heeded Gierek's nervous appeal for national unity. But the party was become a poor thing, and as Ascherson finely puts it, "The Pope's attitude at the Belvedere was that of a person speaking from a position of strength so overwhelming that he could afford magnanimity and concession towards an authority that was, in comparison, so insecure."[13]

At the eighth party congress early in 1980 Gierek stood up and confessed to the desperate straits of the national economy. There would have to be price rises, belt-tightening, and so forth. A rally was held in Gdansk commemorating the successful 1970 resistance to such moves and remembering the dead of that time. The authorities, thoroughly apprehensive, arrested various participants, who lost their jobs. A committee was set up to fight for their reinstatement, particularly that of a local heroine, one of the crane drivers. She got her job back, but things remained tense.

On 1 July the price rises on meat came into force. The condition of the economy was such that the government had to do *something*. But meat is the heart not only of Polish cuisine but of its domestic culture: pork and the crucifix are the potent symbols of the home, and at home the government and its loathsome secret police can't get you. Stoppages at work started all over the country, then ended, always with a pay rise.

The government's only policy was to buy off dissent. But neither pay nor the price of meat was the issue; it was the ineptitude and irrelevance of the party itself.

Gierek, looking over his shoulder worriedly toward Moscow, tolerated the mounting opposition. The Committee for the Defense of Workers' Rights, run by the intelligentsia, went about openly coordinating that opposition, publishing widely and illegally. There were more strikes. And then, on the bright, fresh morning of 14 August 1980, the main Gdansk shipyard, the Lenin, stopped work. A short, stout electrician with an untidy Zapeta moustache, Lech Walesa, shouted to the driver of a heavy lifting machine to hoist him over the steel railings into the yard. Holding onto the back of a big bulldozer, he asked the great throng of his comrades to occupy the whole place. With one roar of assent, they agreed. The yard was theirs.

As soon as the news was out, workers in Gdansk, Gdynia, and Szczecin also went on strike. Whereas the party had created no political institutions capable of either calling on collective creativity or bearing blame, the occupying workers, in contrast, proved to be remarkable and rapid innovators. They circulated decisions from factory to shipyard and back at great speed. They stood by what they wanted—free trade unions—and without being dogmatic, nonetheless never compromised. They also kept strict discipline; any occupying worker found with a bottle of spirits had his name broadcast over the loudspeakers, and the bottle was emptied down the drain.

Walesa was the undisputed leader of the moment. He was a good Catholic, he had been through the 1970 workers' insurrection as an organizer and was deeply marked by the shooting of his friends, he had limitless stamina for debate, and in Ascherson's words, "he possessed a telepathic grasp of workers' emotions and a deep-seated, driving anger."[14]

The strikes and occupations set and hardened. The Inter-Factory Strike Committee published twenty-one demands, a manifesto for a decent future, including access to the mass media, the release of political prisoners, the abolition of censorship, the right to strike (for the whole action was illegal), and, unwaveringly, free trade unions with no interference from the infernal party hacks. They gave, it is pleasant to observe, renewed life to the name they eventually chose for their union: "solidarity," in the lexicon of socialism, is socialism's key bequest to the future. Rejecting the socialism they had suffered under, the Poles kept faith with its traditional values: solidarity, equality, and justice. They called for equal pay raises and no differentials; they looked for natural justice and common law.

Gierek sent a senior minister, Jagielski, to negotiate at Gdansk. He did so with Walesa and the committee crouched in cheap chairs around a canteen table, all in full view of the occupying crowds of workers and with every word broadcast on the public address system. The strikers hit him hard with their arguments, while back in the capital the tough subjects on the central committee spoke grimly of martial law and punitive jail sentences as they waited for the Soviets to come. Gierek held them off, while Jagielski refused to concede party-free trade unions in the endless negotiations.

But the Soviets never came. Solidarity held in factory and shipyard. At Gdansk, the center of the world's attention, the print and television journalists packed in alongside the listening workers, clambering with their long wires and cameras over the jammed limbs and shoulders of the sleeping, singing, arguing men and women waiting for their future. A handful of typists clattered out the press releases without benefit of photocopiers. The workers' families came to the locked gates with flasks of coffee and clean clothes and, in token of their unquenchable hopefulness, fastened bunches of propitiatory flowers and faded pictures of Karol Wojtyla to the gates.

Mass was regularly and piously celebrated in the sun-baked yard, where

Walesa made his daily reports. The government was at its wits' end when the Silesian miners came out. Jagielski had to get a settlement. But how could a Marxist minister and theorist agree to sever the workers from the party?

Agreement, as Walesa must have known, would mean the effective end of the party in Poland, even though it took the country another eight years to clinch the matter. There was nothing else to do. When the toughs in the politburo called for troops to break the strike, the navy chief of staff refused to give his sailors any such orders. Wojciech Jaruzelski, the defense minister, stood by him, in his forbidding dark glasses as well as by Gierek.

Gdansk, the Lenin Shipyard, was the talisman of the whole show. But each strike committee moved at a marvelously coordinated pace toward settlement of the heart of the matter—free trade unions, which would effectually amount to the formation of an official opposition, the first in the Soviet bloc. Walesa dominated the public debate, catching the microphone as it trailed smoothly by its long cable from speaker to speaker, cajoling, rousing to laughter and to tremendous applause, coaxing, encouraging, bringing his people by his passionate and disinterested energy to their famous victory. On 31 August he won, when the movement's discipline was on the very edge of disintegration, almost all they had asked for.

Gierek's heart gave out, just as Gomulka's had, and he retired ill. The Polish debt continued to climb vertically off the graph. In December there were large Soviet troop movements on the borders of Poland. *Solidarnosc* ignored them, ignored the economy, and went right on pressing new political demands. The NATO allies were wringing their hands with anxiety, and the Warsaw Pact nations, celebrating in Moscow by coincidence the twenty-fifth anniversary of their alliance, brooded in secret about what to do, and four Soviet divisions were conspicuously relieved the following week.

The threat subsided. As the humiliation of the party proceeded through 1981, the peasants formed Rural Solidarity, even the annual party congress was turned upside down by opposition, and the Soviets wrote alternately severe and pleading letters. Then in December Jaruzelski, the new first secretary, startled everyone by suspending all trade union activity and ordering Polish troops to assume martial law. In so doing, as the old joke has it, he tried to dismiss the nation and appoint a new one. All he did, however, was, lingeringly, to dismiss the party.

The Brezhnev doctrine had lapsed. The Soviets had moved their troops, maybe threateningly, maybe not. But most important, when one of their satellites had spun into an obvious nosedive, they had not administered the traditional medicine. Andropov's new men in the Kremlin argued against it, and Gorbachev would contend that the Soviet tank had had its day as an encouragement to Communist solidarity. The Poles were persuaded to use their own tanks.

It was a break in the weather, the first clear sign that there were at last limits to Soviet defensiveness. It was just a pity that it happened when across the Atlantic a boisterous if elderly new administration was coming to power convinced that nothing had changed since Joe McCarthy was at work and Stalin was on his throne, and intending to prove it.

IV

The years 1980 and 1981 provided juicy events for someone with Ronald Reagan's simple view of international politics and compulsion to renew and win the cold war. With an effrontery that commands one's clenched-teeth admiration, Reagan and his costar Margaret Thatcher, two of the most vigorous trade-union bashers of the epoch, shed crocodile tears over the hard labor pains at the birth of Solidarity.

Reagan's real contribution to the new coalition of sunbelt rich conservatives was his talent as a spokesman for American patriotism, which he conceived of as the anticommunism he had learned as a young man and consolidated as president of the Screen Actors Guild, combined with the clean and unrealizable ethics of "The Waltons."

Reagan was a simple-minded politician who sets knotty problems of understanding. During his election campaign he came on strong, as all challengers have to, about the incumbent's failure to ward off the Soviet threat—specifically, about the "window of vulnerability" that Carter had purportedly left open. And he strummed the kitsch chords of his signature tune:

> They say the United States has passed its zenith. They tell you the American people no longer have the will to cope, that the future will be one of sacrifice and few opportunities. I utterly reject this view. I will not stand by and watch this country destroy itself under mediocre leadership that drifts from one crisis to the next, eroding our national will. . . . Isn't it time to renew our compact of freedom, to pledge to each other all that is best in our lives, for the sake of this, our beloved and blessed land? Together, let us make a new beginning.[15]

Reagan's point, and the point of Reagan, was to be anticommunist and to revive the thrills of danger and self-righteousness in a country that had been chased out of Vietnam by dinks with small arms and was now being jeered at by a lot of ragheads keeping fifty-odd honest Americans cooped up in the U.S. embassy in Tehran.

Carter had, of course, tried to get the hostages out for nearly a year. He had even authorized a doomed commando rescue raid of quite crazy logistics and battle plan. When at last Reagan began in 1986 to seem the chump he was, harsher voices among the many who called him fool claimed there had been crooked deals behind the scenes of the 1980 election campaign to ensure that the hostages were not released so early that Jimmy Carter got the credit.

During his first term, however, Reagan moved forward on the well-upholstered rhetoric that a very wealthy seventy-year-old who had rarely left the United States would be expected to declaim. In his very first press conference in early 1981 he spoke of Soviet leaders who "reserve unto themselves the right to commit any crime, to lie, to cheat,"[16] and in March he told an interviewer, in a hazily remembered, misunderstood recollection of cold-war game theory, "Let's not delude ourselves. The Soviet Union underlies all the unrest that's going on. If they weren't engaged in this game of dominoes, there wouldn't be any hot spots in the world."[17]

The president had risen on a spume of the moral and moralizing passion always bubbling in the United States. His team used the extraordinary tools of market manipulation to harmonize in a sufficient vote the diverse currents of resentment, chiliastic ecstasy, and ire then flowing through the country. Reagan's team struck old patriotic chords as middle America spluttered over the hostages in Tehran; it revived the ghosts of state socialism with its dilations on welfare and the wretchedness of overtaxation, and promised to exorcise them; it joined hands with big business and the military in an anticommunist chorus to rearm America against the dreamed-up renewals of the reliable enemy.

The orchestration of these thrilling melodies swept Reagan into the White House. Faithfully believing in his well-rehearsed songs, he sang them with conviction. When middle America bowed before him in pentecostal gibberings, Regan was dubbed the prophet of the "moral majority." His followers looked forward to the last days, when perilous times shall come, the Soviet Union will invade Israel, and Christians will be whisked bodily to heaven before the bomb goes off. As Jerry Falwell explained, "Well, nuclear war and the Second Coming of Christ, Armageddon, and the coming war with Russia, what does this have to do and say to me? . . . None of this should bring fear to your hearts, because we are all going up in the Rapture before any of it occurs."[18]

This was an America sliding steadily downward after its very recent ascent to world power. In a nation whose average age was rising steeply, in contrast to a wider world crammed with distressful children, belief in the power of its currency, its technology, its soldiery, may have been shaken but was still integral to the national self-image. A president was needed who would restore faith in these absolutes, and for a few years, the nation found one. It found a man who could sincerely combine the religiose ravings of Jerry Falwell, the foreign policy idiom of John Foster Dulles, the schoolboy economics of Barry Goldwater, and the cowboy individualism of John Wayne.

The president then surrounded himself with people whose worldviews were based on such reckless gibberish. Alexander Haig as, amazingly, secretary of state, the saurian Caspar Weinberger at the helm of the Defense Department, with his sidekick Richard Perle, Jeanne Kirkpatrick at the United Nations, all were loose-mouthed Red-baiters who never brought their imaginations to bear on the weapons they spoke of with such confident ignorance. The Soviets, stiff and scared, listened with incredulity. The colossal and lunatic machinery of American politics had hoisted this man and his wild bunch to power, and his allies and enemies alike—always excepting Mrs. Thatcher, who thrilled to him like a schoolgirl—trod for four more years as on nuclear eggshells.

23

EVENTS XI

The Evil Empire, Star Wars, and Central America, 1983

WHEN Ronald Reagan came to power and spoke so easily of an enemy who would "commit any crime . . . lie . . . [and] cheat," whose only goal "must be the promotion of world revolution and a one-world socialist or Communist state, whichever word you want to use," he was cheerfully sincere.[1] A large nationwide audience responded by rocking and clapping and shouting while Ronnie smiled his famous craggy grin and Nancy gazed at him with her famous gaze. When he announced on 8 March 1983 that the Soviets had "the aggressive impulses of an evil empire," and his opposite numbers were "the focus of all evil," he spoke to and for the viscera of many Americans.[2]

Speeches that go over well with audiences for whom, morally and emotionally, there is no doubt about who is right and who is wrong cannot be analytical or reflective; they must only be sincere. Sincerity is the master value, and only someone with Reagan's *absences*—his lack of irony, intelligence, duplicity—could speak with such confident sincerity in the world of cold war diplomacy. He combined the armor of sincerity with the sword of conviction. Thus armed, Reagan strode off to wage the cold war again. He could ignore domestic politics, about the economics of which he understood nothing, and focus on swamping the ideological enemy with waves of rhetoric, enormous extravagance in defense expenditures, and trivial but spectacular military gestures.

Utterly a product of the film and television industry, Reagan turned the resurrected cold war into the politics of spectacle. As never before, he made the national fantasy into executive action, licensing on a quite new scale the hiring of hit men and anticommunist mercenaries around the world. In the basement of the White House an ideologically fervent and slightly cuckoo young colonel named Oliver North dealt in weapons and hostages in Iran and in gunmen in Nicaragua. When he was finally booked by justice in 1987, even the hysterical show of national support for him could not hide the fact that he all but implicated the chief executive in illegal and murderous deals.

It hardly mattered to Reagan and his boys; they were waging cold war, a good cause. Reagan made a fuss of detecting the "window of vulnerability"; since his purpose was to restore American greatness by rebuilding its superiority in nuclear weapons, he hugely increased defense spending. Doing so felt right to him, and therefore to all Americans who voted for him because he felt as they did. Impulsively, he imposed sanctions against Poland for the same reason after martial law was declared there. From the beginning the Reagan administration and its clients was breezily quick on the draw; they were paying for wars in Afghanistan, Nicaragua, Kampuchea, Ethiopia, Angola, and Mozambique, and giving advice on how to fight them.

Reagan was the ideological light that shone down on this bloodshed, turning the wasteful deaths not into terrible beauty so much as into those gallant, almost bloodless deaths of World War II movies. This is the irresponsibility of ideology: it gives benediction to cruelty when cruelty is disguised as a crusader, and ideology, which has by definition no imagination, cannot see cruelty for what it is.

For all his power, Reagan was hardly his own man in this. In a chilling comparison of the world of 1911 to the world of 1983, Geoffrey Barraclough wrote in 1982:

Modern conflicts begin, not because perverse or ambitious individuals foment them, but because economic and political conditions generate the basis for conflict. No one in his right mind supposes that unrest in the Third World is fabricated in Moscow, though the Russians might cautiously (but so far not very successfully) seek to profit from it. The neurotic fear of the Soviet Union and of Communist subversion which appeared to inspire American policy under Reagan and Haig, the belief, real or pretended, that it has a finger in every pie, stirring up disaffection in Africa, Latin America and the Middle East, corresponds less to political reality than to a gathering sense of frustration, as the United States watches a world built in the American image falling apart about its ears. It is no less dangerous because of that. In 1911 Grey and his advisers in the Foreign Office in London feared that the Triple Entente upon which, they believed, their whole security depended, was threatened with collapse. Seventy years later Washington and Moscow were assailed by similar fears about the stability of the security systems they had built up; and which, like Grey, Nicolson and Crowe, they were determined to maintain at all costs. The difference today is that the cost could be prohibitive.[3]

Reagan and his crew set to work rebuilding American security systems around the world under their peaceful militancy. But the world went on unrepentently tearing them down.

By and large, he rather thought that the new doctrine of "limited nuclear war" was plausible. Or as he so lucidly put it when asked,

> I don't honestly know. I think again, until some place . . . all over the world this is being research going on, to try and find the defensive weapon. There never has been a weapon that someone hasn't come up with a defense. But it could . . . and the only defense is, well, you shoot yours and we'll shoot ours [*sic*, all the way]. . . . And if you still had that kind of a stalemate, I could see where you could have the exchange of tactical weapons against troops in the field without it bringing either one of the major powers to pushing the button.[4]

But in 1981, having sent to Geneva the tough old runt of arms negotiation, Paul Nitze, who they were confident would continue to put a stony face on things, Reagan and his henchmen were shocked to learn that the old cold warrior very nearly came up with a spectacular agreement after he and his opposite number, Yuli Kvitsinsky, had padded away for, as they amiably put it, "a walk in the woods."[5]

Oddly enough, Nitze's chance came from one of Reagan's very singular moments of dash when, on impulse, he broke free of his script and spoke on his own, without thinking whether people would follow up on what he said.

He "outflanked the peaceniks" with his impromptu "zero option," according to which all the new cruise missiles would be canceled if only the Soviets would take to pieces both their new SS-20s and the monoliths they were there to replace.[6] Whether or not Reagan meant it is neither here nor there; but his henchmen made much of the clever slogan "zero option," thereby winning a little time for Helmut Schmidt to overcome the incomprehensible resistance to cruise missiles on the part of millions of West German voters. They all knew that the Soviets would have none of it: why should they trade anything of theirs for weapons that were so unpopular on the other side already.

But in Geneva, to universal amazement, Nitze, who was well known for his bargaining subtlety and stoniness, went for his walk and agreed with the urbane Kvitsinsky to something like the zero option: no cruise missiles and only 150 SS-20s, most of them in Siberia. It was perhaps the first utterly serious and entirely far-reaching arms reduction since such negotiations began. The European governments yammered vaguely when, too late, they got the news; Weinberger and Perle killed the scheme while the president nodded off; and once more the people of Europe, the only ground zero for these flighty little weapons, got out their banners and set off for their capitals.

By the end of 1982, and in spite of Nitze's lapse, the Reagan administration had decisively reversed Carter's uncertain course and set the ship of state confidently back toward the old arctic of cold war. Defense expenditures for fiscal year 1983 were set a cool 17 percent higher than the year before, while Reagan spoke of rebuilding America's strength and putting to rights the "disrepair" in which he had found the nation's armory. He handed out an extra 25 percent of CIA subvention, all to be leveled at terrorist states and Soviet proxies.[7] The National Security Council set its tame strategists to devise a nuclear war plan that instead of resting on the fulcrum of the balance of terror would assume the feasibility of a six-month war, which the United States would win.

This insane scheme made all rational men and women in Europe quake. What it clearly meant was that this group of ideological *ingénus* was looking, not for war—much in their formulas turned on the habitual deliberateness and slow caution of the enemy—but for the resumption of incontestable American military superiority. The United States had the weapons, the strategists said, to "decapitate" Soviet command and control centers as well as the defenses to ensure that if war came the country would be able to keep its own head on its shoulders.

On such evidence, it would hardly matter where the U.S. head was since the country would long since have proved incapable of using it. But the quotidian measures of reasonable prudence were being abandoned in Washington, and in a way that caused people nightmares, given that similar frenzies had twice preceded world war. Looking back at 1983 after only a short time, it is hardly credible that such risks could be taken—as they had been taken at several of the peaks on the long contour of cold war crisis—merely because a governing class and the mass of its supporters thrilled to the shudder of danger and domination, of patriotism and power.

Reagan, well into his seventies, was the still boyish product of the era that had set America on the throne of the world, and he simply assumed the magnanimity and justice of his country's purposes. What it wanted was right and could be had for a price. He threw his weight behind the MX missile, whose fate had been left hanging by Carter, and also the B-1 bomber, which Carter had canceled. These would nicely supplement the monster new submarine Trident. Reagan authorized a return to the manufacture of chemical weapons, the first since Nixon stopped them in 1969, and gave the navy money to buy 150 new fighting ships. Making it clear that the point of all this was to find the weakest points in Soviet military and economic planning, he sought to compel the "evil empire" and, in Caspar Weinberger's phrase, its "totalitarian dictatorship," by force of arms and dollars, to contract and make a public act of political contrition.

Finally, in March 1983, Reagan announced the Strategic Defense Initiative (SDI) during an address to the nation on national security. So far as one can tell, it was all his own idea. If that is so, the celebrated "Star Wars" initiative is an object lesson in the workings of the aura as well as of the facts of power.

Back in 1966, when Reagan first became governor of California, he visited the Lawrence Livermore Laboratory, run by the still dedicated cold warrior Edward Teller. Teller, it will be recalled, testified against Robert Oppenheimer during the anticommunist purges of the 1950s, and the scientific community, whose inclinations are always internationalist, had never forgiven him. The Livermore laboratory was itself a child of the cold war; from its beginnings in 1952 it has been the design center for the development of nuclear weapons.

Teller was its moving spirit, making it intensely competitive with Los Alamos and devoted to blending science and ideology in one inventive triumph after another. A generation later, his laboratory still hires the cleverest young scientists it can find to keep this ceaseless custodianship of American freedoms going. It has plentiful bursaries from Hertz rentals and Coors beer.

Of late, recruitment has proved a more tangled affair. The best and brightest scientists do not all beat a path to the director's door, and for those who do the womanless isolation of the place intensifies "an extreme defense ideology." The laboratory's own development plan regrets that its recruitment has suffered from "nuclear-freeze ballots . . . demonstrations . . . [and] general public concern over radiation effects."[8] But when Reagan met Teller in 1966, each recognized in the other a usable resource, and they kept in touch.

Sometime in 1982 Peter Hagelstein, one of Teller's very brightest young men, completed the mathematics that would make possible a nuclear X-ray laser that could produce a powerful beam of radiation. Teller duly reported this to the president. In the meantime, in spite of the president's keenness for the new MX missile—a keenness attributable only to Reagan's unquestioned assumption that the United States possessing more and more weapons was an uncomplicatedly good thing—it had been rebuffed by Congress. A special commission was appointed to sort things out, with Kissinger and Haig as members, but its conclusions were obscure and contested.

In February, at a Joint Chiefs of Staff meeting, one of them aphoristically voiced a goal against which two generations of nuclear disarmers had chafed in Europe: "We should protect the American people not just avenge them."[9] Reagan, with his actor-politician's ear for the winning phrase, pounced on it: "Don't lose those words," he said.

It seemed that over the years his conversations with Teller had worked on an imagination shaped by the endearing and dramatic simplicities of the Hollywood view of politics. Since the advent of nuclear weapons, there has been a stream of movies in which one side clinches a weapon that will defeat all other weapons forever. Hitchcock's *Torn Curtain* (1966) is probably the best version of such a plot. The hero-scientist who invents the new weapon is an idealist; he intends that his invention shall free the world from being in thrall to nuclear terror.

Reagan, ignorant of science and simplemindedly committed to a foreign policy that accounted for only two forces, U.S. good and Soviet evil, believed that Hagelstein's laser could be this ultimate weapon, a weapon that would work not by eradicating a hemisphere but by shielding the United States from anything nuclear that might fly its way.

The new laser bomb was the product of the theory and technology behind the so-called third generation of nuclear weapons. The first had been the atomic generation, and the second, for which Teller had been the most vehement advocate, the hydrogen. The third generation was to be differently conceptualized altogether. The first two generations of weapons simply went off in an appalling explosion and wiped out everything; the new idea was to give very high intensity and focus to the weapon so that its powers of devastation could be precisely located.

Such weaponry found a place in the dangerous new thinking spreading through the United States as the cold war temperatures were so righteously lowered in the early 1980s. Fiercely accurate and powerful weapons might win wars instead of simply retaliating against attack. They might also give their makers immunity to attack, and this was the dream Reagan had his speech-writer put in his State of the Union address that year. Teller's mad scientists were to devise a weapon that, as Broad put it, "will fire radiation over thousands of miles of space at the speed of light to destroy hundreds of enemy missiles. As the bomb at the core of an X-ray laser-beam battle-station explodes, multiple beams will flash out to strike multiple targets before the whole thing consumes itself in a ball of nuclear fire."[10]

These battle stations were to circle the earth outside its atmosphere and provide an impenetrable shield for the United States against all enemy missile strikes. In 1983 Reagan assigned $26 billion over the next five years to determine the feasibility of the weapon, and the certain-to-be-underestimated estimates of the costs of research and development veered wildly upward into sums that would have bankrupted the nation (up to $1 trillion) to pay for a safety unattainable for at least twenty years. There were those who hoped the cold war might not last so long.

There were also those in Europe who angrily pointed out that any such shield might be fine for the United States but would still leave the Old World to be incinerated. There were many more—the best scientists among them— who thought the scheme was technically impossible, let alone expensive beyond human dreaming. There were those who thought the weapons might be useful additions to the armory in any case, and there were others who licked their lips at the thought of all that research money, even if the thing never worked.

Some witty journalist dubbed the scheme "Star Wars," and it is by that name that everyone, however vaguely, understands the idea. In the movie of the same name, it will be recalled that the forces of evil faced the forces of good, that mass could move at the speed of light without disintegrating, and that Alec Guinness, the magus of the good and happy prince, wielded an indestructible sword of light. Reagan himself appropriated the bit of folk liturgy that the movie made popular: "the Force be with you."

There was an uproar in the White House when Reagan's speech was circulated, with very little notice, before being broadcast. George Schultz had by this date taken over at the State Department from Haig, who had made one intemperate resignation threat too many. Schultz was a heavy, worthy, stolid sort of guy with moral qualities to match, and he was astounded by the president's move. Feasible or not, it changed all the assumptions upon which foreign policy was made—the Soviet policy as much as American policy, let alone that of the European allies. There was a little frenzy of unattributable criticism and private reaction in assorted capitals, but the power of the presidential office was such that no one could organize the consternation and incredulity into a resistance capable of stopping the president's enthusiasm in its tracks. The best anyone could think of was to go along with the idea but pussyfoot about in the hope of so delaying the start of the scheme that it wouldn't survive Reagan's term in office. At this time, he still had five full years to go.

Star Wars was more fantasy than lie. But the Soviets, terrified as always of U.S. technology, and seeing the initiative as a flouting of Nixon's 1972 antiballistic missile treaty, made an enormous to-do, thereby making the project seem the more believable, whatever the scientists said. However it was, Star Wars dominated the new crisis in the arms race and the hapless efforts to control it still plodding on under Nitze and company, so that in 1983 the feel-good president and his ineffable confidence in the white magic of science shone down on the American people, and they reflected back his sunniness to the chief executive.

They were much helped in this mutual radiation by the dependable nature of Soviet arbitrariness. On 31 August a South Korean airplane, Korean Airlines (KAL) flight number 007, veered a very long way off course, flew deep into Soviet airspace immediately over the big submarine base at Petropavlosk, and was shot down by Soviet fighters with the loss of 269 lives.

Any discussion of the incident must bring straight out that whatever the Soviets thought the aircraft was doing, their action was, like so much of their military-political conduct throughout the cold war, frightened and abrupt. If they had known the offending aircraft was a civil airliner that had got lost, their cruel reaction would be not only beyond bounds but also quite inconsistent with their policy over the previous years, since Brezhnev had come to power. But it seems likely that they lost all judgment, and further, that they were prevented from correctly identifying the plane by its technology. The Soviets thought KAL 007 was a spy plane. Why?

Here it must do to summarize R. W. Johnson's remarkably faithful inquiry, which lists the sequence of mistakes the Korean pilot, himself the airline's most experienced veteran, would have to have made if he were simply lost, as opposed to doing a little bit of espionage on the quiet for the United States.[11] Flying so very deep into Soviet space, his aircraft would have brought on bright avenues of electronic detection systems tracking the alien for all of its 365 miles of deviation. He would have been able to take back a map of the surveillance alert network for hundreds of miles, the lines of detectors winking on his copy like the streetlights of a great city. Would the greed of the international intelligence systems stop at taking such risks with a civilian airliner of a client state and squaring its ideologically soundest and steeliest pilot to do the dirty work?

This pilot scrapped his computerized flight plan, took on dangerously too much fuel, manually programmed a new flight plan, made a ten-degree mistake in course, which the two partner computers declined to identify, and switched to automatic pilot without checking that he was on course, thus breaking the most basic of civil airline regulations. Then he lied to ground control about his whereabouts, put latitude but not longitude figures into his computer, and left his cabin for *five hours* without explaining why to his officers (and they in turn had to stay mum about all these known errors). When the aircraft was hit by a Soviet rocket and flew on for twenty minutes mortally wounded, all three officers had to agree not to send out the international Mayday signal. The pilot then flew, while at an inexplicably low altitude, at speeds that might have torn the wings off his craft, and he went on doing this after his plane had been hit by a rocket from a fighter that flew all around him, buzzing him with threats that he ignored for several minutes. When he did take evasive action, he also lied about his maneuver to ground control.

The list of mistakes is too long to be credited as mistakes. Yet the honorable and persistent Seymour Hersh, writing for the other side, as it were, says that he could find no trace of KAL 007 in what the espionage companies—those Washingtonians known as the intelligence community—were prepared to tell him.[12]

Hersh believes his government. But it didn't even believe itself. Both Johnson and Hersh agree that the Soviets on the spot, in a panic at having allowed the alien craft so far into their electronic labyrinth, and certainly not of a mind to check with their seniors in Moscow, believed that they were attacking a spy plane. Moreover, both authors agree that the CIA and its naval colleagues came to the same conclusion almost at once. Even the Brookings Institution historian writes that "the United States (as evidenced in bi-partisan congressional expression as well as administration statements) seemed almost to welcome in the tragedy an opportunity to belabor the Soviet Union with hasty charges of savage barbarity."[13]

The president's team, especially through the capacious mouth of Jeanne Kirkpatrick at the United Nations, had a sanctimonious field day. Yuri Andropov, an impassive man in poor health, in return accused the United States of bad faith, describing with some justice all the Reagan administration's policy as a "militarist course that represents a serious threat to peace," and accusing it of "trying to dominate the world . . . without regard for the interests of other states and peoples."[14]

When, in the dark over Petropavlosk, the Soviet fighters locked finally and lethally onto the tail of KAL 007, the Korean Airlines pilot—a war veteran as well as the legendary "human computer" of the airline—dived suddenly and violently. The Soviet pilot got his order from the ground and fired his heat-seeking rocket, which hit the rear of the 007 fuselage, ripping a huge hole in the passenger deck and the hold. Then the aircraft settled for twenty full minutes at a breathable height, and in the savage cold the survivors hung onto some faint hope as desperately as they clutched the sides of their seats. Then they crashed, and every one of them was killed. Nobody knows who found the black boxes with the flight officers' taped voices on them. Or if they did, they only want them forgotten.

II

At the end of 1983 the Soviets withdrew from the arms limitation talks; cold war relations were freezing harder than at any time since 1962. A month after

KAL 007 went down, there were more nuclear disarmers out in Europe than ever before. England witnessed its biggest political demonstrations of all time.[15] Before Christmas the United States began to uncrate the new weapons that the disarmers had vowed to send back.

Some of these, as I said, were to be set up in little nests in Sicily, others in Holland, and West Germany, and some in an airfield outside Newbury, a modest English market town with its own smart racecourse. The airfield was sited on a sandy, wooded stretch of country known as Greenham Common, and as all admirers of that historical chestnut, the freeborn Englishman, will know, common land is exactly that: land commonly owned by the people of England.

So it came to pass that a group of freeborn English *women* came to camp at Greenham Common beside the grim wire and its high concrete posts. They, too, had come to bear witness against the evil weaponry being uncrated inside the big green bunkers, but they came speaking a strange idiom. They contended that the filthy weapons were the product of the politics of *men*, and they spoke for a womanly politics that refused the steel phallus and its poisonous ejaculation of old militarism. They lived by a communal, even, for their enemies, communist sisterhood of values of which love, mutuality, and a welcome to strangers would be the signatures.

The women built temporary bivouacs, long polyethylene sheets slung over the plentiful hazel and larch of the common. They fastened tokens to the wire—flowers, broom, dolls, baby clothes—and from time to time they implored the guards and the police to join them in order to renounce their guardianship of death. Their tongue was ancient, magical, pentecostal. Occasionally the law-abiding burghers of Newbury dispatched bailiffs to evict the women for trespass or to hose them down for obscure offenses; but the women of Greenham Common took their stand on ancient statutes and legal customs and always came back.

1983 was their big year, too.[16] The numbers around the fires of their little camps multiplied, and some came for a night or for a week, and some left their children in the care of friends or loving husbands and stayed for months. They were called gypsies and much worse by Britain's revolting daily press, as visitors came to declare sisterly solidarity from all over the world. At Easter on All Fools' Day that year the whole Campaign for Nuclear Disarmament came to join them, and in bitter cold weather the disarmers joined hands in an enormous ten-mile circle all around the airfield circumference. Then they blew on chilled fingers, ate their sandwiches, and went home by bus, leaving behind the always vigilant women.

Just before the day-trippers left, three tall figures appeared, marching, trampling at a great pace up the hill and along the flat toward a crowd of disarmers milling around the buses. Three 6-foot-tall women, handsome and stone-faced, staring straight ahead, marched at a cracking pace, their six hands gripping at waist height a board with the bare legend, "Whores Against Wars." The crowd fell back in awestruck silence and watched them dwindle down the long road.

|||
_____

The world was a playground and Reagan was its king, especially that part of the world that North Americans so inexcusably call their own backyard. Nothing angers Washington more than things going wrong among its Latin American neighbors, particularly when the neighbors harbor thoughts of overturning one or another of their strikingly horrible governments or take on other such Jacobin projects.

Cuba, Grenada, Nicaragua, El Salvador, Panama: barely ten million people altogether, and those people living in conditions of such poverty that you'd think all the United States should want to do is distribute dollars and food from buckets in the market squares. But no. The Cuban question seethes in the hot nights of Miami; Grenada was made over by Reagan's men one day in 1983. And the self-styled Marxists of Nicaragua and El Salvador were an open challenge from over the backyard fence to the new Crusade. The rest of the world had hardly heard of either country, but in concert with Alexander Haig, Reagan rhapsodized about the danger to the biggest and richest country ever known from a handful of Communists who could hardly feed themselves and were not so much Marxists as followers of a dauntless Robin Hood from the 1920s, August Cesar Sandino.

Sandino was a Nicaraguan general who back in 1926 refused to accept that the North Americans should be running his country. All the other leaders of civil war factions had obediently deferred to Calvin Coolidge's Henry Stimson. But Sandino took to the hills, and the U.S. Marines followed in hot pursuit.

They never caught him. The Marines trapped him up on this plateau or that, but when they arrived, Sandino had always departed down a secret path, leaving nothing behind him. A *Nation* journalist found him in 1928 high up in the rain forest, in every one of his sixty-five inches a guerrilla leader, with polished puttees, well-cut breeches, and dark service dress jacket with deep skirts, a red-and-black foulard, a stetson, and sad, unfathomable eyes. But the first Somoza, then head of the army, caught Sandino off his guard dining with President Juan Batista Sacasa in 1934; Somoza lined him up in front of a firing squad while he himself went to hear some poetry, a nice instance of politics' consanguinity with culture.

In 1928 Sandino spoke in the pages of *The Nation* straight to the heart of the leftist. In 1983 he might have loaned the same words to Ronald Reagan.

> If the American public had not become calloused to justice and to the elemental rights of mankind, it would not so easily forget its own past when a handful of ragged soldiers marched through the snow leaving blood tracks behind them to win liberty and independence. If their consciences had not become dulled by their scramble for wealth, Americans would not so easily forget the lesson that, sooner or later, every nation, however weak, achieves freedom, and that every abuse of power hastens the destruction of the one who wields it. We march to the clear light of the sun or to death. But if we die, the *patria* lives on, indestructible. Others will succeed us.[17]

It is not hard to see why Nicaraguan rebel patriots took to heart the rhetoric of Marxism and the name of Sandino with which to oppose the rule of the three Somozas. In the 1930s the first Anastasio Somoza, corpulent and cruel, given to unspeakable orgies, favored a Nazi victory in the world. After twenty-six years of Somoza's uninhibited cruelty and indulgence, a poet had the poetic flourish to shoot him dead in a grand hotel. His two sons succeeded him one after the other. The second was as fat as his father. When he went to meet his sumptuous mistress Dinorah at the airport, his wife drove out of the country in a huff.

In 1972 an earthquake demolished the capital city of Managua, and the third Somoza did nothing about it except to steal all the goods-in-aid and sell them off, while his countrymen starved and caught cholera from the poisoned water.[18] But Somoza was just fine. His enemies, especially if they edited critical newspapers like *Le Prensa*, died quite often and suddenly. The Sandinista rebels were in the hills, they were in the suburbs; Somoza was America's man, but Carter thought he was due for replacement because he was bad on human rights; meanwhile Somoza lived in a bulletproof cubicle.

His National Guard had, by 1978, become tyrants. Ill-advisedly, they executed the ABC journalist Bill Stewart by making him lie on the red earth while they shot the back of his head off. This was filmed by ABC, and Somoza was distressed. His embarrassment was compounded by Sandinista victories, and the open support of the people. On 17 July 1979 Somoza took a very large, discreet car at hell's pace to a personal aircraft that departed from Nicaragua carrying him, Dinorah, the embalmed bodies of predecessor Somozas, and the country's liquidity realized as several millions of dollars in crackling new rolls. A year later some sportsman blew Somoza to pieces with a bazooka.

You would think this was the happy ending: the last Somoza was dead, and the Sandinistas could lead their liberated people into a poor but honest future. But this was Ronald Reagan's short sweet hour of history; if there were Marxists in Nicaragua and the backyard of the United States, they had to be eliminated.

The Sandinistas did put their foot in things. They taught an unadulteratedly Latin and revolutionary language, they taught that the revolution would bring the good society and the happy future. They made avid supporters and deadly enemies. The enemies took to the hills and became counterrevolutionaries, *los contras*. They were trained by the CIA and U.S. advisers and paid for by slush money, by private subscription, by Congress itself. Reagan fed the contras out of the back of his hand, put a trade embargo on the country, cut aid by two-thirds in two years,[19] and brought the country's income below that of Haiti, the heart of Caribbean darkness.

The Soviet Union helped the Sandinistas very gingerly, the Cubans more vigorously. They both sent tractors, ambulances, a few doctors, some weapons, pharmaceuticals, and lots of copies of *The Iliad* and *The Odyssey*, which were sold in the supermarkets. Understanding sieges as they did, the Nicaraguans would only buy *The Iliad*. The United States sent a good, quiet American named Tony Quainton, to be ambassador in Nicaragua. He was tidy and honorable. He told his seniors that Daniel Ortega, leader of both the Sandinistas and the country, was a decent man, that U.S. conduct toward Nicaragua had been unforgivably harsh, punitive, and mean, and that the best thing for his country to do was to be generous and kind with the new government as its great traditions enjoined.

In 1983 Reagan appointed the Kissinger Commission to inquire into "Communist encroachment" in Central America, Kissinger, who in 1973 told Allende's secretary of state that South America was a nothing. The ambassador gave his opinion to Kissinger, who disliked it and him and Daniel Ortega. The ambassador was sacked as part of Kissinger's comeback.

Before the year was over, the United States mined all the harbors in Nicaragua and sent a CIA gunboat sailing into the offshore port of Corinto to rocket the oil depot so that three million gallons of fuel went up in an orange ball of fire.

Daniel Ortega and his beautiful partner Rosario Munillo held the loyalty-to-death of the best of the young men, in this case, the secretary-general of Ortega's foreign ministry:

> That there be no SS-9 missiles here pointing at Oklahoma is a reasonable request. That there be no foreign bases here, no breach of international conduct. That we live and let live, as we expect the United States to. We

challenge the traditional view, now sustained by Henry Kissinger, that we are a backyard. We are not, and we are also no threat to anyone. Our revolution was in the first instance a creature of bad American policy toward Nicaragua. But that revolution has happened now, and we are better for it. Our independence is our own business. Will the United States recognize our right to live? This is what peace in Central America depends on. Even now, even in this moment of terror between our two countries, the contacts are strong between the United States and Nicaragua. We approach a fork. We are closer to war and closer to peace than we have ever been before.[20]

But after Kissinger, after the mining and bombing of the ports, after the U.S. trade embargo, after experiencing little Communist solidarity, the Sandinistas hadn't much of a chance, even after the Senate voted in 1984 to cut off funds to the most flagrantly overt of the covert operations assisting the contras. Finally in 1990 the Sandinistas lost an election, and to everyone's relief socialism was put away and normal deference to the United States was resumed. There were still plenty of leftists to kill in El Salvador, some of them Catholics also, but Nicaragua was another small but definite win for Reagan's cold war. After Guatemala and Chile, Reagan had his very own Grenada and Nicaragua. The continent would be safe pretty soon.

24

Mr. Gorbachev Goes to Reykjavík, 1986

T HE foreign policy of the Reagan years turned out to be an unappetizing mixture of grudging hypercaution at the arms limitation negotiating tables, reckless and unendearing braggartry in front of the microphones, and minor acts of war that combined bullying and cowardice in about equal proportions.

For some time all the officers of state and security had sustained the rhetoric of cold war by attacking Soviet clients and "proxies" for their commitment to the forces of progressive revolution (in the Leninist idiom). Thus the most baffling and repellent and yet somehow ignorable phenomenon of international politics in the era, the advent of a world structure of terrorism, was taken as another sign of Soviet fiendishness and ingenuity, for wherever terrorists struck, you could be sure there were Soviet rubles and bullets in their pockets.

For many years and in many cases, there certainly were. Soviet support for insurgency was a policy as consistent as U.S. and British support for counterinsurgency. What is more, Reagan turned domestic moral passion into fuel for a new cold war so completely because the Soviet Union was playing such a buoyant march. It had its new weapons, the SS-18 and SS-20, it was providing the rebels in Africa and elsewhere with ample ammunition, and it was parent and model to the new nations of the Third World. Brezhnev could happily endorse the proud claim of the *History of Soviet Foreign Policy* (1981) that "the Soviet Union is one of the greatest world powers without whose participation not a single international problem can be solved."[1] There was less economic inequality in his country than in the rival camp, the streets were safe and clean, foreign policy had a new adaptability, and Afghanistan was only backyard stuff.

Such a view barely lasted the decade. But it gave the competition its bounce as well as its frustration. In the 1980s the United States wanted to reassert its imperial reach and authority; what better target than terrorism?

On the scale of human nastiness, the U.S. bombing of Tripoli and Benghazi in April 1986 was a small viciousness; at that moment Soviet bombing and strafing of rebel Afghans was killing uncountably more innocents every day. But the bombing of Libya provides a measure of what the cold war had become by 1986. It was an act of cruel, mean-minded oppression, not unlike U.S. conduct toward Nicaragua, El Salvador, and Cuba, conduct of which the liberal superpower will one day be ashamed.

The Libyan bombings were a response to a terrorist act in Germany in which a bar was bombed and an American serviceman killed. A Turkish woman was also killed, but her body was not weighed in the balance of reprisal. A Libyan hand was detected in the murder, though no evidence was ever produced. It was enough that Col. Muammar Qadaffi, the volatile, offensive, handsome Libyan leader, had called irresponsibly for terrorist acts against Israel and its American mentor, and that he had a negligible air force of his own.

The terrorist is the opposite of the demonstrator in contemporary politics. Demonstrators bear their bodies as witness of their helplessness and as signature to their public disobedience.[2] As the police understand, they contradict this meaning the moment they are swept into an act of violence. The terrorist is well named. He is masked. His actions, especially his bombs, take place in his absence and are so broad as to be meaningless. (The people killed can do nothing for his grievance.) His actions simply deny meaning; sudden meaninglessness is the precondition of terror.

What terrorism causes, and is intended to cause, is vengeful rage. The offended state taking revenge is thereby shown to be brutal, as its detractors claim. Thus terrorists recruit support to their cause.

The United States had borne with little patience the capture of its embassy in Iran, the suicide bomb that killed the Marines in Beirut, the several hostages taken by nameless and incomprehensible groups in the Middle East. The Palestine Liberation Organization (PLO) had for years terrorized the outskirts of the state of Israel, well beloved of so many Americans, Jews and otherwise. The Libyan demagogue Qadaffi was doubly to be excoriated as a professed Marxist and a radical Muslim. He hated the United States and said so. In 1984 he helped the striking British miners, socialist enemies of Mrs. Thatcher's benign revolution of the right. It was time to make an example of him, to make somebody pay for all this mayhem. And in any case, the Soviets were sure to be behind it all.

So the Reaganites worked themselves up into a fine moral stew and telephoned Mrs. Thatcher to ask her permission to use British bases from which to fly the sorties. Alternatively, the White House called to say that was what it was going to do anyway, so perhaps she would be good enough to agree. It called socialist France and socialist Spain as well, but those two nation-states declined the use of their airspace for the mission. So the F-111 bombers took off from Fairford in Gloucestershire and flew down the European seaboard, turned east past Gibraltar, refueling constantly from the thirty massive KC-135 tankers that flew with them, and bombed Tripoli.

The bombers were intended to make the sort of "surgical" strike supposedly perfected in Vietnam whereby small targets are "taken out" accurately. In Libya, Qadaffi's house, the Muslim League barracks, a couple of airfields, and what the CIA reckoned were terrorist headquarters in the posh end of the palm-lined Tripoli boulevards were to be cleanly excised.

The surgicality, as the Vietnamese had found, was relative. Sixty-three people were killed, most of them harmless Tripolitans, one of them Qadaffi's adopted baby daughter. The honest colonel was asleep in a tent elsewhere. Only Britain, Canada, and Israel commended the United States. The episode caused collective feeling in Europe to be put off by American leadership faster and farther than ever before. It became obscurely apparent that the whole operation had as much to do with boosting American public opinion and showing the Europeans who was in charge, as it had to do with disciplining Qadaffi's sillier excesses.

Only the English right blew the old Atlanticist trumpet. In the mother of parliament's, parliament's mother-in-law was at her rhetoric again. Mrs. Thatcher said on 17 April, "Terrorism has to be defeated. Terrorism exploits the natural reluctance of a free society to defend itself in the last resort with arms. Terrorism thrives on appeasement."[3] Like all her Conservative predecessors, she invoked appeasement because appeasement, by calling up the moment of Munich in 1938, condemns all those who are reluctant to indulge petty acts of war to losing the Second World War all over again.

The Soviets put it all down to U.S. impatience with its loss of world power. By 1986 Reagan's casual overdraft on his country's production was beginning to run unmanageably away, and the United States was beginning to assume the imposing role of the biggest debtor in history. So the Soviets had a point. But far more to the greater historical point was the Soviets' own economic catastrophe, long and slow in its accumulation and now piling its debris higher and higher upon the utter irrelevance of their economic plans.

II

Time was running out faster for them. Brezhnev died in 1982. His economic program had long since settled into immobility, and his last years, though they included quiet, stolid, and traditional foreign policy, were characterized by corrupt and wholly useless domestic policies. Andropov, his successor, knew how matters stood and made some small moves. During his party career he had been head of the KGB back in 1967 when he met the regional first secretary to the party in one of the prettiest, most fertile regions of the country in Stavropol. Mikhail Gorbachev, then thirty-six, was formidably intelligent and, as everyone has reported ever since, formidably charming. He had a degree in law from Moscow and another one in agronomy from Stavropol; he had highly respectable family connections in the party and a beautiful wife with a first-class degree. Lastly he held a strikingly clean record during his rapid promotion, with no taint of the corruption that poisoned almost the whole Brezhnev bureaucracy and that Andropov himself was committed to expunge.

Gorbachev was successful all the way. He used unorthodox methods of incentive payments and got marvelous results from his region's farmers. He supplanted this method, when it became ideologically doubtful, with mass cultivation teams of tractors and harvesters like those of the American Midwest, stately squadrons of the great mastodons advancing in line across the steppes. He was promoted to national minister for agriculture in 1977, aged forty-six. His first national harvest was the biggest in history. Even when the subsequent harvests went wrong, Gorbachev remained on top of the job, driving for the reforms in roads, transport, and storage that would enable the people to at last be decently and regularly fed.

When Andropov succeeded Brezhnev's inert rule, Gorbachev began reforming and reinvigorating the collectivized farming system with teams of autonomous, free-bargaining workers under leaders they elected for themselves. But the results were disappointing. At every turn his reforms came up against the incompetence, obduracy, and vile inefficiency of the party itself. He saw that national reform must wait on party reform. It was a point sharply made when Andropov died of kidney failure after less than a year in office and was succeeded by Konstantin Chernenko, frail, wheezing with emphysema, and at seventy-two the youngest of a group of geriatric party leaders ranging up to the age of eighty-three, all of whom were deeply marked by the mad fearfulness and iron caution that Stalin and cold war had brought to their lives. If Gorbachev was to make the good society out of the Soviet Union, he had to replace these strong, grim old men and the worldview they had imposed upon their people.

No doubt, the new class of which Mikhail and Raisa Gorbachev were such glowing examples owed the old men much. They were safe from any foreign enemy, adequately fed, decently educated. Their country had fine athletes, noble orchestras, and brilliant dancers. Stalin was a dim memory, the KGB only a dim nuisance. For the Gorbachevs and the people like them, the Soviet Union was ready to be set free, and they were the people to do it.

They were products of the party's successes, and the postwar surge of economic growth that followed Stalin's ruthless reindustrialization. They had seen what Soviet technology could do when applied with dedication to a single area of production: Gorbachev was twenty-six when *Sputnik* went into orbit. They also had a less constricted, suppler sense of the world being more mobile than the previous generation could ever feel. For the old ones, for Brezhnev, Chernenko, Gromyko, the frontiers of the Soviet Union were bare, wind-swept, dangerous places, places to be patrolled and wired. For the new class, frontiers were crossed by international airlines, television broadcasts, by fash-ion, by money.

These intelligent, literate, and undogmatic new people also knew that something stagnant had settled on Soviet life. Heavy industry could not shift itself into robotic or computerized gear; there were food shortages, fewer babies, more drunks, more illness. It was mad that the army should have been making consumer durables like refrigerators for years, madder still that noth-ing could be mended at home for lack of spare parts, and that jeans were exchanged as black market currency. It was maddest of all that the whole country should be forced into deprivation to pay for weapons systems nobody would ever bring themselves to use in order to defend against an enemy with the will and the cash to go on forever.

Reason and revolution were good words for these new managers of old socialism; they would retool the old system without having to defeat capital-ism. They would do it, the new men picked out by Andropov during his KGB years—Ligachev, Gorbachev, Aliev, Chebrikov, Shevardnadze, and Gorba-chev's own recruit Yeltsin—by combining personal charm with Marxism-Leninism, domestic probity with party faithfulness. They would plan the economy and keep people working, but not in the old brutal way. They were faced with very high staffing levels and low productivity, shockingly bad workmanship, often as a result of trying to speed up production, and, of course, a theoretical refusal to acknowledge the benefits of market energy.

When Gorbachev beat his rivals to the leadership in 1985, he forswore

some of the central axioms of Marxist economics. He declared, what was plain
enough, that despite its economic difficulties with debt and its moral difficulties with injustice and unemployment, capitalism still had a lot going for it. In
a famous speech at the twenty-seventh party congress in 1986 he told his
audience, in the same hall in which Khrushchev had attacked Stalin thirty
years before, that the Soviet political economy had to let prices float to market
music, that farms had to be allowed to compete, that wages should reflect the
amount and quality of work done, and that bankruptcy should be the just
consequence of inefficiency.

The crowded hall heard him out to the end. They sat there in complete
silence. Gorbachev stared at them. Then he said simply, "There is no other
way."[4]

As everyone in the West sensed, the man was a hero, as momentous for the
future as Stalin, and the tyrant's moral opposite in every way. Gorbachev saw
the way forward with such calm and clarity, but he also saw behind the
present conditions the set of causes and connections stretching interminably
back. Therefore, change the farms in a region, and you must change the
working rules; change the rules, and you must change the local party; change
the local party, and you must change the national party; change the party, and
you must change the nation's whole political economy, its theory of life;
change that, and you must stop the cold war, singlehanded.

III

So Gorbachev came to the summit meeting in Reykjavík calm but knowing
of his nation's desperate plight. Reagan came to Reykjavík in his usual jaunty
spirit, not briefed for anything more than the usual photo opportunities.

The Iceland summit took place 11–12 October, only a month before the
news broke about the White House's diversion to the contras of funds secured
by the sale of arms to Iran. All kinds of scandal blew about the White House,
much of it politically unimportant, but all of it breathing out a halitosis of the
political spirit, worse, an incompetence, rottenness, and sheer pointlessness
that bore comparison with the most stagnant years of Brezhnev's rule.

The two leaders came to the little island in mid-October, when the winter
was beginning to close mildly in upon the square stones of its granite wharves
and the sea was turning inky black under the heavy clouds. Above the little
capital and its small gothic town hall stands the tidy Hofdi House hotel with
its battlemented facade, dull rooms, and cruce concrete steps in front of its
porch. It was a humble enough spot, halfway between the great powers on a
historically neutral island.

Garry Wills, Reagan's half-grudging half-admirer checks in the contradictory baggage of the Reagan team on 11 and 12 October 1986.

> The summit had all the major elements of the Reagan presidency encoded
> within it—a public relations life of its own, maintained apart from what
> really went on; a confusion, even among those most central to the event,
> about what *had* gone on; a fight by Reagan's attendents for the prize of his
> attention, and a horror at what he finally attended to; contradictory urges,
> afterward, to deny and to defend what happened; a juggling of prior secrets
> in the race to catch up with breaking troubles. This was the culmination of
> Reagan's entire time in office, the summit toward which all the conflicting
> tendencies of his administration had been laboring upward from different
> directions.[5]

The purpose of the meeting was, in the view of the president's posse, to
reassure a jumpy Europe and to make Americans feel good about the new
Soviet leader, about whom Reagan so sincerely felt good himself. This mush
of amiability would surely produce enough topics for a brisker meeting under
sharper eyes in Washington later on.

The action itself was hardly reassuring to *anybody*. There is an eloquent
photograph of the summit's closing moments. Reagan looks every one of his
seventy-five years, his lips drawn grimly inward on shrunken gums, the skin
of his neck slack and lined, his expression plain *dumb*, as though he has
understood nothing of what has gone on. Gorbachev is smiling with a desperate sadness, his eyes fixed on the distant vision of what they have both lost.

It is hard not to be partial. The Soviets had come ready to bid so much,
for behind them so much was poised to be lost. One day it may seem that the
mutual failure to grasp, not the generosity of the Soviets—they had everything
to lose—but the lavishness of the opportunity that fortune had turned up was
what doomed the Soviet Union to its dark 1990s. A quick, sure agreement at
Reykjavík might just have given its economy the boost it needed. The United
States had, after all, openly committed itself to achieving nuclear superiority
and then aimed even beyond that to Reagan's Star Wars scheme. The latter
was what menaced the changed Soviet Union so absolutely, as Gorbachev
knew.

The Soviets were, it transpired, ready to bargain away all their strategic
weaponry if only the Americans would call off Star Wars. Oddly enough, even
though Gorbachev and his advisers obviously did not believe that the laser
technology could ever work within a price that would make it worthwhile,
they took the scheme seriously and foresaw their own weapons systems being
drawn inexorably into a new arms competition. Presumably across Gorba-

chev's pate fell the shadow of his military. They insisted that, whatever the cost, new weapons must be found to match the new weapons of the enemy.

To begin with, the discussions took up themes from the Nitze and Kvitsinsky double act in Geneva. The easiest question was going to be settling the future of the in-between class of missiles known as intermediate-range nuclear forces (INF), which included anything that couldn't cross half the world or that only launched itself from a tank. The INF class also included the objectionable cruise missiles when they'd only been in their dens for a year or so.

Reagan, Schultz, and Nitze were the U.S. top three; Gorbachev was supported by Shevardnadze and Admiral Akhromeyer, each of whom Nitze strongly approved. But the Americans had not rehearsed much. Reagan did a brief and flatfooted speech about Soviet disregard for human rights, and then, after a certain amount of the technical stuff, which is usually enough to fill up official communiqués on these occasions, Gorbachev, in his first direct encounter with the American president, presented a short, breathtaking paper.[6]

To begin, he proposed, with perfect seriousness, halving strategic weapons. Gorbachev knew, of course, that a quartet of the elite unemployed of former administrations—George Kennan, Robert McNamara, George Ball, and McGeorge Bundy—had canvassed such an idea vigorously in the recent past. He further proposed the *complete* eradication of the middle-range weapons as a preliminary to a new and comprehensive nuclear test ban, which would even include the little bangs Kennedy and Khrushchev had still allowed themselves after the 1963 treaty. All these astounding proposals were conditional upon Star Wars research being confined strictly to the laboratory.

Reagan was a man for whom the immediacy of his responses and intuitions as he met and touched other people directed his primary judgments. He liked Gorbachev, and seeing Gorbachev's solid frame and manly good humor, Reagan moved spontaneously to meet the other's openness. He wanted America to rule the world, sure, but that could be done on the strength of the country's wealth, freedom, confidence, its sheer merit. In the meantime, let's use our huge political power to do away with these weapons that do nothing except terrify us all. Let's do it. But, of course, I'll hold on to my dream of a perfect defense canopy, the magnificent inheritance under which American children may be forever happy and untroubled.

Reagan's assistants were caught up in the quiet, intense excitement of the moment. Hardball player Richard Perle gave a ten-year timetable for eliminating all ballistic missiles. The Soviets looked doubtful. Their strength lay in the big stuff; what about cruise, only minutes from East Germany?

The subordinates talked all night. Everyone was tired to death, and the meeting was due to end soon. It was late afternoon on Sunday, 12 October.

Gorbachev proposed the destruction of all nuclear weapons. Reagan came gallantly up to scratch.

"All nuclear weapons? Well, Mikhail, that's exactly what I've been talking about all along . . . get rid of all nuclear weapons. That's always been my goal."
"Then why don't we agree on it?" Gorbachev asked.
"We should," Reagan said. "That's what I've been trying to tell you."[7]

It didn't happen, of course, Reagan had no authority to do such a thing; Congress would have forbidden it, the nation itself would have forbidden it. But as the light failed on an autumn afternoon in the Arctic Circle, there shone out the possibility of an end to the diplomacy of terror, an end to the cold war itself.

IV

I cannot doubt that it made a difference to that impalpable presence, the feeling of the time. If the two most powerful men in the world could hold in their hands even for a moment, the reality of nuclear disarmament, then the world could indeed change itself.

Narrowness, bigotry, cant, and misrule quickly piled on the event, transforming it from a moment of revelation into a farce. There was a relief in this bathos. The vision had gleamed before them for a moment, then gone out. The foundations of political diplomacy had shifted momentously, and the people on the spot felt a mixture of shame and exhilaration. In this state of overexcitement, the hack negotiators began to rewrite and represent the meeting for the public's comprehension.

They rewrote it in the first instance, however, for senior colleagues at home. It had been, after all, a narrow squeak. What if the Reagan party had come home saying they had agreed with the Soviets to eradicate nuclear weapons within ten years? *What if they had?* Imagine Gorbachev in Reykjavík. He is quick, intelligent, opportunistic. He takes risks with a bright eye and an unusual openness of disposition for a man who has been a politician all his life. He is both shrewd and sympathetic in judging others. He knows something, not everything, of the economic disaster in front of his country. At home is his political father, the patient, impassive Gromyko, a patriot and a party man. Gorbachev is still a little in awe of him. Gromyko would never take such a gamble. Gorbachev does, the President's face brightens and opens; he believes with all his heart in abolishing nuclear weapons. He agrees.

Reagan's national security adviser, John Poindexter, had been watching the baffled television journalists prating into their cameras, each report contradicting and obscuring the last. As soon as the summit team arrived back home, he seized the president for a moment alone.

"We've got to clear up this business about you agreeing to get rid of all nuclear weapons."
"But, John," replied Reagan, "I did agree to that."
"No," persisted Poindexter, "you couldn't have."
"John," said the President, "I was there, and I did."[8]

Incredulity was the standard response of the professionals, agitated evasion the response of those assigned to cover up for Reagan's goofs. In the middle of it all, the president was smiling benignly, believing sincerely he had done his best but had been turned down by the men behind Gorby who mysteriously didn't want his America or anybody else in the world to have a nuclear-safe home.

The European prime ministers and heads of state watched the garbled news with consternation. The NATO generals—one imagines them crowding around a TV in the mess, shouting at the set in unbelief—were very shaken. Faithful cold warriors like Mrs. Thatcher did not know what to say. The eponymous Larry Speakes, Reagan's press spokesman, first told the press corps that the president "misspoke," second, that he had been "misquoted," and finally, that he had been "misunderstood."[9]

After all the yammering, the Reagan team pulled itself together and translated the confusion as a diplomatic success and a propaganda victory. The presidential team had kept cool heads by holding on to Star Wars, which had terrified the Soviets into offering *any* concession for the sake of killing it off.

It was about this time that the baseball metaphor "spin control" got out. Spin control means throwing out political messages with just the right degree of curve on them to exclude the possibility of a clean hit by the wrong bat but to ensure they land smack in the mitt. Spin doctors are the managers of falsehood and the pilots of ambiguity; once a political message goes astray, everything about it may be "plausibly denied."

After the shock of Reykjavík, the NATO allies bore down on the rogue president and settled the status quo and its iron curtain massively back in place. The French and the British, whose silly little nuclear forces Gorbachev had blandly counted out of the reckonings, were the most vociferous about the impossibility of contemplating such a huge change. The NATO generals wriggled in discreet discomfort; their defense ministers met on the grand old Scottish golf links in Gleneagles and tried to soothe all the alarmed bosoms.

Gorbachev kept after his goals. He said repeatedly that the ideas expressed at Reykjavík were indeed "unorthodox," but that his own concessions were considerable. He expressed surprise that even the West Germans were rattled at his proposal to clear their country of middle-range weapons. But he nailed on Reagan and his most devoted followers in NATO the responsibility for losing the chance to change the world. The peoples of Europe were as divided and confused as they well might be after such a public commotion, after so much contradiction and revision. The peace movements were confidently reproachful of their national leaders.

But the disarmers must have sensed that even if the chance had been missed, after so many years of dour immobility, of piling up mountains of weapons, of millimeter yieldings at the negotiating tables, something tremendous was on the move. Whatever Gorbachev was doing, the epoch's frame of feeling was widening and lightening with every passing moment. The people who cast that frame of feeling and set it in place, who lived firmly within its space, had become old and crabbed. Another way of seeing and feeling was coming in like a tide, filling people's veins and clearing their sight. That tide would melt the icy fields and flow unregardingly past the iron curtain, leaving it stranded and rusty, the new waves bustling past to the shore.

25

The Secretary-General—Lord Peter Carrington

SOMETIME in early August 1944, not long after Frank Thompson's execution, while his brother was battling up the spine of Italy in a tank and George Kennan was in Washington wondering what on earth to do about the Soviets when the Nazis surrendered, Peter, sixth Lord Carrington since shortly before the war, borrowed his commanding officer's car and, with three boisterous friends, went swanning off, via Chartres and Versailles, along the Champs Elysées themselves.

Peter Carrington was, at the time, twenty-five, an Old Etonian, and a spry young Grenadier Guards subaltern with a good tank war behind him and in front of him, and plenty of champagne to drink at the Ritz.

He was, is, an officer and a gentleman. But like all his generation who went to war, he was deeply affected by the brave civility of Britain's civilian army, its calm competence so painfully acquired, its detestation of much of the old ruling-class tarradiddle about good form and silly protocols. That army would return the new Labour government in 1945. Peter Carrington not only had learned its excellently egalitarian lessons even in the Grenadier Guards but he stood in his own family's decently liberal line: the dynasty included some Whiggish members of parliament, and his father had been educated by accident in the deplorably leveling, even Jacobin surroundings of Melbourne Grammar School.

The civic tradition he inherited, his lively sensibility, and a comradely militia, all made the young Carrington as much a European in his class formation as the Thompson brothers were in theirs. He saw, with keen and sympathetic eye, the gray-faced survivors creep slowly from their hovels and bombed-out cellars in the ruins of Cologne. He also saw the helpless Soviet prisoners of war, sometime slaves of Hitler, desperate and terrified at the prospect of certain liquidation when they were sent home. He saw his brother officers, hardly less helpless, forcing these wretches into repatriation with a mixture of compassion and ruthlessness.

> When I look back and recall the twin steeples of the mighty cathedral of Cologne rising from the rubble of a city in which it appeared to be the only building standing; when I remember the desolation of Europe forty-three years ago, the populations starved, demoralized and apathetic on the one hand, filled with hatred and the desire for revenge on the other; when I remind myself of the vicious wounds inflicted by the nations of Europe upon each other in those appalling years, I marvel at what has since been achieved.[1]

Forty-six years later, he said to me: "It's hard to remember it now: the muddle and the great multitudes of refugees, people starving, the ruined cities. Certainly nobody was seeking *more* confrontation. The Russians came on with such outright hostility, they refused any cooperation at the United Nations. We might call it paranoia on their part now, but they *did* mean it, you know—or Stalin did; he thought, he spoke in terms of ultimate Soviet, or Marxist, domination.

"In the Army, and at home as well, I think, there was this great residual feeling in 1945 that the Russians had carried the brunt of the land war, but there was also the feeling—well, the reality, plenty of people in uniform experienced it directly—that the Russians were all woodenly hostile. We just hoped it would go away.

"Of course, it didn't. But we were all war-weary. I suppose it wasn't very admirable, but the *prospect* of standing up to the Soviet Union in 1945 was just too much. It wasn't on. So the cold war was a relief. You thought, well, let that hold the line while things settle down. And I don't think the British people would have *trusted* the Tory party to hold off the Soviets. It took Ernie Bevin and Attlee—the most *sensible* man I ever knew—to work with the Americans and make the new defense arrangements work. And work they did."[2] As he has written:

> It is probably simplistic to say, "NATO has preserved peace"; that makes a lot of assumptions about Soviet intentions at particular times which are

equally difficult to prove or to deny. But what I find demonstrably true is that NATO, by its combined efforts, has produced a certain balance, a certain Western strength and confidence, which cannot have failed to inhibit any temptation to adventurist Soviet policies at least in a westerly direction. I think it was Palmerston who said that Russian expansionism is like water running downhill. Checked, it may find another way; but it can be checked. To check it requires adequate strength and evident will. NATO provided both, and continues to do so.[3]

Peter Carrington is a chip off a very ancient English block, but for all that, an accessible, warm, and friendly chip, with bags of charm, a shrewd eye, and the sort of workmanlike, no-nonsense, and bluffly confident intelligence conferred by the goodness and the grace which on his birth have smiled. He has a crisp, class way with approbation and disapproval: this man is "one of the ablest I have ever known," that one "unassuming and unambitious" (high praise), this other, however, "a man of emotional prejudices and a taste for abrasive language and behavior." But he honors honor and speaks without affectation of personal sincerity, dignity, and uprightness.

He took this firm, sane, and easy speech with him into the public offices that his inherited title promised him. *Noblesse* is as *noblesse* does, and Carrington had his nobility as well as his presumption of power. A junior minister under Churchill when the old, old man returned to peacetime power in 1951, Carrington rose steadily from farming office (he was, naturally, a great landowner) to the military. Joining the Defense Ministry in 1954, he slipped into the career in which he found his fulfillment for the better part of the next forty years.

Seen from a little way off—Boston harbor perhaps—Carrington is an unbelievable trace of the old England defied by the Founding Fathers. From much closer up, however, he is a product of a worthy Whiggishness and the sort of liberal public spirit and calm assumption of rational command that qualified him and his like to keep the cold war obediently in place, as enjoined by the benignant rulers over the ocean. When in 1970 Carrington became defense secretary, he sat down at the desk of the outgoing Labour Secretary, Denis Healey; details of style and party obligation apart, his slim frame fitted the ample space left by his portly predecessor perfectly.

The pair of them, both with a keen zest for human peculiarity, were completely in and of their world—that enclosed, busy world of international policy elites, a world sometimes desperately important even if routine, puffing more than a little with its own self-importance, rightly persuaded that the future of the globe lay in its well-manicured hands.

It is a new class that has replaced the portentous older class of ambassadors in their sashes, orders, pumps, and stockings. It travels ceaselessly and frenziedly but, like the kings of Bali, remains calm, almost motionless, faintly smiling at the center of international bustle. It was the class that, whether genially (Lord Carrington), woodenly (Andrei Gromyko), with a vulpine intensity (Paul Nitze), fixed the limits and compounded the intricate protocols of the cold war.

As Carrington said, "People like certainty. You know, the steady continuity of the Warsaw Pact against NATO. Certainty and fright. The cold war was kept in place by fright, the military alliance held together by *fright*. Every three years or so the Soviet Union managed to frighten everybody, Berlin, Cuba, SS-20s, and so forth, and then as the fear diminished and there were no sudden jolts, the bonds began to soften. By 1980 we hadn't been frightened for a long time.

"After all, if five years ago you'd told me that five years hence the Soviet Union would begin to break up, I'd have shouted, 'Hooray!' But now? After the relief of the end of the cold war, there's a new, strong uncertainty, and nobody likes it. I don't. It was that continuity and fearfulness which kept British policy solidly bipartisan for the whole cold war."

Carrington moved steadily and easily through the defense establishment and in the company of his seniors in the United States for years. According to the sonorous British liturgy of arms, he had been First Lord of the Admiralty between 1959 and 1963, had held diplomatic office in Australia when the empire was still almost an empire, and had led the House of Lords in its ancient and creaking half-subservience to the commoners. Throughout this long passage, from making ceremonial salutes on the quarterdeck of British warships in the 1950s to occupying the massive desk of NATO's secretary-general in Brussels in 1984, Carrington kept a dry, sharp eye on the interests of his beloved Britain as his great nation passed through "its extraordinary transition from offshore archipelago to maritime and commercial empire and back again all within three centuries."[4] As Britain's power fell steeply away, he watched it go without sanctimony and sought, according to the lights of a Tory landowner, politician, and soldier, without unseemly fuss or passion, to meet the country's always heavy but welcome military duties. While dealing honorably with treasured friends in Washington, he did what he could to nudge his graceless nation nearer to the European cousins that were its natural allies.

So NATO was for him a house with many mansions. He had been foreign secretary in Margaret Thatcher's government from 1979 and resigned in 1982 in a thoroughly principled way when his department was taken by surprise by the Argentinian invasion of the piddling and useless Falkland Islands. He worked a couple of years for Henry Kissinger's outfit advising the first world on the precarious dangers of the third, and then was invited by his grandee class to "serve the Alliance to which I'd devoted a good deal of time and thought and in which I strongly believed."

His time as foreign secretary coincided with the advent of the last heavy freeze-up in the cold war before the picnic in Reykjavík in 1986 started its global warming. Characteristically, he took with due tolerance the increase in megaphone diplomacy when Ronald Reagan came to power. "Beyond doubt Carter allowed U.S. defenses to fall into a poor state. Cap Weinberger was quite right to do something about the U.S. forces. And Reagan was a simple man; he was really obsessed with all that stuff about standing on your own two feet, walking tall, motherhood, and the evil empire."

Carrington is no ideologist; he holds his beliefs, as it were, metaphysically. They underpin a world rather than fill a mouth and are all the plainer to see for that. He spoke with cold detestation of the Soviet invasion of Afghanistan, with a lighter dislike of the Labour party's strenuous efforts to defend a kind of state socialism. He was a steadfast, not a rabid cold warrior, and he embodies, with his informal dignity, cheerful charm, resolution, flair, and sheer bloody nerve, the liberal and civic values of the best of old England's upper class.

And so, "When Mrs. Thatcher wanted to stop me going to Moscow as foreign secretary, I simply went as president of the European Council, which Britain happened to occupy at the time. Even if you disagree with people as fervently as I did with the Russians, there's every reason for talking to them. Both Reagan and Margaret Thatcher wouldn't even talk to begin with."

Nevertheless, there was never any doubt for him who the enemy was, nor that it could only be kept securely penned in by the NATO alliance. As defense secretary he befriended, as did Denis Healey, his Labour predecessor, the sort-of-Labour chancellor of West Germany, Helmut Schmidt (though he thought Willy Brandt no great shakes). Carrington was quite certain that "Britain needed to stay in the strategic nuclear business," not only to keep shiny that old chestnut, "U.S. credibility," but also—to the beat of an antique drum—so that France would not possess the only indigenous nuclear weapons on the Continent.

Helmut Schmidt was quite flat about NATO. In his view, an Anglo-French nuclear force could not possibly replace an American one. "The Federal Republic is in NATO because the United States is in NATO. And if the United States forces, and above all their nuclear forces, were not here the Federal Republic would probably soon go neutral."[5] Some people might even have thought around 1984 or so that such neutrality would have been no bad thing. But Carrington's faith, like Schmidt's, was placed in the plain fact of the weapons and their ungainsayable presence. "People confuse intentions with facts. You have to plan for the *fact* of the weapons' existence, and that remains true after the end of the cold war; no doubt, in the uncertainty of things, you have then to plan even more carefully. Gorbachev won't last forever. Khrushchev disappeared, and this chap may disappear. The peace movements were a bloody nuisance; what they did was counterproductive. I don't actually believe that Gorbachev did all he did for any other reason than to put the Soviet economy to rights and thereby to retain superpower status. I don't believe he left Afghanistan because the invasion was so unpopular with the rest of the world but simply in order to get rid of the problem and maintain the Soviet Union as a great power. It's a great mistake to believe he's a softie. He *couldn't* have expected the Soviet Union to break up.

"So you need NATO, but it must evolve into something different. What the United States wants is a say in the destiny of Europe and sufficient information about the European Community's economic policy. People say that NATO can't change into such a body. Why not? Transatlantic politics are crucial to the world's future."

Carrington's most admired friend and colleague among the foreign policy elite was Henry Kissinger.

I got to know Kissinger very well indeed. He was—indeed is—a remarkable human being. His love and devotion for the United States, his deep and intelligent patriotism, were utterly sincere; but his attitudes of mind and his understanding were thoroughly European, deriving from a German background and immense learning. I found him at all times a joy to work with because of the speed and originality of his mind, his flexibility and his enchantingly funny sense of humour. Henry Kissinger, too, had imagination. He thought big and he thought profoundly. He could be adept and tireless in negotiations but it was always against the background of a carefully reasoned and long-term political calculation. He was a great public servant.[6]

He added, in our conversation, that "Kissinger had a broader vision of the world than anyone I ever knew. He wasn't always right, of course, but always fascinating. He had vision—not *a* vision, but vision. I don't, because I'm not as clever as him. And he was well matched in Nixon, you know, however unsavory a figure he was. Nixon had an astonishing feel for foreign affairs, and Kissinger would agree with that, although they drifted apart at the end." But Carrington, at once unassuming and utterly assured, did have his vision. He remained a relaxed but (in his own word) passionate European. The line of his unhurried, undogmatic, deeply conservative politics was aimed at an "increasing harmony of European voices."

With Kennan and McNamara, Adenauer and Brandt, Bevin and Healey, Peter Carrington was one of our true guardians for the period of cold war, whether we wanted or needed guarding in any case. Nearly half a century of settled fearfulness, of the massive continuity of terror and the dread facts of the weapons, mutated Ernie Bevin's elephantine and peppery ruggedness into Carrington's smooth, likeable aplomb. Each bore the scars, the medals, the honors and dishonors of a gallant cold warrior.

26

The Dissenter—Edward Thompson

MORE than any other single figure, Edward Thompson focused and expressed the principled dissent of Europeans to the direction taken by the march of time.

His dissent was well known in the United States, but he remained so intransigently an English European and a romantic socialist that his ideas, his prose, the strength and corrugations of his position, make him difficult to transplant and translate into the frames of mind and feeling of American culture. And yet he spoke and speaks a libertarian tongue, learned from an American mother, an almost-Buddhist, anti-imperialist father, and a socialist brother.

Frank Thompson's younger brother Edward was born in 1924 and grew up, a little awestruck, in the shade of Frank's remarkable distinction. Being brought up in that loving household, with its vigorous commitment to living by liberal values, Edward Thompson learned his family's devotion to the causes of freedom and justice, and learned also that those noble causes are inextricable from the actualities of organization, prison, exhausting travel without enough money, trying to collect enough money to keep organization going, digging into one's own never very full pocket, writing late into the night. He saw the many Indians who came to his father's house with gifts in tribute to his father's efforts on behalf of Indian independence. He read the letters sent to his father from prison by Jawaharlal Nehru.

So when Frank went to war, Edward followed as a volunteer only three years behind. As a twenty-year-old subaltern he was commanding a troop of tanks in the hard campaign against the Germans up the mainland of Italy. It was a hellishly slow advance against a well-armed foe that had its supplies close by and was dug in on the reverse slopes of every pleated valley. The army edged forward up narrow roads and in the Italian heat; although the Renaissance cities more or less escaped, a large number of British and American soldiers—many of the latter with Italian names—did not.

For the third convention of European Nuclear Disarmament, held in Perugia in 1984, Edward Thompson wrote a short memoir of his previous visit to the city on 19 June 1944.

At 6:30 P.M. his tank was trundling and squealing toward the suburb of Fontioeggi, second in a troop of tanks reduced by battle damage and engine failure to two. He might have been in the first tank but for army protocol about the troop commander not risking himself in the lead. It was blazingly hot. When the front tank was directly hit, two of its men got out and ran to safe cover in the narrow street with its high-walled gardens; the other three were killed. Thompson broke protocol by running forward to the Sherman under rifle fire, to call down into the turret. He received no answer.

It was a tiny exploit in that formal operation, the liberation of Perugia, itself part of the larger victory in Italy and, ultimately, the victory in Europe. But Thompson's memoir goes beyond the leggy boy running through bullets to check on his men and finding half of one burned to ashes in the driver's seat. It includes the letters sent to him when, in spite of the certainty of the troopers' death, the army had only posted them as "missing" and their families were being put through the mutual strain of hope and anguish for impossibly long. Thompson quotes their letters; they break one's heart.

"My brother was all I had in the world we had lived together the last ten years since losing our mother and I would like to know a few more details, if the spot where he is buried is marked or did an explosion make this impossible?"

"Please did one of his friends pick some wild flowers and place them on his grave?" And always, whenever I sent the least scrap of news, this was received with pitiful expressions of gratitude from the kin of these troopers. They seemed to be astonished to receive any attention from anyone in authority: "Hope I am not asking too much of your time . . . ," "I am so sorry to trouble you. . . ."

"Will you try and do me a kindness and see that his personal belongings are sent back to me and if you really could get a photo of his grave it would set my wife's mind at rest as she is greatly grieved."[1]

Forty years later, addressing a convention trying to unmake the wretched aftermath of that bit of war in Perugia and the world war it was a part of, Thompson turned his recollection of those dead troopers and their grieving relatives into a restatement of what became his imperial theme as a historian.

He noted that much of the experience of modern warfare is both as exemplary and as inconsequential as that forgotten skirmish. The young troopers in the tank wanted to get safely home. But they also wanted to finish the war and to fulfill certain necessary duties and loyalties. "They were democrats and anti-Fascists. They knew what they fought for, and it was not the division of Europe, nor was it the domination of our continent by two arrogant superpowers."[2]

It is a note that recurs in all of Thompson's writing. Throughout his work and his life he tries to write history that speaks of the best purposes and the best people of the past. Both vision and action rewritten as history, teach that the massive inertia of the past is only given direction as we rediscover within it the best present that it might have made, that it might still make if only we would collectively renew our efforts and reconnect with the intentions and energies of the dead. The dead go on before us, larger than in life they seemed, and as we heed them, in our history books and in our lives, their energies flow again down the reopened channels of feeling and imagination; and they fill us with new possibilities.[3]

Thompson's capacity for making himself a conduit in this way was a piece of good moral luck. It was a gift. His experience, given his gifts, led him from his devotion to his brother to those Eighth Army tanks to the British Communist party. Maybe this last step is what made it so difficult for the American intelligentsia to admit Thompson to the arcana of what is a deeply conformist elite culture. Thompson's oddity and awkwardness made him an equally uncomfortable friend and ally in England. His writings are so personal as well as resonantly public. He calls fools, traitors, timeservers, and lickspittles by exactly those names.

He meant his communism. After he left the army and prepared with his mother the little collection of his brother's diaries, letters, and poems, he went with her to the dedication of his brother's memorial at Frank Thompson Halt in Litakovo, Bulgaria. From there he went on to Yugoslavia in 1947 to work beside his future wife, Dorothy, building a railway in the heterodox new Communist republic whose many national factions, so long at one another's throats, had committed themselves to the long, new, and difficult revolution of economic cooperation.

The icy air of the cold war hardened around Thompson and his wife when they came home.

In Britain the small number of communist intellectuals belonged to a defeated and discredited tradition—or so it was the business of every orthodoxy in our culture to assure us. . . . Our solidarity was given not to communist states in their existence, but in their potential—not for what they were but for what—given a diminution in the cold war—they might become.[4]

And so when 1956 came along and brought crises of allegiance and conscience to West European Communists, Thompson was unequivocally on the side of Leszek Kolakowski in Poland—on the side, that is, of those who held to the promise of happiness made in the name of a truthful socialism and fought the crude assertion of socialism as power-without-value made by the Stalinists.

When power-without-value was so irresistibly asserted with the aid of tanks in Budapest, "ten thousand people walked out of the British Communist Party and I can think of not one who took on the accepted role . . . of Public Confessor and Renegade," least of all Thompson himself who, exiled from a party that had so murderously betrayed its humanism, continued to speak in both public and scholarly life for socialist ideals that honored the name of William Morris as much as that of Karl Marx.[5]

Thompson took Morris's name well. In 1955 he published a full-dress critical biography of Morris that, to the puzzlement of both art and literary historians, featured Morris's commitment to the Socialist League and to writing the broadsheet *Commonweal* every bit as carefully as his wallpapers and verse romances.[6] Thompson describes Morris stumping the country to address the countless public meetings of working people that eventually tired him to death, and he praises Morris for pursuing politics in the midst of his art, for his public-spiritedness as well as his designs. Morris carried a banner for turning the work of life into a work of art, an ideal that Thompson himself would live by—writing, fighting in and through the word, for a cause greater than ourselves and only alive in such lives. I suppose it is what we mean by saying of a life that it is an example to us; there are few enough to go by.

Throughout the 1950s Thompson worked with his wife in the adult education branch of Leeds university, one of those offices beyond the university walls where grown-up students, in England, the men and women cut off by social class from educational opportunity, may come to ask and answer questions raised by their own experience never even heard of back on the campus. Out of these densely woven, homespun curricula came in 1963 his Tolstoyan history, *The Making of the English Working Class*.

His great book respects the inevitable creativeness of ordinary, everyday life and the local knowledge those living such lives so painfully acquire. Thinking and feeling thus, and with 1956 behind him, Thompson threw himself into the work of the British Campaign for Nuclear Disarmament, whose early days foreshadowed exactly the reformation of popular consciousness of which he had written.

Not that there needed to be (or was) anything too theoretical about the CND. The horrible weapons were there and had menaced the Koreans (all of them). Let's get rid of them (why the hell not?). The movement was born, swelled, lapsed, as such movements will, and disappeared into a few bookish, leftish households like the Thompsons' own, while its drum was hung waiting to be struck when a later armada of missiles was sighted off the Atlantic coast in 1979.

Thompson went briefly and by invitation to a conventional academic post at the University of Warwick, and then, repelled by its dealing with local business and finance, he noisily left his job and turned full-time to the lonely, ill-paid craft of writer.

He had a living to make. Through the 1970s, as a consequence, he wrote a good deal of polemical journalism in the few hardy weeklies of the left, *The New Statesman* and *New Society* on one side of the water, *The Nation* on the other. He turned the scholarship of his legal history in *Whigs and Hunters* to new account by arraigning the legal system of Britain, nay the very judges themselves, for cravenness, *bien-pensant* obedience to the powers that be for "taking liberties which were once ours," for seeing sedition in honest dissent and a traitor in every socialist.[7]

Behind all this committed, exhausting, and necessary engagement, his imagination was dominated by the grim old scenery of the iron curtain and the nuclear bomb. During his years of journalism Thompson forged a prose style equaled by no other commentator or historian of the period whom I have read. In a bad time, when the English language as spoken in public turned rancid with cant or hard with crisis management, Thompson did much by himself to keep its critical idiom clean and sharp and magnanimous.

But he paid the price of exposure. An eloquent writer of great heart will be rebuffed woundingly; he was, and was much hurt in it. A generous man with cordial impulses may also be made touchy by misunderstanding and angry by even admiring criticism. An open and impulsive man will be bruised by political bullying. This man and his prose lived through these vicissitudes and became known as the first wordsmith of the peace movements, as much of a presence as Bertrand Russell had been in the same place on the platform before him. He was miscalled by cold warriors on every side as knave and fool, blamed by Muscovites and Washingtonians alike for threatening the cause of peace-through-lots-of-nuclear-weapons with his naive demands. He thereby greatly heartened hundreds of thousands of demonstrators trudging through dust or mud in Sicily or the Netherlands or over Greenham Common.

His single most important statement was no doubt the pamphlet *Protest and Survive*, noticed in these pages. His long apprenticeship in opposition to cold war, the mastery and range of a prose style drawn from the fighting tradition of English and Irish dissent—from Milton, Swift, and Morris—his tall, craggy, white-maned good looks, earned him his place alongside Joan Ruddock and Bruce Kent as one star in the trinity leading the European peace movements. The interdependence of mass media and power politics today demands the creation of stars. It was a piece of tremendous good fortune that in 1980 the British found these three, just as in 1967 the antiwar movement had thrown up its own remarkable cast—Chomsky, Hoffmann, Fonda, the Berrigans, Sloane Coffin, Ellsberg, Mailer—on the steps of the Pentagon.

Thompson was made for the moment. He could draw fire and return it masterfully and swiftly. He proved an adroit negotiator. When Reagan came up with his "zero option" in 1982, Thompson, while arguing that it should be accepted vigorously pointed out that it would look like a trick designed, as George Ball said two years later, "to make the administration look generous in offering to give up any deployment of missiles, while at the same time assuring that the negotiations got nowhere"—unless NATO could offer very much more than was on the table. On the terms of the zero option the Americans only held back the cruise missile if the Soviets dismantled everything.[8]

Thompson was much vilified for this response, and for exposing the fraud of the Soviet Peace Committee, a propaganda machine, and the machinations of its stooge, the happily named Dr. Popov. Thompson was subsequently described as "a notorious anti-communist working for the CIA" in *Rude Pravo* (the Prague daily) and as a "mouthpiece of anti-Soviet conceptions" by the Soviet Peace Committee itself; he was widely called disgraced and disgraceful by one Wieseltier in *Nuclear War, Nuclear Peace* (prescribed in that order, one presumes) and by George Ball, no less, in the *New York Review of Books*.

Such an array of enemies greatly cheered his host of admirers, supporters, and friends. What is less remembered from the seven or so years he and his wife spent in the limelight, is their tireless, tired-out responsiveness to an impossibly demanding schedule of appearances, statements, writing, speaking, going to conferences, and plain marching and sitting and being trodden on by cops. World politicians wield their influence from large offices equipped with telephone exchanges, wordprocessors, and young men going ahead of

them saying "hold two elevators." The Thompsons did everything from a
house overlooking Worcester, on one regular salary, and with an effort of
personal will strong enough to fuel a rocket to the moon.

Behind it all, behind the labor of body and soul, lay Thompson's piercing
recollection of the chance lost in 1945 to release the "spirit abroad in Europe"
on behalf of all people fighting to put down the grim past, and his sense of
outrage that forty incredible years later the occupying troops of the two
superpowers still came and went as they pleased over the ancient battleground
and littered it with their deadly monsters. Thompson's most abiding contribution to the cause was therefore the Appeal for European Nuclear Disarmament launched on 28 April 1980 and thereafter the keystone of all his argument against the continuation of the cold war.

The appeal was substantially of his drafting.

We are now in great danger. Generations have been born beneath the
shadow of nuclear war, and have become habituated to the threat. Concern
has given way to apathy. Meanwhile, in a world living always under menace, fear extends through both halves of the European continent. The
powers of the military and of internal security forces are enlarged, limitations are placed upon free exchanges of ideas and between persons, and civil
rights of independent-minded individuals are threatened, in the West as
well as the East.[9]

The document went on to ask Soviets and Americans not to deploy their
latest weapons intended for "medium-range" effect, begged for continentwide
response and coalition, and acknowledged that there is a world outside
Europe, that Europe is not the only place in the world, and that the old
continent had twice "disgraced its claims to civilisation by engendering world
war. This time we must repay our debts by engendering peace."

Did the appeal lose? Were all Thompson's heroic labors of the next decade
and more—stopping his scholarship in its tracks—unavailing? The cold war
stopped all of a sudden, not down the slow gradient that was the best Thompson and his cheerful supporters, myself included, had hoped for.

As it stopped, Thompson said early in 1990, when struck suddenly by grave
illness (from which he recovered),

It is profoundly moving to see the forms of the cold war dissolving before
one's eyes, but dissolving most of all on the other side. The cold war has not
been an heroic episode, an occasion for triumphs, but the most futile,
wasteful, humanly destructive, no-through-road in history. It has led to
inconceivable investment in weapons with inconceivable destructive pow-

ers, which have—and which still do—threatened the very survival of the human species, and of other species perhaps more worthy of survival. It has nourished and reproduced reciprocal paranoias. It has enlarged authoritarian powers and the licence of overmighty security services. It has deadened imagination with a language of worst-case analysis, and a definition of half the human race as an enemy Other."

Looking back on his own efforts within popular protests, he said,

At certain moments history turns on the hinge of new ideas. I think there was such a moment between 1982 and 1985 in which peace and human rights movements together broke the cold war field-of-force and gave history a new hinge. In any case, the non-aligned peace movement made a large contribution to the end of the cold war and is one of the only traditions to emerge from it with any honour.

On this note I should like to make my parting salute to Edward Thompson, a name that, in these days, we should peculiarly honor.

27

The Novelist—Joan Didion

SHE was born in 1934 in Sacramento, California. So she came from the basin edge on the other side, the shore mysterious to Europeans who cross the stormy Atlantic and go only as far as New York, familiar from a lifetime's television, but rarely reach the Pacific.

The Pacific Ocean is Joan Didion's stretch of wide water, and it holds her. She is pulled out toward Hawaii, beyond the frontier, the last state of the West and the first islands of the archipelago that leads into the East. She travels down the California seaboard to the wretched semistates of the countries that live at the edge of her country's enormous and casual wealth. Her imagination works in the dense heat and thick air of Honolulu, Guam, San Salvador, Los Angeles, Melanesian and Spanish names. When she goes to Miami she finds a Hispanic-Caribbean city that mirrors the form and structure of cities the other side of the equator—gleaming, obsidian skyscrapers hoarding their cool wealth, the mad rout of the poor in the stinking streets below.

The battlefields of the cold war are a long way from these hot gates: Vienna, Budapest, Prague, Europe with its wide boulevards and rivers. But the business of hating communism goes avidly forward in Miami, leaving eyeless corpses rotting in dumps and cars burned out black beside the roads;

it equips countless hard-faced young men with M-16s and gives dark and beautiful women seated at glossy dining tables their plump fervor, their creamy dresses and heavy gold bangles.

The political artist works in such images—corpse dump and dinner table—and yet Joan Didion is a political novelist only because, as she says, "it's all politics, isn't it?" Even so, she works obliquely on the topics of power and her nation. The question for her, as for all novelists, is to identify "specificity of character, of milieu, of the apparently insignificant detail." The geography that holds her is the deep blue Pacific and the narrow seaboard stretching south into the continent's other half. So specificity, milieu, and detail, the writer's trouble-and-joy, take her well beyond the weather. Though weather matters: "I happen to like weather, but weather is easy. . . . Anyone can do palms in the afternoon trades. . . . I mean more than weather."[1] She means, I would guess, the *gap*, the gap between minute, pleasurably subtle observation (which she's so good at) and the content or essence of action.

Measuring the gap is the hardest thing for the political novelist to do. It is all the harder in a time of cold war, when the point of politics is to blot out the connections between purpose and consequence, to blur the distinction between threat and desire, to stretch for as long as possible the moments between apprehension and expectation, hope and resignation.

Standing in this murk, anxious to discover how in her nation public figures might lead worthy lives one could admire and even emulate, Joan Didion has to feel with her fingertips for what she can find.

Her only light in the political murk is the feeble glow of journalist's prose. This is what illumines the world of affairs. It doesn't show us much, but it is all we have. Despising communiqués and judging politicians to be mostly horrible, Didion tells whatever stories may yield up their little bit of human meaning in the narrow beam of that same prose.

Thus, in her novel *Democracy* (1984) she finds the details of democracy in the mean lives of the rich and successful family in which the father just misses the Democratic nomination and one grandfather murders his daughter. But the subject of the novel is the gap in the heroine's life between her vivid hope for "eccentricity . . . secretiveness . . . emotional solitude," and an everyday routine made up of photo opportunities.[2]

Like no one else, Didion captures the U.S. version of cold war politics as lived by public figures who have no private lives. But her point is not, of course, to shed crocodile tears, with the yellow press journalists, over this lack; it is to follow those individuals who, promptly and excusably, remove themselves from public view and make what settlement they can with the seat of a 747, with a lover seen ten times in ten years, with a TV news clip of themselves dancing ten years before.

Didion's line toward this bit of personal algebra was, as one would expect, oblique. She went to Berkeley, graduating in 1956 as all upright and idealistic young women and men of her generation did, with a degree in English. In those days there was plenty of work in journalism for a tiny, bird-boned, waiflike woman with a good degree, a fey sensibility, a way with words, and a strong dose of *nostalgie de la boue*. She worked on *Vogue* at a time when neither women nor journalists were very political. But "after my husband [the writer John Gregory Dunne] came back from several weeks spent in 1963 for *Time* covering Vietnam, he had a strong sense that the war wasn't winnable. Now I'd found a very visible politics going on around Berkeley and the Free Speech Movement—it stayed that way for years—and there weren't many people reporting it. So I wrote *Slouching towards Bethlehem* [1968] because I was (and am) flatly certain that things are falling apart, although more slowly than I believed then.

"At that time I couldn't see the politics. I had to wait for them to come through. I still thought, for instance, that Vietnam was all just part of the responsibilities of being a great power, that a border war was better than World War III. Until one morning in 1968 it came in a flash as if I'd been asleep all that time that stuff just didn't add up, that *everybody* was lying."

All the time American foreign policy bubbled and seethed below the surface of the Pacific, and the flights went out from Washington to Guatemala, and Ecuador, and Uruguay (Philip Agee on them). And the lines ran back from the Philippines and beyond, from Vietnam and Cambodia, to touch the lives sunbathing in the green Californian gardens and stir them up. Writers among them had to find a quite different pitch and tone in order to write about their country.

Being a writer is a rum business, rummest of all in relation to what preoccupies you while something else is going on. Joan Didion went to a film festival in Colombia, where films cannot be wilder than life, and "I began to wonder how on earth the South had gone one way and the North, the United States, a completely different way. So when I started *A Book of Common Prayer* [1977], I knew there was a subject there in the dark.

"That subject—if subject is the word, that whatever-it-was I wanted to write about, *was going to write about*—started from people sitting around in good clothes in beautiful Colombian gardens which were linked all the way to the countries of Central America. Somewhere along those lines the money travels, and the guns and drugs as well."[3]

She had to find a way of catching the great American omnipresence, and both her disposition and her aesthetics impelled her to find it in a few, frugal images. The heroine of *Democracy*, Inez Victor, wife of a Democratic senator who misses the presidential nomination, "said the 3:45 A.M. flight from

Honolulu to Hong Kong was exactly the way she hoped dying would be. Dawn all the way."[4]

Didion had collected news clippings from 1975 when Saigon was falling, including Hunter Thompson putting a cable through to his editor at *Rolling Stone* saying that it was going to take a lot of money to get him out of Saigon. She smiles. "He said, 'After Saturday it's big greens on the barrelhead.' " She lit upon the image of the giant green helicopter lifting off the roof of the American embassy.

As she read a few of the official policy statements published by the Government Printing office, one called *Base Development in Vietnam*, "it came to me that it was about *contracts*." Behind the action of *Democracy*, miniature and potent as transistors, are the contracts that take Inez's lover Jack Lovett hither and yon across the Pacific and Indian oceans and the China Sea. The contracts do not *cause* the action; the purposes of all the characters, except the man who wants the nomination, her husband, are as obscure to them as they are to us. Purposes inconsequentially work out, except for someone's being shot. The plans of individuals, "if the convergence of yearning and rumor and isolation on which [they were] operating could be called a plan," bind them to life, give the giant helicopter its thrust to lift off the embassy roof.[5]

The plan in question is that of Jessie, Inez's daughter who, in March 1975, took herself off from King Crab in Seattle by way of an Air Force transport plane to a job serving in the American Legion's burger station outside Saigon. The weird girl made the trip with no more credentials than a breezy manner, honey-colored hair, and a faked press card, plus her well-grounded belief, "that whatever went on in Vietnam was only politics and that politics was for assholes." She provides Joan Didion with another strong image to play off against the moment of the helicopter.[6]

Perhaps the victims of democracy in *Democracy* are those ignored and exploited by those eager to assume democratic office. Certainly all the (considerable) animus in the book is directed against the failed presidential nominee, who "was not an insensitive man but who had the obtuse confidence, the implacable ethnocentricity, of many people who have spent time in Washington." The novel's meager sympathy is reserved, so far as one can tell, for the heroine and her heroin-shooting daughter, for whom the costs of public life are, Inez tells a reporter, in a typical metaphor, memory damage—*"as if* you'd had shock treatment. I mean you drop fuel. You jettison cargo. Eject the crew."[7]

That remark catches Joan Didion as she alights on the gap between love and politics. She doesn't even want to talk politics in her novels, not because it is unimportant but because the political conversation of mankind so garbles the facts and is so fixed and false a set of exchanges that it provides neither a point of entry nor a point in making an entry. "Things that might or might not be true get repeated in the clips until you can't tell the difference," Inez says.[8]

So Joan Didion flies from one horrible spot to another, not at all looking *for* something, but looking *at* things, appalling things in the face of which, repressing her shudder by an act of will, she has braced her frail, tiny body and terse speech.

Democracy is a bleak, sparse book on an exigent subject. It was on the *New York Times* best-seller list for nine weeks. It has its news all right; the news that we can not expect democracy, however much we may want it, from a world like *that*.

So it is natural for her novels to sound like a peculiar sort of journalism, and her journalism reads like a queer novel that is also true. Joan Didion speaks in a true American voice that is all her own. She reports that there is no incredulous tone adequate to the monumentality of American self-deception.

She went to El Salvador and found her image. It is a body-dump. "Terror is the given of the place," and so she went to the Puerta del Diablo, to look at the numb aftermath of terror: "Bodies . . . often broken into unnatural positions and the faces to which the bodies are attached (when they are attached) . . . equally unnatural, sometimes unrecognizable as human faces, obliterated by acid or beaten to a mash of misplaced ears and teeth or slashed ear to ear and invaded by insects."[9]

She holds horror in place by the tight-lipped reticence of a style which assigns its author, like all good styles, a moral position in a political world. At the Puerta del Diablo she found a man teaching a woman to drive a Toyota pickup along the very rim of the steep and dreadful dump. Three children were with them, playing in the grass above the vertiginous cliff littered with "pecked and maggotty masks of flesh, bone, hair." It was only later that she wondered why they should choose a corpse dump for a driving lesson.

The cold war and the fear of communism are at the heart of El Salvador. It could not be such an unspeakable place without hearty and lavish American efforts to make it resemble the best of Miami. Her sense of this discrepancy puts acute strain on Joan Didion's manner, in both her physical presence and the presence of her prose. She was in El Salvador with her husband in 1982; she reports faithfully "the mechanism of terror" that held her tight in its grip while she walked carefully and courageously about, *Heart of Darkness*, textbook of America's Pacific, in her pocket. Alongside terror in a car casually boxed in by expressionless toughs armed with G-3s ("I studied my hands"), or terror on a porch at a candlelit dinner she quotes the unyielding cheeriness, "the studiedly casual, can-do, sheer cool," of U.S. statements about this awful place.

"There are no issues here," a smart Salvadoran told her, "only ambitions,"[10] and that seemed about right to her. If you drop the usual political analysis—leftist, rightist—and you see American and Soviet weapons in the hands of state killers as part of the intricate machinery of terror, then you can only conclude that a shopping mall crammed with gilt goodies and bright bathing towels, like the image of an oasis, is hardly a rampart against communism. If you conclude, as Joan Didion did in 1982 after her visit to the San Salvador morgue (easily arranged, the door stands open), that life there is untranslatable into Washington's language, then a quite new story must be found with which to understand the Third World and its old facts of filth and murder.

Joan Didion's response to Ronald Reagan's or John Kennedy's tall tales is open contempt. She went to Miami, the capital of Cuban Florida, in 1985 and 1986 to add substance to her horrified sense of the *irrelevance* of received wisdom, received facts, and received ambassadors to the nightmare life-in-death of Central America, to the pratings about democracy and elections, to the busy plotting of the left.

What sends Miami mad and red-eyed, both among the harbor drug and gun dealers and the jeweled, white-suited rich in the Omni Hotel, is Castro's Cuba first, and gringo perfidy about Cuba second. Kennedy had promised to help, had accepted the flag, and then had failed them at the Bay of Pigs; twenty-five years later, Reagan had sincerely mouthed the lines written for him, "We cannot have the United States walk away from one of the greatest moral challenges in postwar history." In the interim, the Miami Cubans had plotted and bought weapons, from time to time murdered one another, and lived the life of nationalist exiles, forever drafted in a war forever on the edge of starting.

Joan Didion found these well-dressed, well-read citizens with overseas interests living off the conviction that they had been traduced by six administrations, which had maneuvered Cuba and Miami in and out of world and Washington policy for strictly presidential ends. Thinking the Miami Cubans right, liking them rather "in spite of their tendency to bomb one another," leaves Joan Didion perched on the view from nowhere, hating American politics and politicians, reporting in her tense, laconic voice the appalling facts and the appalling people, with nothing much to bet on except a few exiguous truths and her tenacious skepticism.

"When I read *Inside the Company,*" she said to me, "it all came together. The CIA and the KGB station chiefs playing golf together every Wednesday."

It is in *Miami* and *Democracy* that she adumbrates a bitten-off sort of theory of cold war politics. The world boils and seethes, and the presidential team organizes world media attention around "issues" and this or that "focus," first putting something front and center, so that it is much-mentioned, then pulling it back when people have had enough. In Miami it may be that the right people will also step forward to dance to their tune when it is played, but then again, maybe not. A good president, played by, rather than playing to, the orchestra, gives everyone a dance, sufficiently on cue.

This is our new politics, the politics of spectacle. Didion, a woman, watches, without wishing, from its edge. Her feelings are intense, but unexpressed. They reside in the many silences of her writing. These attributes make her an exemplary novelist for the honest citizen of a nation at cold war.

28

Mistrust—*Missing, All the President's Men, Gorky Park, Edge of Darkness,* and the Lexicon of Apocalypse

JOHN Locke, whose liberal theory of society underlies the contribution of the United States to the way we live now, put the unfashionable value of trust at the center of a society's constitution.[1] Trust is not in itself a noticeably liberal value—it doesn't, that is, depend strictly on those freedoms and rights which are by now almost the only agreed values left in the bare political arena. Everybody agrees that trust has a place in practical, everyday exchanges; but we seldom acknowledge its place in our political life, nor even in the emotional self-analysis we now commonly use to validate our private lives.

Trust is individually given as a consequence of custom, ceremony, habituation, exchange. It is a value woven into the roles each individual carries out in his or her daily doings. To be wife, lover, mother, husband, friend, customer, client, adviser, attorney, is to follow (and improvise) the rules of conduct that define the roles, and thereby to win trust. Break the rules and you lose the trust.

But trust is also given to social institutions. Being a member of an institution entails assuming and living at least some of the values of that institution. A trade unionist makes a commitment to solidarity, an intellectual to criticality, a friend to friendship itself, a Catholic to obedience, a socialist to equality and justice.

The key moral casualty of the cold war was the ancient, necessary value of trust.

"You can never trust the enemy," "Remember what they did that time, look at what they're doing now. . . ." Each appeal for ideological support was built upon the contradictory (and shaky) foundation of the other's untrustworthiness. Each shady action—spying, lying, treaty breaking—was justified as the only way to do business with such perfidy. The Soviets found plenty of ideological ammunition in capitalism's avowed cynicism about human altruism, its official preference for the ruthless hand of competition to sort out the weak and inadequate. The West found no less historical evidence in the record of treachery left by Stalin. Each side could rightly accuse the other of lack of trust and trustworthiness; these arguments were used to feed the dark and mutual suspiciousness of decades of arm's-length diplomatic mountaineering.

Nobody can afford to be an innocent about the place of trust in our political beliefs and actions. Trust in other days was learned by lived habit and in local association. The kind of trust that holds society together—trust in its currency, its clocks, its officers, its news—is nowadays dependent on how people believe their society works. But very few people trust their national political system, still less the international one.

This loss of trust during the cold war epoch may be detected in countless ways and can be found in millions of hearts. The most regular evidence for it transpires in our stories—our anecdotes, our murmured autobiographies, our novels and movies. But the loss of trust, the growth of mistrust, has also been a loss of substance, of *volume,* in the great names and master symbols of the polity. Trust, after all, is transitive; you trust *in* something—in others or in ideals. If you cannot trust strangers, or work with fellow citizens whom you do not know, and if trust in your polity is confined to its paper money, its delivery systems, and, warily, a few of its kindlier officers, is it worth defending against the enemy? What are all those weapons for?

The answer from the heart is still, for a surprising number of people, the flag of their country. But as Alan Bennett has reminded us, the alternative, treason, has lost a good deal of its appeal because nobody has any idea which ideals are worth committing treason for.[2] The nation-state is and will remain the unit of world politics, but outside radical Islamic states, not many nation-states can count on their people's unquestioning loyalty. The worldwide collapse of communism makes it very plain that the Hammer and Sickle is an even less plausible rallying point than Old Glory. And the only battlefield on which the Union Jack will rouse men to shed blood for it is a football stadium.

II

After *1984* and *Animal Farm,* mistrust of socialism was a commonplace. In that painfully comic film *I Married a Communist* (1950) the video blurb asks, "What happens when politics invades private life?" The handsome, home-loving American discovers that his new wife is a Communist agent for whom love and marriage are legitimate instruments of political infiltration. The honest and unpolitical hero learns that a leftie doesn't know what love is and may even put public ideals before private tenderness. She, in her turn, learns better. In the cold war movies and novels there is a gap between the world of politics, full of untrustworthy politicians, and national values, which are what politics should really be about if only those blasted politicians were not always jostling for power.

The hinge of the epoch turned when American movies began to mistrust, not power politics, but the values men and women really live by. This slow shift may be read in a dozen movies and several dozen novels, but Constantin Costa-Gavras's film *Missing* (1982) is the work of art that requires its audience to live the process through most piercingly.

Missing is set in Santiago during the Chilean coup d'état that killed Salvador Allende and Victor Jara in 1973. The missing young man is the son of a prominent and incorruptible Mormon businessman. The son had taken his happy marriage and blithe wife to live in Chile in a vaguely radical, vaguely bohemian, post-1968 sort of way. The story of the film is the search for him by his wife (Sissy Spacek) and his father (Jack Lemmon). The story is a true one. Costa-Gavras, a leftist and a libertarian, has no need to point its moral.

The missing man is found to have been arrested at home while his wife, trapped in the streets by the curfew and rightly terrified of the patrols, dozed the night away crouched in an entry. In the dark there were sudden bursts of automatic gunfire. Two truckloads of noisy soldiers drove through the streets. Two terrified people ran pell-mell down the street. A body lay on the sidewalk, heavy, commonplace, the blood long since drained away into the shirt and trousers. A white horse, of splendid lines and muscularity, galloped at full tilt along the broad boulevard, its silver mane and tail flying. These spurts of life and death we see from the young woman's hiding place.

She pieces the story of her husband's arrest together from neighbors' observations. Her husband was taken away by soldiers, as were two friends of his. Her father-in-law flies in, heavy in manner, exasperated with his son, anxious, but not yet afraid. He knows the ropes but is not heavy-handed. He has contacted the embassy and stirred up senior officials through contacts in his church and in Washington. The ambassador's stooge has no news of the young man. Maybe he's up-country? He had one or two compromising friends. . . . The father becomes even more exasperated at his son's playing at journalism, at being a writer, at being a househusband.

Things begin to look grimmer. The embassy still knows nothing, not even the grizzled, handsome CIA officer who once made a pass at the young wife. Father and wife go on a dreadful visit to the morgue, and there, among the rows and rows of crammed corpses, the blood dried around the purple bullet wounds, is one of the two American friends arrested with the missing man.

Jack Lemmon's hot-tempered, decent faith in American rights keeps him going, going to the embassy, keeps him believing the smooth young spokesman's evasions and the ambassador's lies. The same faith—this is the strength of the movie—keeps him pressing for admission of complicity from these practiced circumlocutors when he knows his son is dead—shot and then buried in the concrete of a road's foundations. He learns what the state will do and tries to hold that state, his United States, to its principles. These are his rights, and the rights of man.

His gallant daughter-in-law understands evil quicker. Her generation has learned the mendacity of its own rulers, and their cruelty. She loved her husband fondly and deeply; her father-in-law learns of that love and also learns the goodness of the children's marriage, their serious trust in truthful journalism, in the declaration of independence they lived in every turn of their phrases and bodies, in their boisterous cookouts and in their unofficial broadsheet run off riskily on an old mimeograph machine.

So the story is not that no politician may be trusted and all power corrupts. It is the more specific story learned so bitterly and incredulously by settled men like Jack Lemmon's well-off Mormon in a porkpie hat. His contemporary generation includes the lying ambassador and the cheerfully knowing killer up-country who has helped get the arms in, spread the dollars about, hold the dissidents, permit a few murders.

The manner of the movie is transparent. The story is assembled retrospectively as one would learn it oneself. "Where was he yesterday? . . . and before that? . . . and before *that?* . . ." It is pieced together in bits of news from those, either hunted out or bumped into, who know the bits: an ex-cop who has taken asylum in the jammed Swiss embassy garden; a neighbor, and another neighbor; a tourist girlfriend of the wife's; a man in a local aid office; mementoes in their devastated apartment. Costa-Gavras underlines the story process not so that we will conclude that *Missing* is just a story but so that we can also weigh its evidence and deduce its circumstances correctly—so that we shall know the truth. It is a trustworthy film.

Months later the young man's body is exhumed and he comes home in a coffin. The end is then our anger. We *ought* to be able to trust a power like the

United States, which ought not to sanction the murder of its nationals. What would it be like to be a citizen of a trustworthy state?

Answering such a question brings the great political theorists, de Tocqueville, Schumpeter, and Rawls, running across the fields in their long coats. They come to the graveside and look down. There is a coffin and inside it the decomposed body of a bright and trusting young man dug out of solid concrete. One of them must be right about how the trustworthy state should have protected the boy, to say nothing of the easy-going country in which he had briefly settled.

III

Probably it was Nixon's abdication and the interminable washing of dirty linen on the television screens that really finished for a long while public trust in the American polity. No doubt, the small-mindedness, the cheap fixing and automatic lying of the main players, helped by making so immediate and audible the casual denigrations and barroom cynicism. But these lapses were gathered into the icy current of the cold war moraine.

The state had been lying for years, pointlessly on the whole. It had lied less to deceive the enemy across the iron curtain and more to deceive its own people and keep them docile. Its incessant propagandizing had made public relations the fifth estate of the realm: news had become no more than a balance set by policy between intention and apprehension. Hannah Arendt notes that the publication of the Pentagon Papers exposed not only governmental lies over Vietnam but also two new kinds of lying in politics, an art in which plausibility had always taken priority over truth. The first is the lying of the advertiser and public relations man concerned about getting a good gloss on the governmental image. The second kind of new mendacity she detects "appeals to much better men" and takes the form of self-deception—fixing the facts so that they fitted the pet theory of policymakers about how things ought to be going. Both kinds of liars were lying to audiences at home, not those in Saigon or Moscow. Arendt reiterates what has become an axiom of contemporary liberalism, almost amounting to a new article of the Constitution: "So long as the press is free and not corrupt, it has an enormously important function to fulfil and can rightly be called the fourth branch of government."[3]

The muckraking journalist has long been an American folk hero, and quite rightly, too. There is now high tension in the courts of Western government between those journalists who wait for the press briefings, the background unattributables, the deceptions of the hired mouths, and the off chance of actually putting a question to the man in charge, and those journalists who are trying to find out what is really going on but who come up, unsurprisingly, with not much more than eyewitness reports. The ambitious journalists are routinely promoted from the battle front to the White House lobby.

Since Ellsberg's and Sheehan's Pentagon Papers, however, and since the investigative coups of Woodward and Bernstein after Watergate, journalists have moved closer to the offices of power and stand in a structurally ambivalent position toward them. On the one hand, they report what power says in the way power wants it. On the other hand, they speak for mistrust toward power and seek always to disconcert it, to tempt it into indiscretions and to confess to error.

This was not Walter Lippmann's way. Lippmann was a friend of the president and of the president's friends. His hearty advice came from outside the Oval Office, no doubt, but was certainly reported to the incumbent. The star journalists of the latter part of the cold war would never have essayed Lippmann's fruity prose; a much dryer, more laconic bunch, they found an idiom and a political space deliberately oblique to the diction of the day. They spoke not for or to America but for an exiguous truthfulness, to an undefined, anonymous, literate public scattered across the country.

Such figures turned up in a number of cold war movies and thrillers. They became the culture's microphone of incredulity and talisman upholding the old freedoms and the First Amendment. They stood for something serious, even when they came on as drunk or venal or just a nuisance. *All the President's Men* (1976) teaches mistrust as a political value, teaches that power habitually lies and that the methods of surveillance and clandestine inquiry had better be turned against the state before the state itself is confounded to decay.

As our two heroes, Bob Woodward and Carl Bernstein of the *Washington Post*, so significantly played by Robert Redford and Dustin Hoffman—Captain America and Jewish wisecracker—follow the Watergate trail by patient telephoning and canvassing doorsteps, they find it signposted by lies, dirty money, and countless denials. Ben Bradlee, their editor, backs them in spite of pressure from the White House and the publisher. By dint of methodical disbelief, incessant hard work, and playing off the double-crossers against one another, truth wins the day, and another American success story is written.

Such journalists can move easily from a presidential press briefing to the shadiest dive in town. They stand outside the ordinary traffic of social structure; they have an unusual mixture of friends on each side of the wavy line between crime and justice, truth and falsehood, that runs through the intelligence community. Such a journalist is one of the few people who might find out what is happening and do so *on our behalf.*

The only person with equivalent access to both the high and low rungs of the social structure is the policeman. But he is hired by the state. To be as free as the journalist he, too, must learn mistrust. In life, as well as in the books and films, he is shocked out of loyalty to the state by a jolt to his private life. But because he is a good cop, he keeps his trust in the possibility of justice. The cold war cop is a figure with a long lineage—the lonely man of principle standing up to the corrupt system.

In journalism the truth-teller is often the photographer: in real life, Don McCullin, Philip Jones Griffiths, the Magnum collective; in movies, the reckless secondary hero of *Salvador,* (1986), the quiet Cambodian of *The Killing Fields* (1984). Their pictures take the measure of the gap between how things really are and what people are told through official channels. The wider the gap, the more political mistrust becomes a mark of the free citizen.

The good policeman plays counterpoint to the truthful journalist's tune. Both anticipate the community of goodwill. The journalist tells truths to shame the state for its devilish activities. The policeman finds out the truth to recall the state to its duty as custodian of justice. Both teach that reasons of state never justify discounting the principles of justice.

As the cold war entered its fourth decade, a sneaky story began to go around that there was nothing much to choose between the American and Soviet states for ruthlessness, corruption, lies, and hypocrisy. *Gorky Park* (1983) is only one sufficiently intelligent thriller showing that, although each state's officers may be complicit in the other's graft, they will stop at nothing to keep the truth from getting out.

At stake is the Soviet monopoly of sables. An American has got some out and back to the United States. Good old capitalism. Three people are found shot and frozen in the snow beside the skating rinks in Gorky Park, their faces and fingertips scraped off. Arkady the good policeman, son of a state grandee, is the homicide specialist.

Book and movie alike combine that harsh knowingness, the specifics of violence, and a quite vivid and particular love of place that together so characterize the high-class formula thriller. Old cold-war theater has its usual place in the long interrogation to which the very badly wounded Arkady is subjected. Narcotized, beaten, shot, and hunted, he is implausibly indestructible, but agreeably so. The point of the story, discernible in spite of the formulaic breathlessness, snappy action, and bafflingly arbitrary plot—is about the complicity of states and the trustworthiness of love of one's country, in this case, Mother Russia.

The KGB and the FBI are in everything up to their necks; they kill indifferently, everyone kills indifferently. Osborne, the American homicidal businessman (played, of course, by Lee Marvin), is shot for justice by Arkady somewhere on Staten Island. Then Arkady leaves his wild-eyed Russian girl to become an American, frees the sables into the snowy afternoon, and goes home. She asks, "Will I ever hear from you?" "No doubt. Messages get through, right? Times change."

IV

Edge of Darkness was made by the BBC and shown in 1985. I take it to be one of the most remarkable works of art made for British television.

It is what Vladimir Nabokov once called "topical trash."[4] That is to say, its subject matter is the most everyday and urgent of fears. It exudes what at first looks like pious concern for superstitious environmentalism. It is very specific about real events: its action includes the very NATO commanders' conference held at the Gleneagles Hotel in Scotland in 1985, hard by the venerable golf links, at which Reagan's negotiating policy for Reykjavík was prepared.

Television is nothing if not topical. But in addition, *Edge of Darkness* settles viewers into the cold war frame of feeling and onto the edge of their seats by hanging out all the usual thriller signs—the smell of duplicity, the secret-service paraphernalia. Then it simply walks away from them, and viewers on the edge of their seats are left there, looking for the wrong climax.

The beloved daughter of Ronnie Craven, a widower policeman, is inexplicably murdered as he brings her home in the rain after a student political meeting. She is blasted away with a shotgun as they walk toward their village house in Yorkshire. (Near the end of this five-hour-long series we learn quite casually that on the spot where she died a spring gushed out where no spring had ever been before.) Craven discovers that she had been an active member of a secret green group that has penetrated hidden underground vaults where nuclear waste is stored and uncovered deals being done between U.S. and British corporations in weapons-grade plutonium.

Through the action runs the leitmotiv of a mysterious freight train carrying a squat ovenful of millions of years of radioactive sludge, trundling and clanking over points, moving as such trains do around us every day.

The police know who the killer is; a hit man from Northern Ireland, where Craven, the mute and anguished hero, was a crack interrogator. Craven baits the killer to his house, again in torrential rain. At the moment when the killer, a shotgun poking into Craven's neck, is about to tell him why he shot the girl, he is shot himself by watcher-police above the house.

The girl, a beautiful, big-eyed, cheerful Yorkshire lass, helps her father as a ghost. Sometimes a little girl, sometimes eerie, more often a reassuring presence as well as a keen blade of loss, she is ghostly in the way the dead are to those who loved them and to whom they were as familiar as breathing or loved music. She disappears only when something in her father's feeling goes wrong.

Craven is joined in his pursuit of the truth by a devil-may-care CIA man, Col. Darius Jedbergh, who is under enigmatic orders to "break into the ball park and steal the ball." Jedbergh has already infiltrated the greens and helped them get into the plutonium store so that he can report to U.S. intelligence the records of its plutonium which the British keep so dark. But now something cheerful and attractive and anarchic breaks out in him, the duplicity in which he thrives subsides, and Jedbergh speaks as simply as Deerslayer of the war between good and evil, the future of the planet. His English ally in this special relationship, rationalist and empirical and domestic to the soles of his caving boots, wants knowledge, not revenge. But in this case, knowledge would be as good as revenge.

They enter the labyrinthine plutonium store together, guided by an old acquaintance of Craven's, a leader of the National Union of Miners who has both secret responsibilities for the underground store and charges of ballot-rigging pending against him. The elderly miner helps them out because, as he says, he "hasn't sold out completely." The trio silhouette a new political alliance ranged against the reckless new freebooters of an economy run wild who will take *any* risks with the future of the world for the sake of the profits and a thin line of advantage. They are the most fearsome soldiers guarding the new world order that will replace the cold war. They are the strategists of what Raymond Williams, a couple of years before *Edge of Darkness* was shown, called "Plan X."

Plan X is sharp politics and high-risk politics. It is easily presented as a version of masculinity. Plan X is a mode of assessing odds and of determining a game plan. . . . To emerge as dominant it has to rid itself, in practice, whatever covering phrases may be retained, of still powerful feelings and habits of mutual concern and responsibility. . . . At the levels at which Plan X is already being played, in nuclear arms strategy, in high-capital advanced technologies, in world-market investment policies, and in anti-union strategies, the mere habits of struggling and competing individuals and families, the mere entertainment of ordinary gambling, the simplicities of local and national loyalties . . . are in quite another world. Plan X, that is to say, is by its nature not for everybody. It is the emerging rationality of self-conscious elites . . . [and] it is its emergence as the open common sense of high-level politics which is really serious. As distinct from mere greedy muddle, and from shuffling day-to-day management, it is a way—a limited but powerful way—of grasping and attempting to control the future.[5]

Plan X in *Edge of Darkness* is enacted by an insolent Tory minister, a cold, hard, youthful English businessman who owns the mines and drowns all trespassers, and a neat, boyish, middle-aged American who is an evangelical of the space-traveling future and the vizier of the plutonium bazaar.

The two heroes find the hot cell by way of tunnels and vaulted rooms hollowed out by miners' hands more than a hundred years before—"Victorian values, Mr. Craven." (They also find a nuclear war retreat and museum store built just after the Cuban crisis, sheltering a whole vintage of chateau-bottled St. Julien and the best concert LPs of the day.) After breaking into the cell, they shoot their way out and, fatally irradiated, steal two blocks of plutonium in a green Harrods bag, and escape to break the news.

Driven by their different recklessnesses, each is now doomed to die. Jedburgh goes off to Gleneagles, leaving a trail of CIA corpses behind him. At the hotel, in an effort to push the story out into the public domain, thereby to save it, he clashes his two plutonium pieces together in the face of his evangelical enemy, and the generals, admirals, and technocrats pile out of the conference hall in undignified terror.

No news escapes. Jedburgh primes his deadly stuff into a crude but serviceable atom bomb. ("Plutonium may not be user-friendly but as a means of restoring one's self-respect it has a lot going for it.") Craven tracks him down and, for Scotland's sake, reports his whereabouts. Craven's ghost-daughter visits him and leaves him a flower of the hardy black alpine perennial that she loved and which lives above the snowline of the unreceded ice age. Jedburgh chooses death in a bloody shoot-out; Craven sits it out with whiskey in the kitchen. In the last shots, the plutonium in its improvised bomb is safely winched from the loch while the dying man watches alone from the top of a rockface. The wind whistles thinly through clumps of the black flower as the credits roll.[6]

Maybe some improbabilities are too much: the news of Jedburgh's conjuring with plutonium would certainly have leaked out. The underground gun battle would not have been quite so crass. The Greenpeace environmentalism is a bit kitsch, and Jedburgh is too much altogether. (In one exquisite touch, we learn that his passion besides golf is watching TV ballroom dancing.) But these faults are trifles in a narrative that so masterfully gathers up the themes of Plan X and connects them with the ordinary details of British political life. *Edge of Darkness* ties together the commonplace murderousness of Northern Ireland, the arrogance of a party too long in power, the always faltering nature of an economy whose leaders will try anything for the sake of profitability, and the lived concerns of a new, intelligent generation convinced that the rich world must revere the authority of the planet before that authority exacts the obedience that is its due.

This is the post–cold war politics now struggling to be born.

V

The last fiction of my book was not a film or a novel, but one of the tallest and best-known stories of the day. Like all tall tales, nobody ever knew whether to believe it. Gradually, however, it became the vehicle of more mistrust than anything else. It was the story of nuclear deterrence itself, as told by governments, and no memoir of the age would be complete without it. It was the heart of ideology; it was policy become narrative.

The case for nuclear weapons starts from the premise that they can be neither annulled by law nor disinvented.

If nations possessing nuclear weapons are attacked, they will use those weapons sooner or later.

If a nation with nuclear weapons wants to attack first in order to prevent a surprise nuclear attack from its enemy (the "preemptive strike"), it will most probably only be deterred from doing so by the knowledge that retribution would be certain, swift, and severe ("massive retaliation").

Thus each nation with its weapons is held in check by fear of the destruction each would wreak on the other, no matter which nation fires its nuclear weapons first ("mutual assured destruction"). This is the balance of terror.

These theories have arisen from the ideological schism of cold war between the superpowers. Each side holds a further thesis enunciated by Zhdanov for the Communists, Truman and Acheson for the liberal capitalists, that the other is an inherently expansionist power bent on world domination and the defeat of ideological rivals. Nuclear weapons are the only guarantee that these ends will be frustrated.

Since the North Atlantic Treaty was signed in 1949, a number of additions have been made. One was a distinction between a nation that launches all-out nuclear attack without warning (the "first strike") and a nation that uses smaller nuclear weapons on the battlefield, perhaps as a warning as to what might happen next ("first use"). This distinction was proposed after the invention of much smaller, "smart" weapons, with their wits stored in their proboscis, which could be targeted much more accurately than the old mastodons.

NATO planners, then, devised a more "flexible response" by which nuclear weapons could mount a graduated ladder from little and smart to enormous and stupid. They thereby adjusted the earlier, simpler strategy of retribution which would follow any invasion or crossing of the "trip-wire" frontiers. The vaunted flexibility to burn up the place a bit at a time also blocked any Soviet impulse to try an old-style invasion with tanks but without nuclear weapons on the wager that the United States would never incinerate Europe if it was not being touched itself.

These refinements led to the convention of distinguishing battlefield or tactical nuclear weapons (which is to say, weapons rather smaller than our sleepy chum, the Fat Man) from strategic weapons, any one of which would obliterate Philadelphia and poison Pennsylvania. The escalator mounted its steps toward the burning-out of the globe, the idea being that at any step the combatants could call a halt before proceeding to the final devastation.

It was not, however, until the early 1980s that the argument about winnable nuclear war really surfaced. Before then, the flexibility of all available responses lay not so much in a supple choice between blowing up the whole of Westphalia or only a little place like Wuppertal. The flexibility was cerebral: the enemy, seeking weak spots in the armory of his opponent without putting itself at terminal risk, must see that there is no move it can make that is not unwarrantably dangerous.

These are the assumptions of thinking about nuclear war. So far as could be determined, they had Soviet equivalents. They are not, contrary to the assertions of some expositors, very difficult to grasp. Although the intricacies of the technology have been used to embellish the basic plans, and flexible responsiveness has encouraged bright young planner-killers to come up with new applications of old weapons (as happened with cruise missiles), especially in the race toward "parity," it is an easy enough game to understand.

The game turns on accurately imagining all the moves your adversary might make (moves motivated by his ambition to conquer and expand, which is a datum). These speculations are then regularized as a series of predictions, each matched by formalized gambits. The protocols of game theory lent themselves to the organization of these giant, imaginary maneuvers, and NATO planning and training centers snatched them up eagerly.

Nuclear war games, however, are not exactly zero-sum in their imaginary outcomes. Each player judges what to do according to the best possible result only for himself. This criterion rules out altruism and turns prudence into the central political virtue. The "prisoner's dilemma" is a frequently used model in the game of superpower relations.

The dilemma that faces the prisoner faces another prisoner in a different cell at the same moment. Each is charged with a crime. Neither can reach the other. Each is advised that there is not enough evidence to convict either of them, but that if both confess, then they will both be sentenced to a period in jail. If, however, both refuse to confess (remembering that neither can know what the other is doing), then both will be released. But if only one decides to confess and the other does not, the confessor will go free *and* be given a reward, while the other will be given an even heavier sentence than if both had confessed and been convicted.

What should the prisoner do? If he could speak to the other prisoner, they could agree to say nothing and go free. But he can't. If they both confess, then they both stay in jail. If he confesses and the other prisoner does not, he goes free with his bounty. If both stay silent, they get out. The lowest risk to the prisoner is to confess; confession might also earn the big prize. It might also, however, bring on a jail sentence.

The disarmament-rearmament spiral that dominated the cold war may be understood not so much by translating its tensions into the choices posed by the prisoner's dilemma as by grasping that these were the devices the cold war managers used to control and rationalize their enormous, insane, and competitive bureaucracies.[7]

Given the appalling dangerousness of defense establishments, the efforts of forty-odd years went into making technique the foolproof control. Technique is thought to be proof against fools because human judgment and cognition have so far as possible been removed from it. Technicians are highly skilled and thoroughly demoralized. Hence, in the ethos of disciplined substitutability essential to all large organizations, but especially military ones, there is always a routine to follow. Stupid people can replace clever ones with no danger of lethal mistakes being made. A skilled technique is a routine; naturally, some people are better at it than others, but to exclude incompetents, the technique can simply be graded.

Even management as methodical as this, however, reaches breakpoints at which judgments must be made. The manager handles judgment within a strictly defined field of authority. He begins by basing judgment as far as possible on calculation. If the jumbled signals of life can be registered on the ordinal scale, assigned a value, and printed out formulaically, then the buggerings-around of human inanity can be canceled out. When, in spite of following this procedure, human fatheadedness continues to intrude, the manager acts to the limit of his authority and then refers the decision upward.

These procedures hold throughout the defense establishment. They are the synapses of a giant social cortex—a cortex, however, whose intelligence is entirely artificial. This system feeds its eager circuits with messages, processes novel information into the preferred sacs, indeed, *learns* to adapt and to reward itself for adapting. But it cannot judge, and it never reflects.

The careful enclosures of task and attainment, of means-end planning, cannot face the unprecedented, cannot deal with shock, cannot act with flair or imagination (it would be too dangerous). Planning for nuclear war must stay within the limits of a safe routine. It truly must. But in dealing with error or disaster, routine can be as lethal as recklessness. All simulations of limited nuclear war have ended up as uncontrolled exchanges of mass destruction exactly because the rules could not stop the machines in time. (To say nothing of the fact—and nothing ever *has* been said—that nuclear explosions blow the guts out of all electronic circuitry.)

In the bureaucracy of nuclear defense planning, this most durable and delicate of world bureaucracies, the thought-forms that dominate it work against all human interests. But it is a human trait that when brain and culture have worked busily together to evolve new thought-forms and to turn them into art, its human practitioners do not easily give it up. Their lives were constituted by these crazy conversations. When a bright boss like Herbert York, the first director of defense research and engineering at the Pentagon, demolished the intellectual ground on which certain defense thought-forms were based, planners instantly created another little platform from which to look out for the enemy.

In Herbert York's famous law of diminishing returns on national security, security stands in a fixed ratio to the increases in weaponry.[8] As one superpower increases its military might with new weapons, so the other superpower hastens to catch up. As both superpowers pursue both policies (getting ahead, catching up), the national safety of each degenerates at speed. Although both superpower planning outfits recognized this law, each continued to play for advantage, in the unkillable hope that the next move would not only make it more powerful but also more secure, and that this lead could be held. Moreover, since U.S. planners were themselves in the grip of the managerial thought-form always aspiring to totalitarianism, it was only predictable that they would come up with their final opus—the Single Integrated Operational Plan (SIOP), the last and tallest tale before the end.[9]

The Plan cannot be discussed without hilarity. It covers all eventualities but starts from the cheery premise that all U.S. weaponry will begin nuclear war in the pink of condition and would not be caught by a clandestine Soviet strike with unlit fuses. This is the best case, which gives its structure to the Plan. In the worst case, U.S. nuclear forces have been, in the inimitably euphemistic habits of the trade, "degraded" but are still operable. It seems that the ultimate worst case (for U.S. nuclear planners), American weapons reaching such a state of degradation that there are none left, is beyond the scope of the Plan.

To the planners, however, this inadequacy is logical, for they do not judge the Plan by its capacity to defend American cities and people. The point of the Plan is to knock out Soviet missiles, preferably before they are launched. They add an evaluative scale for different stages of nuclear engagement, ranging from the probability of kill (that is, knocking out targets) to what might be called (but isn't) the probability of *being* killed but instead "pre-launch survivability." The top-heavy hyphenation of those remorselessly abstract compounds cuts them off from real life and human ends. Yet the end of all humans is what they unabstractly project.

There is an odd counterpoint between the cumbrousness of the special concepts devised in strategic policy (a phrase itself redolent of megalomania) and the flavor of rather nineteenth-century romance in the nomenclature of the weapons themselves. The conceptual style is aggressive-technical, stripped of colloquial overtone, demoralized as by hard realists. Myself, I relish "hard target kill capability" (dumping on the missile silos before the missiles slope off) for its tight-lipped precision; others prefer "circular error probable" (guessing by how much you might miss the target), "throw-weight" (means just what it says), and the mighty MIRVs—for nothing is more characteristic of nuclear weapons discourse than its acronyms—which are multiple independently targeted reentry vehicles (rockets with willful baby rockets on their backs). These resonant chords are linked by the melodious lilt of "capability," "facility," "proliferation," "posture," "counterforce," and a clutch of no-nonsense metaphors—"firebreak," "bean-counting," "bargaining chip," and the like.

These terms have accrued meaning and thickness by protracted use, so that we can entrust our lives to the planners, even though they daily compute the eradication of millions of our lives. Actually, it is only at the insistence of such high-IQ morons as Herman Kahn[10] that anybody talked much about dead bodies before such associations as Scientists against Nuclear Armaments came along; the death toll is not usually mentioned by either the strategists or, of course, the politicians. What was at stake for them all—and there is no question of how much it mattered and how fast it got through into popular consciousness—was sheer bulk: the bulk of U.S. weapons on land and sea and in the air, as opposed to Soviet bulk.

When plans turn into weapons, perhaps because such matters were in the hands of scientific killer-visionaries instead of prosaic killer-strategists, they were given names very different from those emanating from the offices of abstract hyphenation. The act of naming has its necessity.

Nuclear weapon names seem to come from schoolroom memory. Thus, the fearful monsters that will shoulder their way up from the deep before bursting into the sea air and soaring off on their mission, spouting orange flame behind, are offspring of the gods: Poseidon, Polaris, Thor, and Trident. Truly American missiles that will be launched from the homeland are Honest John and Minuteman. Weapons to be launched from man-sized trucks, with merely local death in their bellies—from one little kiloton up to ten or so, a trifle, a trifle—have been christened Lance and Tomahawk in an amiably archaic way.

Such were the plots and protagonists of the greatest story ever told. It had an audience and a cast of billions. The hardest judgment of all is deciding whether or not it was sheer luck that prevented the final take.

PART V

Finale

29

The Final Chorus, 1989

IN 1989 the new tide fairly flooded through Central Europe, touched China and Southern Africa, swirled fiercely into the Soviet Union, and left the Middle East and both Koreas high and dry but peering anxiously out the window. For years the political commentators had assumed the fixity of cold war politics and its ideological iceworks. They had declared themselves mightily impressed by the solidity and weight of the political institutions on each side of the ideological divide. They identified the difficulty of changing all this as, in the jargon, the problem of agency. *Who* could do anything about it? In the meantime, the cold war had continued for nearly fifty years, and there seemed no reason why it shouldn't go comfortably on for another fifty.

Agency was indeed a problem. The grand headquarters of NATO and the Warsaw Pact, the vast submarines far below the ice pack, the unobtrusive, home-loving men going to work every day with the keys of missile silos ready to insert in consoles in their offices, the summitry-jamborees and the defense correspondents, the silent factories and the shining weapons being wheeled below great hangars on quiet trolleys, it all looked so permanent, so beautiful, so organized and ordinary. Who on earth could stop it? Who would want to?

To everyone's amazement, the first agent of the momentous changes of 1989 was that old historical standby, the people. The way had been prepared by one man, Mikhail Gorbachev, who had shown that the massive immobility of the way of the world could be shifted if there were a will to do so. But the sweeping changes were driven forward by crowds and crowds of people, nearly all of them peaceable, not so very many hurt by their spontaneous and wholly reasonable disobedience of the prohibitions that had kept them back so long.

To be sure, those prohibitions would have continued to hold firmly if they had been backed by tanks and machine guns as they were in 1956 and 1968. Here, too, Mr. Gorbachev made all the difference: it is said that, if not for his intervention, East German hospitals would have been told by the bosses to expect 100,000 casualties when the crowds were cleared. That simple moral was brought home in June by the Chinese government, which released tanks against a few thousand students in tiny tents and left the tank commanders to clear Beijing's Tiananmen Square by whatever method suited them best.

Elsewhere, however, the tanks stayed in their barracks, although it was a near thing in Berlin and Prague. In Bucharest they appeared briefly and uncertainly, first for the state, and then, in a surge of sympathetic conversion that was one of the best moments of the year, for the people.

But the most striking thing about 1989 was the visibility and efficacy of the crowds. Each country had its different order of crowd, each crowd its distinctive, national character. The Romanian crowd that jeered the appalling Nicolae Ceausescu right off his perch on the balcony was packed and static, a slow crowd. The crowds that piled out of East Germany into the West by way of the Hungarian border in September was hurried, absorbed, excited. It was a quick crowd. The crowd that massed on either side of the Brandenburg Gate, poured through the checkpoint in both directions, and chipped off bits of the Berlin Wall for mementos was a happy, festive, party-going sort of crowd that had come to see history for itself and turned out to *be* that history in the making. The breaching of the wall was the sign that the crowd had made history. Witness had turned into action. The serpentine files of Polish believers threading through the long grass to the huge cross Karol Wojtyla had erected in an empty field was a thin, gentle, eloquent crowd, though there were hundreds of thousands in it. When he returned as Pope John Paul II to bless his people, the gentle sound of their applause was like the pattering of rain, rising and rippling endlessly until it so saturated the policemen ranged against them that they gave up and, soaked in benignity, began to go home.

It would be forgivably innocent to suppose that these unarmed crowds did everything for themselves, just as it would be to thank Mr. Gorbachev and give him his Nobel Prize for stopping the cold war all by himself. Behind and among the crowds, however, were the men and women who had kept resistance going from small, ill-lit basements and in illegal broadsheets for so long, who had sustained the dissident organizations at the cost of peace of mind and, at times, spouse, home, children, and who now were at the forefront of improvised coalitions with nonparty names.

Behind them stood the ruins of state socialist economies.[1] In 1989 the Soviet budget deficit was 20 percent of net material production; four years after Reykjavík, military spending still accounted for 15 percent of that product. The Soviet Union had beggared itself trying to meet the challenge of U.S. weaponry. The corresponding extravagance of the NATO countries had not yet hit home so heavily.

The Soviet Union had kept production on a war footing that never faltered. Their satellites were obliged to follow suit; the manufacture of steel and armaments and heavy industry took absolute precedence. These priorities were backed by strict controls on civic expression, even on ordinary domestic encounters in cafés or shops. Social development from East Berlin to Bucharest stopped dead in 1949. The national leaders were "Stalinized," each turned into the simplified image of the poster; Big Brother was born. Whatever improvements in the lives of the East European poor were effected between 1945 and 1989—and they were considerable—all experience was dominated by the grim forms of life invented during wartime.

The system seized up. A new generation, some of whom in East Germany, Czechoslovakia, and Hungary had seen something of Western television, and all of whom had thrilled to Western rock music, refused to believe in the capitalist bogeyman or in the need for a war economy.

There can be no doubt that the Soviet central committee called the cold war armistice. But the change in feeling, the tide coming in, had already been making itself felt for a few years. As it did so, as liberalizations popped up all over the East, and as the tanks with cold engines sat locked up in their marshaling yards, the flood gathered that would go quite out of anyone's control.

It might go to the good or the bad; once power is released there are plenty of gangsters waiting to snatch it up. It might sweep away its authors, including Mikhail Gorbachev. But it would sweep on its way to an end that only began in 1989.

II

The year began in Poland. The Poles had been a slow crowd, an interminable, thin line of pilgrims trooping from Gdansk in August 1980 through to Solidarity's eighth-anniversary strikes. During those years the union had been illegal, gone underground, surfaced again, and kept up its indomitable style of public debate, including an hours-long broadcast on state television between Lech Walesa and the head of the state trade unions. All this while Andrezj Wajda, Poland's great filmmaker, was putting his considerable cultural authority behind Solidarity, particularly in an enormous documentary about 1980 and in an open parable about the French Revolution, *Danton* (1982), in which Jaruzelski was cast as Robespierre and Walesa as Danton.

But Jaruzelski had held off the Soviets in 1980, and he now gambled with resignation of his post in order to win official recognition of Solidarity in January 1989.

The central committee, in turn, gambled *its* authority on early elections, the first in Poland against any but stool-pigeon opposition since the Communists won hands down in 1946. Solidarity, for its part, wanted Jaruzelski as president and a free hand for themselves to organize as a trade union. Before the union knew where it was, the agreement settled by the central committee on 5 April gave it virtual status as a political party.

Solidarity was headed by its citizens' committee. It put together a slate of candidates in a hurry, often forgoing electoral procedure and simply asking them over the phone to run for office. The election was to be the preliminary to several years—many Poles put it at four—of preparation for full-blown multiparty democracy.

Nobody was ready for what happened. Solidarity ran a simple campaign, counting on Walesa's unique authority and its own unimpeachable moral position. Its badges and flags all carried the now world-famous scribbled signature, red on white, *Solidarnosc*. Its slogan was "We must win." The Communist party ran on behalf of a muffled coalition with neither policy nor leaders.

Yet except for Walesa the Solidarity candidates were in a state of nail-chewing suspense about the result. The voting papers were long and complicated, requiring voters to cross out all those on a list of thirty-odd candidates to whom they did not wish to give their vote.

It must have been comically plain to the officials in the polling stations what was going on as the long lines formed to vote on 4 June. They must have heard the repeated buzzing of the ballpens echoing on the narrow deal shelves as every voter crossed out every Communist name for seat after seat, right across the country.

Sixty-two percent of the electorate voted, which can be read as a certain skepticism about any politician's ability to rescue a modest economy $32 billion in debt and with no prospect at all of paying it off. But almost all the vote went to Solidarity, where it could. In the curious electoral college prepared for 4 June, it won its maximum one-third in the lower house of the parliament, the *Sejm*, and every seat in the Senate except the president's.

Walesa remained the uncontested, unofficial leader of Poland, still indomitable, still with his astonishing stamina, increasingly confident and dogmatic. His unofficial cabinet, the Citizens Parliamentary Club, was stacked with the intellectuals who had remained at the forefront of the struggle. Jaruzelski scraped home as temporary president on a narrow Senate vote in late July, and the Communists and the Solidaritarians together bent to the tasks of tripling prices, cutting wages, and asking the International Monetary Fund for more.[2]

III

Hungarians run a long revolution. Since 1956 János Kádar, his signature forever warm on the death warrants of Imre Nagy and his men, had coaxed his country toward low-level comfort. He had kept out of Czechoslovakia in 1968. His dissident intelligentsia had on the whole stayed out of jail. In 1989 Kádar fell ill at just the right moment and lay dying as the body of Nagy was at last reburied with proper reverence in front of the whole capital on 16 June.

In early 1989 the students had been in the streets again, wild lads shouting at troubled memories. They called themselves by an old name from 1848, the Young Democrats, and they had an appropriately dashing and magniloquent leader, Viktor Orban, who called up the ghosts of student leaders dead for thirty years and demanded at a Budapest demonstration in March the withdrawal of the Soviets.

When Kádar fell ill, he was replaced, unusually, by a presidium of four men. An opposition coalition—Young Democrats, Free Democrats, trade unions—pressed so successfully for talks on a new constitution that the presidium agreed. The groups met through the summer until the end of September and produced a Magna Carta with new laws, new political parties, and a date for multiparty elections.

In May some Hungarian official found the nerve and the power to open the border with Austria. As the democratic future of Hungary was being drawn up without acrimony, thousands of East Germans poured through in a motley caravan of rattling old Ladas, across the tip of Czechoslovakia, into Hungary, over the border into Austria, and thence to West Germany. They came to stay, bringing their families, the portable bits of their homes, and the artisan skills with which East German apprenticeship programs had carefully equipped them. The news photographers of the West waved them across the frontier under the barrier arm, and they shouted, waved, and sang their way into the consumer society.

By the time the constitutional deliberations were done, communism had been polished off in Hungary. On 23 October, the thirty-third anniversary of 1956, the president declared the new republic open, and the people below clapped politely and went quietly away. The thing was done.

IV

The Berlin Wall was built in August 1961 to stop the steady departure for the West of 1,000–2,000 people per month. When Hungary raised the small door in the iron curtain in May 1989, 50,000 East Germans left for good before the end of October.

The East Germans had watched the developments in Poland and Hungary on their West German television sets beamed from West German stations, and they became restless. Their skilled workers were leaving home forever, and looking around at the remaining seventeen or so million, they saw how many of them were elderly. They weighed up the low level of their prosperity, the showpiece of socialist political economy, they heard the old lies from their leader, Erich Honecker, and they remembered his predecessor, the Stalinist Walter Ulbricht, and their parents' brief disobedience in 1953, with its roll call of the dead after a summary introduction to the T-34.

So, at last, they became angry once again. A few small objections flared up, and the police, sons of the poor, beat the slightly less poor back into subjection. On Saturday, 7 October, the state officially celebrated its fortieth anniversary.

There were protests at the celebration, and arrests and beatings as usual. Most of the upset was in Leipzig, and after the cops had done their habitually brutish best on Saturday, the Leipzigers turned out on Monday in tens of thousands to bear, once again, silent witness to their sense of outrage. Some of their number were founders of the still illegal New Forum, a self-consciously civic movement formed to recall a sclerotic Communist party to the best principles of citizens' socialism.

The crowd in Leipzig was either dense, purposeful, completely pacific, or heeding broadcast appeals from three leading New Forum members—a priest, a conductor, and a comedian—for Gandhian disobedience, to repudiate violence. The troops on alert were never called out. By demonstrations, public prayers, and vigils (in a country with few Catholics), the implacable, peaceful dissent moved across the country like an anti-cyclone.

Honecker, last of the old tough subjects, was voted out by his peers and a conciliatory, temporizing successor, substituted. He reopened the frontier to Hungary, which had been closed to stop the exodus; Honecker's stooges resigned in droves. Finally, the people of East Germany came in vast numbers to the center of Berlin on 4 November and with a few speeches and their mere, enormous presence—a still crowd with a fixed, pacific purpose—transformed their state and its beliefs in a flash. The offices of the secret police were opened, the council of ministers resigned as one, and after twenty-eight years the Berlin Wall was opened and its impending demolition announced.

What was left of the government was as good as its word. Over the weekend of 11–12 November, two million East Berliners crossed to the western half of their hometown to buy a few goodies and savor the pleasure of sauntering past the impassable barriers. The balls and chains smashed into the top portions of the wall while a crane dumped great lumps of concrete on one side. From all over Europe students poured into Berlin for whom the Wall was the symbol of the mess their parents and grandparents had made of a Europe which was become their playground. Broke, scruffy, unquenchably happy, they came in their jeans and T-shirts, to gouge out a bit of wall, to go through the forbidden gaps, to share *Traminer* and *Dunkelbeer* from the wiped neck of the bottle. The generation that would have to atone for the sins of the fathers restored to the great city a heart-rendering, instantaneous camaraderie.

V

In 1989 the students of Eastern Europe were as bold and brilliant as the students of 1968. In Poland their professors were working alongside Walesa; in Hungary the students were the Young Democrats; in Prague they started everything off again.

The Czechs arrived slightly later at the year's theater. Czech students wanted to change everything, to throw out the awful old bullies and liars who treated them like dogs and fools, to stop suffocating, to speak out argumentatively, *to hear the truth*. On Friday, 17 November, to commemorate a student martyred by the Nazis, they revived the symbols and ceremonies of 1968 and marched that evening straight into the police truncheons, boots, shields and all, singing and holding aloft candles, broom, and branches of oak flowers still on the trees.

They went down before the heavy clubbings with broken wrists, splintered cheekbones, and punctured kidneys, they went down or they ran away, and the world's television cameras recorded it all, honestly.

The next morning, the students called with one voice for an occupation of all the universities and polytechnics in the country. Not long out of prison, Vaclav Havel, the intellectual hero of 1968 and the prime mover behind Charter 77, the twelve-year-old manifesto for the people's freedom, hurried to the theater where he was so well known, "the Magic Lantern." The students were there, as were members of the many underground and unoffi-

Figure 29.1　The demolished Berlin Wall, 1989. (The Hulton Picture Company)

cial resistance groups. On Sunday, 19 November, they constituted themselves the Civic Forum and became the government-in-waiting and unacknowledged legislators of their nation. They published demands, rattled out from desktop word processors: resignation of the loathsome President Husák, a creature of Moscow; many other resignations; an inquiry into the secret police; and liberation of all political prisoners.

Wenceslas Square was both theater of the crowds waiting on the tiny stage at the Magic Lantern and their auditorium. There they heard the demands repeated endlessly, above all by the guileless, awkward, eloquent Havel, the moment's hero. Good King Wenceslas listens also from his plinth, and the message from the Kremlin was that Husák was on his own; there would be no tanks from Moscow or Warsaw this time.

The Velvet Revolution was a revolution of the cultural class. Havel was at the center, exhausted, animated, gentle; around him were actors, writers, journalists, students, professors, teachers, many of them exiled into manual labor by Husák, all of them up and boiling to avenge their books.

The movement was of a piece and very pure. Its roughly produced photolithographed paper was called *The Free Word;* it sold out daily before breakfast. The movement knew it had the enemy on the run, but its leaders seemed never to have calculated anything. Havel always looked surprised; he was the opposite of a Leninist, which was why he was elected to the presidency so triumphantly. He was pleased and surprised to hear that Alexander Dubcek was returning from his cottage-in-exile, and on 24 November Dubcek arrived.

He only served that evening in November to repudiate the enemy; his benign remarks did not match the moment of 1989. Dubcek was still a socialist, and good luck to him. But he was first and foremost a Slovak and a victim of the old order. The crowd hailed him that night in triple, smashing shouts of his name, and the tears poured down his candid face.

By the second weekend of the revolution, Czechoslovakia was effectually being ruled by the amazing unanimity of its mass meeting. Hundreds of thousands gathered in the square, and Havel and his faithful Forum told the crowd that the Forum was only the transition to free elections. The crowd trusted him absolutely. As names of the existing central committee were read out over the loudspeakers, they were derisively echoed by 5,000 throats. By Monday morning, 27 November, the day of the general strike, the central committee members said they would resign.

The old men did so slowly and with ugly reluctance over the next two weeks. All that time the Magic Lantern and the Civic Forum were the nervous system and brains of the revolution, the playwright and his intellectuals— writers, actors, students, blithely planning the good society, bringing the theory of the seminar to a highly practical revolution.

The ruined government promised a new coalition of members adequate to the national crisis, then delivered the old has-beens. The Forum called another national assembly to protest this ignorant cynicism. It was held in the fierce cold in Wenceslas Square. Once more the crowd hooted and catcalled at the reading of the central committee names, and one by one the members shuffled out of office, finally leaving only Husák. Then he went as well, turning out the lights.

VI

By the time Havel was elected president and Dubcek chairman of the federal assembly in Czechoslovakia, the most murderous old man left in the old Stalinist system had been routed out of office. By the end of 1989, Nicolae Ceausescu had been Romania's dictator for forty years and had turned, from never very promising beginnings, into a tyrant of Byzantine proportions. He left behind an unfinished and grandiose palace with gold-leaf embellishments when he and his wife set off for an exile supported by what were sure to be enormous private bank accounts embezzled out of state funds. But their people caught up with the Ceausescus and dealt with them, as the CIA says, "with ultimate prejudice."

It is hard not to feel a hideous relish at their brutal end. During his last year in office, Ceausescu had ordered the destruction of all the houses in two out of every three villages of his country. He *what?* He ordered the destruction of two-thirds of the villages in Romania.[3]

He explained that it was part of modernization, or industrialization, or some other-ization. Romania is the most backward of the old Balkans. Its valleys are green and lush in the spring, and the villages share the ancient vernacular of the Mediterranean—whitewashed stone, slate, and mottled terra-cotta tiles. Bony olive trees and tough vines cover the hillsides, and the hedgeless roads are still trodden by donkeys and the occasional oxcart.

The bulldozers came in the spring of 1989. They gouged deep ditches behind the little streets and pushed the houses into them, "like a mass grave," as one villager said. Then they ground down the rubble and flattened the site. The villagers were encouraged to destroy their own homes, and those that did, hoping perhaps to salvage a little for their smallholdings, were pointed out to Ceausescu as virtuously supporting his policy and its vision of a future in the best and grayest of high-rise tenement blocks.

You hear such a tale, a tale of bullying hypocrisy and utter disregard of natural feelings for home, and you feel dull anger boil over. It was such a feeling that overcame the tyrant. There he was on his balcony with his hired stooges cheering him as he recited the triumphant progress of his destruction program, when quite suddenly the strange spirit of disobedience abroad in Europe rippled through the crowd and turned it from sullen obedience to anger.

The scene was broadcast all over Europe as part of television's revolutionary spectacular in 1989. The crowd began to mutter and then to roar in derision at the dictator. Ceausescu stepped back in anger and consternation and tried to quiet the crowd by wagging his outstretched hands soothingly. But the roar of hate went on, until he hurried away, the city was in the streets, the secret police started firing, and the army wavered between duty and mutiny.

Romania was the only European country in that amazing year to see revolution (and counterrevolution) in military forms. The secret police fought fiercely, and there was bloody street fighting. The Securitate opened heavy fire on a crowd in the northern city of Timisoara, and several hundred were killed. At that point the army turned against the tyrant and joined the people. There were assaults by combined units of army troops and armed civilians that gradually sought out and shot the sniper-policemen holed up in the blocks of flats. Once the army had turned, however, it was quickly plain to the hated Securitate that it was doomed, and it judiciously surrendered all over the country.

Many of the secret police were shot when they did so. A lot of vengeance had been exacted by the time the new year came. Its most summary moment was the execution of the tyrant himself and his wretched wife after a short, hammering kind of trial. The dead bodies were displayed to grim effect on television, and the leaderless counterrevolution died away.

Thousands were killed, and without the focus of a Stalinist government, the country had no political structure from which to run itself. But there can hardly have been an honest human being anywhere who did not rejoice at the overthrow of a regime frozen in the darkest moments of Stalin's 1950.

Just before the wholly unexpected revolution came, the Romanian poet Mercea Dinescu wrote hopelessly to other writers in the West about how inconceivably awful life in Romania was. Poland, he wrote, was free by comparison; Havel in prison with his laptop was beyond a Romanian poet's imagining: "disinfect your door handle thoroughly; it may be poisoned." And he wrote, "in actual fact there are twenty million protesters in Romania, unpublished dissidents who live their lives gagged."[4]

On Christmas Eve, the nation removed its gags, and with one voice removed the enemy.

VII

On a savage tangent to the European jamboree, the Chinese government served up a reminder that the self-congratulatory Westerners could usefully keep their caricatures of communism in good working order. When several thousand students piled into Tiananmen Square in Beijing in early June, built themselves a plaster-of-paris Statue of Liberty, and settled down in little bivouacs to wait for the government to resign, the politburo responded tardily.

The Western news and camera teams poured in. Since the Carter and Brzezinski rapproachement with China, the Chinese had been exempted from cold war strictures. They had their peculiarities, no doubt, especially with regard to the Khmer Rouge in Cambodia, but then, so did the United States. They looked as if they were going to tread their inscrutably Oriental path toward the market economy, and in the meantime, American business would sell them what it could.

So the crunch of bones under tank wheels in Beijing came as a shock. There were upheavals and deaths elsewhere in China, but the telerevolution was booked for the capital. So the world could watch the action in Tiananmen Square, even if the rest of the Chinese could not.

Not everyone in the Square was a student.[5] There were workers from nearby factories, piratical with their red cloths tied around their heads. People had brought their little children to stare at the sudden, temporary settlement of untidy campers. At night the darkness was speckled by little fires all over the square. There were no public lavatories and an acrid smell of urine.

Small groups of riot dispersal units from the army had tried already to clear the square, but their shields and batons were no match for the thousands of students with bricks and bottles, so they went away. The square was packed and busy and throbbing with excitement. John Simpson from the BBC, roaming around and about after losing touch with his film crew, suddenly saw hundreds of figures, thousands even, standing stockstill around a lampost. Out of an old-fashioned loudspeaker bolted to it came these words: "Go home and save your life. You will fail. You are not behaving in the correct Chinese manner. This is not the West, it is China. You should behave like a good Chinese. Go home and save your life. Go home and save your life."[6] The people listened. They did not move.

The tanks came sporadically, as though they had no coherent orders. One vehicle was only a light armored carrier with a small squad of soldiers in it. A corner of the crowd turned into a lynch mob as the carrier ran down several people, grinding limbs under its tracks. It stuck on a block of concrete. The crowd yelled and swarmed onto the helpless vehicle, recklessly pouring gasoline everywhere, dropping Coke bottles of blazing gas down its flues. The vehicle caught fire. Simpson goes on, unforgettably:

> The screaming around me rose even louder: the handle of the door at the rear of the vehicle had turned a little, and the door began to open. A soldier pushed the barrel of a gun out, but it was snatched from his hands, and then everyone started grabbing his arms, pulling and wrenching until finally he came free, and then he was gone: I saw the arms of the mob, flailing, raised above their heads as they fought to get their blows in. He was dead within seconds, and his body was dragged away in triumph. A second soldier showed his head through the door and was then immediately pulled out by his hair and ears and the skin on his face. This soldier I could see: his eyes were rolling, and his mouth was open, and he was covered in blood where the skin had been ripped off. Only his eyes remained—white and clear— but then someone was trying to get them as well, and someone else began beating his skull until the skull came apart, and there was blood all over the ground, and his brains, and still they kept on beating and beating what was left.
>
> Then the horrible sight passed away, and the ground was wet where he had been.[7]

They say the lynch mob was not students; the students sang the "Internationale" and stayed in the open spaces of the great square and vowed to observe Taoist nonviolence. But the smell of blood was in the air, and the soldiers were enraged at their comrades being torn to death. They came back in a massed phalanx with bayonets and machine guns and squadrons of tanks in battle order, and the students and the mob died together, crushed or shot. Once again, communism provided the televisions of the world with living images of its valueless power, power that came down like a fist on its children and smashed their lives out, all because they didn't feel the same about things as their fathers, and wanted to say so.

VIII

Warsaw, Budapest, Prague, Bucharest, Sofia, Berlin, are the names of the end of the cold war, as they were of its beginning. Europe is a complacent continent, never more so than in its wealthy cities. When it has prospered, it has done so at the expense of the rest of the world, and when it has fought with itself, it has drawn the rest of the world into its uncivil wars. The ruling class of the United States is the progeny of Europe, and it has learned the parents' ways.

The mighty and puissant United States would still try to dominate the great area of its imperium and the tiny area of its backyard. The Soviet republics would quarrel more or less violently among themselves. But the two great powers would have to do without the cold war, and the tale of its two cities, Washington and Moscow. That story had frozen history stiff with fear, stayed governments in a condition of perfect righteousness, armed and mutilated the wretched of the earth far beyond their means or their desserts, and driven honest citizens out of the company of friends and deep into their burrows. Rumor and its many tongues would now need new mouths to speak from.

30

BIOGRAPHY X

The Peace Woman—Joan Ruddock

HOWEVER rarely sentimentality can be cited as historical cause, it was surely a change of heart as well as economic collapse that prevented the tanks from rolling in Poland and East Germany, and it was certainly a change of heart that brought many millions onto the streets and boulevards of Western Europe to protest so politely against the politics of incineration.

The demonstration march is a curious thing, but a potent phenomenon. It has a long history, not without martyrs and corpses; but in the West since 1968 it has had a quiet enough record and a strikingly formal, modest, and good-tempered standard of conduct. These people with their strollers, scruffy dignity, and their conversational cheerfulness have come to bear witness as individuals in a crowd of millions to their own quiet helplessness, their conviction that they and their small children trotting beside them are ill served by the dread, shimmering lines of missiles with their trip-wire sensitivities.

Across Europe and, in glimpses, across the United States, Joan Ruddock became one of the key spokeswomen for the generation that would finally refuse to ratify the declarations of cold war adhered to for forty-five years. The inchoate surge of resistance and resentment issuing forth as the 1980s peace movements did not of its nature take kindly to the leader principle, and still less to adding on its own account to the galaxy of international stardom. But those movements were keenly conscious, as never before, of the public media. They were certain that media attention had to be captured if their convictions were ever to bear fruit. With Joan Ruddock as their very own celebrity, the peace movements could pursue the strategy of judiciously balancing the colossal spectacle of the mass march with the image of their individual representative talking persuasive good sense on the TV screens of millions of homes.

Given the change of heart, it had been inevitable that the representative would be a woman. At least since 1968, if not from the beginning of the cold war, it has been the hearts of women that have been changing the fastest, by devoutly recommending to the human race a diet of the milk of human kindness, and its companion dishes of egalitarianism and emancipation.

A good society in the future can only be realized proleptically by living according to the best values of today. The women's movements, as protean and uncoordinated as the peace movements they have dominated, have claimed those values for themselves, and therefore for all of us. Their prime target had to be the nuclear war with which men had threatened them for so long.

Some women fought off that war with stinging mockery, by pointing out the resemblance of the shining missiles to each warmonger's cherished phallus. Others resisted with the pure, speechless, dogged intransigence of womanly disobedience; as they did, encamped at the gates of Greenham, where young wives and old, lovers and daughters, each refused to do as they were told by the men arranging for everybody's complicity in their own destruction.

But nuclear war also had to be resisted in the television forums of democracy. The women's movements and peace movements alike needed a paragon: a good-looking, well-dressed, scientifically articulate, resolute, and gallant heroine, young enough to speak for their hungry generation, old enough to command the allegiance of the older people who had struggled for so long and so little. Joan Ruddock met each criterion squarely.

Strikingly handsome, she was born into a municipally leased house and to Welsh working-class parents in 1943; she inherited from her nation a fluent Welsh tongue and, contrary to her parents' predilections, its radical politics. At London's Imperial College of Science in the 1960s she anticipated the radical actions of the 1990s by her presidency of race relations and international student associations. When she graduated she quit her scientific discipline to join one of the earliest single-issue pressure groups to mobilize professionally in Britain.

The issue in question was housing—housing for the poor, the homeless, the destitute. The reference point was the twentieth anniversary—in 1968—of the Declaration of Human Rights. In fighting this battle Joan Ruddock learned early the skilled maneuvers of pressure-group politics; the careful academic scholarship behind calm rebuttals of government figures; the courteous, firmly oppositional TV debates; the accurate press releases; the fund-raising among genteel social consciences; the big platform and public meetings; the seventy-hour week.

This was her hard, exhilarating preparation for life with the peace movements. She ran for parliamentary election for the Labour party in the pretty little conservative town of Newbury where she lived. It was December 1979, and Mrs. Thatcher's new government had just obediently agreed to accept the natty cruise missiles from the United States, to be parked in some anonymous spot.

"I'd been at a meeting and went into the station, bought a local paper, and saw the huge headlines about the cruise missiles. They were going to Greenham Common, two miles up the road. I sank into my seat and said, 'Oh, no,' because I knew that that would be a campaign in which I'd be totally committed. I saw my holidays and a restorative period at home I'd planned just melt away. These weapons were *really* in our backyard now. There was no knowing where the political campaign would end."[1]

There is an excitement like terror at such moments, exactly because there will be such a struggle, and one's heart must be right in it. But then there is the tiredness. "I said to myself, I'll get supper and I won't do anything until I can think constructively. And then, when I got home, the local Labour party secretary phoned and said simply, 'Well, Joan, what are we going to do?' So four people set out to campaign against cruise missiles at Greenham Common. From then on, as they say, it's history."

The peace movement in Britain roused itself from its patient somnambulism; it was joined, in Holland, Italy, Germany, by all the Europeans disturbed by their own governments' weary acceptance of this latest nuclear Thing, and the marches began to roll.

"I don't know how often I'd be called late at night because ABC would have a film crew ready at dawn on the Common, and I'd be there to say what had to be said in front of a lovely pink sunrise with the mists clearing through the trees."

She was by then working for the Citizens Advice Bureau, a state-funded counseling service. The British government, ever assiduous in prosecuting the cold war on whatever terms suited the boss in Washington, was indisposed to allow its own employees, especially clever and beautiful ones, to object. Not surprisingly, then, a little squirt of a government minister traduced Joan Ruddock as using its office to subvert the state's purposes. "Of course, I kept to office hours, I never cut corners." It is a deep satisfaction to record that one independent inquiry that he hired to fire her exonerated her completely and recommended lavish new moneys for the impoverished bureau.

In 1980 Joan Ruddock was voted into the chairmanship of the Campaign for Nuclear Disarmament and became a household name. The cobweb of coalition politics needed a strand spun linking CND and the Greenham peace camp, symbol of opposition to all hot or cold wars.

"I'd been helping organize supplies to the camp from the first, but it only became an all-women's camp after a few months. When it did, could such a traditionally male organization as CND—remember how much it was based in the Labour party and the British Communist party—could it give a feminist camp its endorsement? It was a horrendously difficult business, but the decision couldn't have been taken with a male leader.

"I never wavered in my utter admiration for the Greenham women and their resolve. They had to give up so much. I understood why they had to be *absolute*. Eschewing all comforts, they *had* to take a purist position. *I* had a different role, and they attacked me for it. I went to the camp one day and in deference to them I didn't wear makeup, but because it was bitterly cold I put some cream on my face. I heard someone say in the most scornful way to a cameraman, 'She's wearing lip gloss,' and it hurt. But I understood.

"There again, I wouldn't be a hypocrite either. I always dressed carefully in order to deny the caricature peace woman. I still do as an MP, and I had a clear duty to accept every media invitation and look the part. I *was* the part. That was a feature of the CND leadership and of its integrity over all the years of its activity. Of course, the burden of work was phenomenal. Every evening, all the weekend, all my leave. I went part-time at my job for a couple of years."

Over the years of Joan Ruddock's celebrity as Europe's peace woman, British ministers always refused to debate with her. Some, it was said, were simply frightened; but all were party to a suppression of debate increasingly characteristic of British political culture as the elite felt its control over assent to the cold war slipping away. The older generation in the United States had reestablished its grip on the management of assent after the deplorable interludes of antiwar festivals on the steps of the Pentagon and the Kent State student insubordination. Its will was still clamped tight on the nuclear handles. But Joan Ruddock spoke ringingly in many British and European ears, and chalk-striped British ministers would not parley with her for fear the Americans would cut them off without even a key to a Polaris code. She stood with the peace movement for what the British ruling class most feared—a Europe that had moved confidently away from the Pax Americana and was led by Germany and France.

"I always thought that if Mrs. Thatcher—or any strong prime minister —had woken up one morning convinced that nuclear weapons were useless, she would have scrapped them. They haven't kept peace in the world since 1945 in any case, and they made a Europe in which there clearly could have been no point to war. . . . The weapons were irrelevant to Europe; they were merely instruments of foreign policy for both superpowers.

"The United States always seeks to be, and is, the dominant force in the world. Its enormous power is all used for its own interest. Britain lives with its supine acceptance of patronage—that's the special relationship, paying the dues when they are called in. It would always have been very difficult to break free. But always possible. Now, more than ever."

But Joan Ruddock is not anti-American, no more than Edward Thompson, English son of an American mother. In both the argumentative ethos of American public debate and the repressed, docile, and inaudible conversations of the British, she speaks with representative clarity for more than the constituency of women and their long fight for their freedom. She speaks for those unorganized millions in the progressive classes of the world who, living their sufficiently comfortable and civilized lives, have come to believe that world peace is a feasible project. Allowing for all the difficulties, acknowledging the certainty of human violence and the facts of technological destructiveness, they have edged toward the view that peace is more than the absence of war, and that war as a way of resolving human disagreement may be made to disappear.

"There wouldn't have been an end to cold war without the peace movements," she says flatly. "There would have been no change in the East unless the activists—including Gorbachev and his set—had shared our goals and seen how many people supported us. Gorbachev was surrounded by people who *thought like us*. The key resistance was to continuing the arms race. That gave the space for rethinking in the East."

Joan Ruddock's story will continue long past the end of my everyday story of cold war. Making peace so forcefully was the first job. What must follow is building the institutions and creating the language that may make for a peace culture. "The corps of people and the body of knowledge must be maintained *now*, after cold war. We mustn't lose that history."

After the years eaten by nuclear locusts, the crops wasted, the lives destroyed, the only good the cold war can bring us is to teach us how not to do things another time.

31

The End—An Interpretation

THE honest NATOpolitan—citizen of the North Atlantic's mighty necropolis of nuclear destruction—says simply that Stalin and his gangsters were so awful that there was nothing to do except put a ring around them and wait for everything to get better. This gallant, a soldier-bureaucrat or adviser-academic, still thought at the end of the epoch as if history stopped with the Long Telegram (as if Kennan had never changed his mind so audibly), and as if the cold war were the one defensive move made by the affronted West, which calmly prosecuted it until it was no longer needed. How wrong is he?

He has a much less appealing superior. A leading politician of one of the NATO democracies, his superior believes not only that the incipient enemy must live to the bitter end within a huge encroachment of nuclear firepower, the graceful missiles exactly lining its frontiers, but also that only this always visible certainty of chaos and old night kept Europe peaceful for the duration. Indeed, this complacent figure also claims, with a fine blindness to the grateful dead, that the weaponry (plus his ruling party's unfailing vigilance) kept the peace for the epoch, that the cold war, being not-war, was the happiest and simplest mechanism of control the world could have devised, and that it did a worthy job for half a century.

The noisiest proponent of this view dominated British politics for exactly the last decade of the cold war. Margaret Thatcher spoke out sharply—while Ronald Reagan, her ally of eight years' standing, spoke out genially—for completeness and consistency in Western cold war policy, praising the West for the moral superiority of its values (ideology), the intellectual coherence of its policy (strategy), the readiness and resourcefulness of its military production (economics), and the caution and judiciousness of its diplomacy (politics). All the allies are thus impregnable, or at least undefeatable, even if reducible to a heap of ashes. If their resolution proved sufficient, however, this unhappy eventuality would never come to pass, and good would triumph. Which so it did.

Margaret Thatcher and Ronald Reagan—both of them narrow, even bigoted, each blessed with luck, money, and a strong will—by the accidents of electoral systems were brought to power at the same moment and with the same vision. They returned with a noisy to-do to the lessons they had learned in the early 1950s, when the machine of Stalinism far exceeded in reach and power the mad old despot's grasp on reality. By invoking these lessons, they reawoke antisocialist fear and loathing for ten years, and in a lot of people. By 1989, when those heroic and bloodless victories in Eastern Europe brought an end to the cold war, Reagan and Thatcher had done much to reinstate the rhetoric of Kennan's "containment" and Dulles's "rollback." Their power over the communications systems of their countries meant that for many people there seemed to be no good reason why the cold war should not go on forever.

For a few years in the 1970s a different story had started to emerge, but it never got very far with the politicians. By its lights, the United States had been at least as responsible as the Soviet Union for putting the cold war show on the road and in fact did more to intensify the conflict by its determination to spread the Pax Americana across a genteelly subservient world committed to open-door trade, liberal capitalism, free elections for all parties except the Communists, and happy consumerism.

Such a story called into question American motives everywhere, and particularly American conduct in Vietnam. Attacking the popular idea that American democracy was essentially decent and that its institutions could readily be transplanted wherever the flag took them, this story flatly opposed the notion that the United States both had the right and was right to intervene in any country that fell short of the standards of the U.S. Constitution.

The heresy upset a great many people, as it was intended to. Postwar reconstruction had been so successful. After the bloody deluge of Fascism, Stalinism, and the firestorms over Japan, there followed in the United States and Western Europe a period of such quiet, energetic, and successful rebuilding that politics came briefly to seem both in the streets and up in the ivory towers, a tradition of complacency. In the famous phrase, everyone supposed that the end of ideology had come, that the lion of capital would lie down with the lamb of labor and nobody would need to worry about class struggle or poverty or exploitation anymore. Domestic affairs were so peaceful, at least for politicians and students of politicians, that political argument was wholly displaced onto foreign affairs and political reputations were made on the basis of cold war rectitude.

Elsewhere, of course, ideological politics held on its bloody course. Fanatics and gallant democrats clashed and killed each other in Warsaw and Budapest, and, as was to be expected, the fanatics won. Millions of Korean fanatics in sneakers rushed in upon the machine guns of the united Western powers and were only discouraged by having their country bombed into the mud.

These excesses served the development of selective vision in the West. The combination of a war so cold that it froze all political initiative in the icy stillness of nuclear terror and an economic boom that turned public-spirited citizens into private consumers, contracted the people's political pupils until they could see only the small details of everyday life, where any sensible person prefers to live in any case. Political questions turned on the extent and quality of individual powerlessness, whether in the teeth of the cold war winds or in the warm currents of consumer capitalism, which so corroded the long traditions of communality, social welfare, and equality.

This was the ethos that formed the generation of senior politicians passed or passing from the world stage in 1990. As motley a bunch as Charles de Gaulle, François Mitterrand, Helmut Schmidt, every American president from Harry Truman to Ronald Reagan, and each British prime minister from Harold Macmillan to Margaret Thatcher has been united in the cold war strife which divided their extremely discrepant intelligences.

A new generation came along that, as new generations will, refused to believe the old story. As the cold war neared its end—and now that it is over, one can see that the end was coming throughout its slow coda—a certain shrillness beset the devout cold war storytellers. They panicked a little at the passing of a tale that had held enough of the audience in thrall and themselves in comfort. Some reaffirmed the righteousness of the cause; others found new agents to blame for the cold war, making Roosevelt as responsible as Stalin for selling Eastern Europe down the river, or rehabilitating Eisenhower for his wise passiveness.[1] But all the old cold warriors were remorselessly pleased with themselves, whether for their prudence or their dynamism in winning the cold war on behalf of capitalism without using nuclear missiles.

II

Then, of a sudden, it was done. Self-righteousness had its sweet moment. One looked on at the television screen in a pleased way, it being a thrill not much experienced to see tyrannies toppled in a week, and in a flash the storytellers were back to say how all this was the result of their advice. Saying that they had always known it would happen and that all's well that ends well, and, naturally that they—senior politicians in power, their hired mouths and zombies, people like that—should be left to carry on the good work.

When the wall fell, the organs of liberal opinion in the United States and its satellite Kingdom responded by simply playing a triumphal march. They wove counterpoints about their leaders' claims that the cold war had been the necessary resistance of naturally peace-loving democracies to the threats of expansionist power, that it had been prosecuted with courage and the display of military power, and that by acting thus the West had so overloaded the economy of the Soviet Union—an economy already made incurably inefficient by its planners' deranged ideology—that it collapsed in ignominy. The outcome was a just victory and a peace from which the capitalists would eventually make lots of money by teaching capitalism to socialists. Understandably, upright left-liberals may have winced at the tone adopted by the more practiced sycophants, but they would have to make, as we shall see, ample concessions to the argument. In the long run, however, what is not at stake are the questions, who won? and, who was to blame for it all anyway? In the long run what matters is what the consequences of the cold war were for the future.

The cold war has been the supreme fiction of the epoch. Different interpretations of it both on our side and on theirs, have competed for dominance. Some have pursued the argument about who started it and who was most to blame. Others have been more interested in the argument about passivity or dynamism: Was Eisenhower a bumbler or a subtle tactician? Was Kennedy a reckless gambler or a cool hero? Was Brezhnev slothful or peacemongering? At the end, the argument was about the cold war's inevitability: Stalin was such a bad lot, if we had no choice but to hold down both him and the consequences of his tyranny at all costs, did we really win?

However you answer these questions, the complex legacy and the frames of mind conferred by the cold war will shape the future. It will have profoundly shaped those who will make the future happen. Consequently, the version of cold war history that we choose to tell will make all the difference to the future we make. There is already a hell of a fight starting about which version to choose, and the word is out in the marketplace that writers, advisers, film and TV producers, aides, and policymakers are all needed to explain our recent past and direct our immediate future. Old theses are being taken down and dusted off; old has-beens—Kissinger, Nixon, Rusk, that crew—are coming forward with brazen bids for office and a terrific judiciousness.

In this vanity fair of opinion, the winning stories, not surprisingly, are going to be those that re-present the United States and its allies to themselves as righteous, peace-loving, brave, and prudent. The trouble is that any effort to beat the devil by telling truths and appealing to standards of objective judgment will have to set up its stand in a marketplace where all the peddlers are paying lip service to the same standards but are naturally pulling the audience their story best fits. Telling the cold war history is already becoming a matter of looking for the most applause, and of shouting rude words at those who find it.

Well, that is how things are. As part of my own reckoning, I shall say that it seems a bit rich to rename the epoch "the long peace" and to fish out all that tired old stuff about the end of ideology once again, much less to crow about winning the cold war, breaking up the Soviet empire without firing a shot, and bankrupting its treasury in readiness for invasion by McDonald's.[2]

The long peace did, after all, accumulate a considerable body count—that happy neologism of cold war. The bullets were largely exported to the Pacific rim or the South American rain forest, where the cold war's victims were mostly small and either very dark or small and olive pale.

Nearly 90,000 G.I.s died in Korea and Vietnam: 33,629 in the first war, 58,132 in the second.

Three million Koreans, northerners and southerners, civilians, soldiers, children, died of cold war in Korea.[3]

Four million Vietnamese, northerners and southerners, died similarly in Vietnam.[4]

Four million Cambodians died at the hands of Pol Pot, the Communist Khmer Rouge leader who was himself a direct product of Nixon's prosecution of the cold war in Indochina.[5]

Three-quarters of a million men, women, and children died in the anticommunist (or pronationalist) civil wars in Malaysia, Borneo, and Indonesia from 1950–1967.[6]

There are no reliable figures out, but there can hardly have been less than two million deaths in Afghanistan since 1980, victims of Soviet and American rockets and ammunition.[7]

The Angolan and Mozambique civil wars, also fought by black and white Africans with Cuban and American advisers in the bush, killed hardly less than three-quarters of a million.[8]

God knows how many Ethiopians and Sudanese have died for the dictatorship of the proletariat.

From the CIA coup in Guatemala in 1953, via Chile, Nicaragua, El Salvador, and Panama, another 750,000 were blown up or starved to death in the name of liberty. Some were soldiers, and others were, as they say, innocent bystanders; most were under the age of fifteen at the moment of their death.[9]

In bloody old Europe, only a few thousands died, shot by Soviet guns or run over by Soviet tanks. Some tens of thousands more—labeled dissidents or enemies of the people, schizophrenics or saboteurs—died in the Gulag after 1946.[10]

If there were cold war memorials—an empty missile silo, perhaps, a roadblock, a burned-out village—these sixteen million names could decorously be recorded on their tablets, if it weren't for the fact that nobody knows the names. The only archivist of most of these corpses was their common memory, and that died in the strafing, the burning, the forced-draft urbanization, the famines, the insurrection, the high jinks of stoned boys with guns.

Cold war; unpeaceful peace; the armies of the dead.

Some would have died in any case. The realists are always quick to remind us that wars, like the poor, have always been with us, and that until very recently going to war has been for most men and not a few women a natural, self-explanatory, and fulfilling way to confirm one's manhood, even if the price of the experience is one's life.[11] Many went to war in 1914 for the joy of it, and plenty of men still remember their soldierly spell between 1939 and 1945 with pleasure as well as with advantages. Short of thermonuclear explosion, which so diminishes the fun as well as the duration of a war, it is likely that this attitude would have killed a lot of people during the epoch quite without the ideological lift given to firing off napalm rockets by cold war poetics.

No doubt. No doubt the family oligarchies of Central America would have continued to fight to the death with one another without the cold war, though they might have run out of ammunition sooner. No doubt Ho Chi Minh would still have fought to throw the French out of Indochina, though we will never know if he might have settled for the quieter life he agreed to with the British in 1945 before they double-crossed him. And no doubt the corrupt, bullying, and repressive regimes of Generals Park Chung Hee and Chun Doo Hwan in South Korea have been preferable to the dismal bleakness of Kim Il-Sung's rule in the North. Pity about Pol Pot, of course, and the Puerto del Diablo, but you can't make an omelet, etc.

But how can we imagine a world without a cold war? Take it away and what then? The old simplicities and certainties have vanished but the weapons remain. What if the Soviet Union breaks up in disorder, and those who can seize hold of the old weapons, brandishing them again in an effort to recover their lost magic? So some say that it's best to keep a version of the cold war going as long as we can.

There are sorely tried concepts in support of this position: "Cold war bipolarity served the cause of peace remarkably well"; "power vacuums are dangerous things"; "nuclear weapons have played a major role in bringing about the evolution from cold war to long peace"; "the purpose of nuclear weapons should be to maintain a healthy fear of incautious action . . . a healthy respect for a major method by which we have achieved the long peace."[12]

So the legacy of the cold war is that the world shall be kept quiet by the United States and its slightly reduced pile of nuclear weaponry, with a bit of help from the diminished Soviet Union, which may remain in irreparable economic difficulty but will also remain imposingly well armed.

This vision is likely to commend itself only in Washington—and is likely to cause bellicose resentment in many parts of the globe for that reason alone. It is the logical conclusion of the cold war only according to the tritest textbooks of international relations and those hospitably self-serving centers and institutes that have invented the lexicon and tall tales of the epoch.

These experts have preached two basic lessons: First, the two superpowers have competed for influence and the expansion of power because that is what great powers do. Second, the superpowers had irreconcilably opposed visions of the good society. Each therefore sought to extend influence and power by persuading the rest of the world to see things its way. Sometimes this persuasion was helped along by guns and bombs, but mostly by propaganda. For the first time, the apologists say, international relations were grounded in ideology.

The second idea assumes certain ideological absolutes and points to another casualty of the cold war: the political imagination. Socialism being forbidden, political argument and vision had to make do with liberalism's rather scanty diction. Individuals and their rights have been exalted and the public good and the powerful fulfillments of the citizen's life have been almost entirely removed from vision.

By the same token, the political argument on the wrong side of the iron curtain has more notoriously eradicated, with the aid of policemen, the taken-for-granted joys of saying and buying what you like. Thus the consumer-hedonist faced the badly dressed, badly housed, and badly fed comrade, who finally, and understandably gave way all at once, and came across eager to share the delights of the shopping mall, the freeway, and the well-tempered dormitory suburb. Their choice, the greatest simple outpost of liberalism, was clearly made.

III

After all is said and done about supply-side economics and free-market competition, the good life as lived or imagined by two-thirds of the populations of North America, the European Community, and the Antipodes is wanted by both those who enjoy it and those who do not. A decent income, a secure job, full employment, glistening shops, affordable medical care, safe cities, and sufficient government are the material grounds of the free society in which individuals may cut what figure they please so long as they do not cut into the free figures of other individuals.

Back in 1945, therefore, the real victory was Roosevelt's, and perhaps a little that of the British Liberal lordship, William Beveridge, who is given the credit for dreaming up the welfare state. The New Deal was clinched for the United States by the Second World War. The war brought the colossal increase in gross national product and productivity that the New Deal inaugurated. It confirmed full employment and bathed the economy in the radiance of ideological self-satisfaction, benefits extended to Europe as well by the Marshall Plan. The wartime record of the socialists and the military experience of working in solidarity in the occupied or beleaguered countries infused Europe with a larger dose of what some Americans might think of as socialism, to be sure; but by 1950 or so the Communists were just nowhere in Western Europe and were reviled in the United States.

Socialism in the West was useful only as a scourge with which to beat dormant ideals into some show of wakefulness. The recrudescence of Marxism in intellectual life after 1968 had little more effect on the political conduct of the Western nations than to call them back to their professed liberal values. The historians who so admirably wrote the critiques of the Vietnam War could hardly point to Hanoi as the celestial city; they had to settle for identifying American violations of the nation's first principles: rights, freedoms, choice.

As a result, the third biggest casualty of cold war was the promise of happiness held out by the visionaries of socialism. Tough cold warriors always had a short way with socialists at home, of course. They said, cursing a little, "If you want socialism so much why don't you go and live in Moscow or Beijing and see how you like it." It was a crude argument and was never met well enough by the objection that Stalinism and Maoism were not true socialisms. But it was also an effective argument because it squelched any effort to apply socialist theory to capitalist society, thereby cutting political imagination off from a key source of renewal and civilization.

No doubt, Marx himself deserves to be a casualty of the cold war for the failure of both his economics and his political predictions. Economic development never came anywhere near bringing about the socialist society, and the proletariat showed little or no interest in its future dictatorship, even supposing the party would ever have given it the chance to assume office.

Nonetheless, Marx is only one of many socialist thinkers to have diagnosed the viciousness of unbridled capitalism, the oppression and exploitation to which it inevitably gives rise, and the repellent class arrogance created in its successful bourgeoisie. Who then supposes that his way of life should be the way of life. Marx might have had exactly in mind the earnest AID official in Managua or Saigon strenuously checking the ballot boxes in yet another round of free elections. Socialism, forever disfigured by the cold war, is still the only rhetoric left with which to berate the delusions and cruelty of horrible old capitalism.

IV

If the cold war cut Karl Marx and William Morris out of the conversation of all but a small corner of the culture, it did worse for the social and sociable virtues. Hating their ideological opposite number led both sides, as everyone noticed, into some entirely unholy alliances. Americans got chummy with such guardians of human rights and the open society as Synghman Rhee and presidents Diem and Marcos; the Soviets installed their own protectors of the proletariat in Czechoslovakia and Afghanistan. Both parties claimed moral superiority, of course, but what with their grasping allies, their state assassins and surveillance hoodlums, their bombers and drunken troops, neither could be believed except by their most credulous citizens.

And so, after the solid reality of the dead and the impalpable losses of political imagination and the generous vision of socialism, the fourth biggest casualty of the cold war was the idea, and the realization, of civil society.

The old-fashioned word *polity*, from the Greek *polis*, has been revived of late to restore this loss. The polity is the daily intercourse of good government in which some citizens represent others; others watch, evaluate, judge, and criticize; still others listen approvingly or disapprovingly; and those remaining look on with anything from mild interest to friendly indifference. This is the common, vague picture of the ideal democracy carried around by most people, if only as a way of belittling the pitiful distortions of democracy in their own societies. The oxygen of the polity is the degree of knowledge and zeal to be found in its citizens; the lifeblood of the polity is its public-spiritedness; the homeostatic regulator of its metabolism is the vote. This is the body politic of the ancients.

In the age of public opinion, the polity has become a quaintness. The success of consumer capitalism and the rigidity of the cold war mortally wounded the mild public-spiritedness with which the European welfare states emerged from the Second World War. In the loose federalism of the United States, socialism was a swearword and even liberalism was tainted by association in the very nation-state that invented the doctrine. As Anthony Giddens says, the common appeal of the nation-state—now as never before the entrenched instrument of world politics—was only to its nationalism.[13] To the degree that this is true, then even after cold war the prospects for peace continue to look bleak.

Nationalism has a horrible record in the twentieth century, but nobody minds that much.[14] As soon as the Soviet Union relaxed its grip on its member states after 1989 and the monolithic party began to break up into its components, nationalists began to kill anybody in their locality with a smaller national membership: Bulgars against Turks Serbs against Albanians, Azerbaijanis against Armenians, Estonians against half their own population.

Even in the West, where creature comforts tended to upholster the more fervent political passions, Mitterrand's France and Thatcher's Britain kept up their repellent nationalisms, caricatured in the latter case by the vainglorious expedition to the tussocks of the Falkland Islands in 1982. And over the whole continent broods the biggest, darkest nationalism of all—Germany, the reunited nation, has hardly more than the deutsche mark and its swaggering anthem to pull together two parts so radically estranged by forty-five years of consumerism and communism.

In the United States the claims of nationalism have divided the country as much as they have united it. The nationalist, flag-waving kitsch of President "Walk Tall" Reagan topped off the decades in which the nation and its public largely assumed its rights of officiousness over more than half the globe. At the same time, however, minority nationalism took on grand nationalism with resounding success: allegiance to black beauty, to Hispanic, Indian, and aboriginal membership, always cast in the legalism of individual rights, filled civic space wrested from the flag.

Nationalisms and rights are cashed in terms of personal identity. The magnetic fields of consumerism and cold war have damaged civic life by splitting personal values from public ones. Institutional values such as loyalty, duty, solidarity, and mutual help have lost out to individual values such as sincerity, honesty, spontaneity, and self-fulfillment.

It is a question of emphasis. Obviously, each of these values names something strong and good in life. But the cold war has tended to strengthen our private values and weaken our civic ones precisely because civility has vanished from the forum of civil life. The best lives of the epoch turn out to have been lived in private.

V

The military theorists and historians say that if we want peace, we must understand war. But if we want a peaceful future, we must also understand what the cold war did to domestic lives. It turned civil discourse on both sides of the iron curtain into nationalist and ideological rant and drove virtuous life into the enclave of the home and family, where the virtues have nearly starved. It did fearful damage to the great ideals of the Enlightenment, which had fueled the sibling doctrines of liberalism and socialism. It made common efforts in the name of human progress and emancipation impossible, even incredible.

The fifth casualty of the cold war was therefore the idea of progress itself. Progress has been poisoned by its association with lies about modernization in the Soviet Union and with rationalization in the old industries of Pennsylvania or Yorkshire; but let me paraphrase Leszek Kolakowski's definition of progress as democratic socialism. It simply *is* "the obstinate will to erode by inches . . . avoidable suffering, oppression, hunger, racial and national hatred, insatiable greed, and vindictive envy."[15]

At best, this transcendental appeal was drowned by propaganda; at worst, it was energetically subverted by the insane overmanufacture of the instruments of world destruction, by the squalor of state surveillance, torture, and lies, by futile deaths and cold murder. For the record, it should be remembered that when the cold war ended so abruptly—and what is more, ended in exactly the way everyone said it would never end—each side possessed *ten thousand* strategic nuclear warheads, that is to say, 10,000 weapons each individually equivalent to at least one megaton (and many much bigger) and therefore capable of doing about ten times as much damage as was done to Hiroshima.[16] This figure does not include the hundreds of smaller nuclear weapons capable of wiping out a village or a city.

The figures have numbed us by now. Some have tried to stir us back up by calculating how many times each human being on earth could be burnt to a crisp, blasted to a shadow, irradiated to death in a trice, or sifted as ashes if the whole infernal pile went up together. But the results are dull. All we can dully say is that the figures about the weapons of mass destruction betoken the nations' long trek away from progress and the Enlightenment dream of men and women in rational and altruistic cooperation gradually making a peaceful world.

The cold war is over, but a strong flavor of cruelty is sure to linger. If, as Michael Howard claims, peace as the natural condition of humankind and war as its inhuman converse is an ideal of recent and liberal provenance, that ideal has been clinched by the invention of nuclear weaponry.[17] But peace itself is still a fragile concept, improvised merely as the relative absence of warfare. For the Enlightenment thinkers, peace was the absence of recourse to warfare in resolving human quarrels, but even they were guilty of bad faith on this score. Europeans and Americans alike imagined the good society in terms of capitalism and industrialization; they entirely neglected the significance of military formations. Market forces trumped armed forces.

Heirs to two centuries of thinking about class and the just and equal humanizing of production, we are left with neither a tradition of political thought nor an interpretative framework for understanding violence and the nation-state.[18]

The modern nation-state and the nationalism it has sedulously encouraged has first centralized and then industrialized the means and control of violence. Liberalism forbade military rule, so modern society developed state surveillance. Socialism, which promised redemption from the iron cage of modern capitalism after the bloodshed of the last revolution, ingenuously took peacefulness for granted. It foundered accordingly from within and without. The nation-states assailed it, and its own economic theory beggared it. The optimistic tradition and political theory of liberalism cannot base peace on anything except goodheartedness—a view of power contemptuously demolished by Machiavelli a very long time ago. Progress into peacefulness has a very long way to go.

VI

If the optimistic and progressive line about the possibility of a rational world has been shattered by the wastefulness and unreason of the cold war, at the same time the professors of hard realism and the heralds of free enterprise have broken down on the road to world domination. Indeed, they have gone almost broke manufacturing weapons unsubjected to either intelligent criticism or the disciplines of competition.

The surprising sixth casualty of cold war has therefore been the possibility of the United States taking on a benign, freely trading, limitlessly productive, and, where necessary, temperately restraining hegemony of the world. The chorus of triumphalism that arose from the voices hired to flatter national elites when European communism gave up the ghost in 1989 could not survive a serious reading of the U.S. balance sheet. Paul Kennedy is only the latest and most accessible economic historian to point out that the United States irretrievably overspent its patrimony in prosecuting the cold war and thereby driving the Soviet Union to the edge of bankruptcy.[19] The United States was in better shape than the old enemy, no doubt, but then the old enemy had long since given up on any Khrushchevite ambition to bury capitalism. Throughout the 1980s Washington, the capital of capitalism, went on blindly spending under President Reagan's inspired direction as though there were no tomorrow.

But tomorrow came, and the ten years of overspending on weaponry turned the United States from the largest global creditor, distributor of the dollar and the workshop of the world, into the biggest debtor nation ever seen.[20] No country had ever before gone into the red in peacetime so comprehensively.

At a time during the 1980s when wisdom and pragmatism alike would surely have reckoned that the landscape of the cold war was everywhere softening and putting up little green shoots, the American defense budget was tripled. The consequence was that, like the great empires that preceded them, the Pax Americana and its grim rival, the Suppressio Soveticus, spent up their authority, whether fiscal or military, and could no longer rule the world or its waves. As was to be expected at a time in which all history is, as never before, immediately world history, in which information and money move thousands of miles in a second—and deadly weapons also—these two empires lasted a bare half-century, whereas their predecessors in Spain and Great Britain had at least delayed decline for four times as long.

Whatever happens to each, the superpowers will continue to be militarily superpowerful. But ultimate authority will be denied them. China, Japan, and a Europe dominated by Germany will all jostle for a sort of victory, and none will any longer do as they are told by the sometime bosses, however warily they treat them.

The sixth casualty of cold war has been the absolute power of superpowers.

VII

The cold war has been an extraordinary show to watch. Whether you take the view that all human life is fatuously self-deluded or the view that it is blindly heroic (or anything in between), there has been plenty of action to bear you out, and the action has been as full of color and suspense as it has been undoubtedly slow and dreary.

Indeed, one could easily say that the cold war's achievement has precisely been its spectacularity. Politics became a public spectacle that we paid our taxes to sustain. Safe in our private lives, we could remain in our seats and let our politician-actors represent us in the great issues of life's narrative.

Taken as an amiable parody,[21] the cold war turns out to have been both high jinks and grand opera. The audience went onstage only at key moments and when invited by the actors, usually just to vote. In an odd corner of the action, now and then, some of the audience joined in so heartily that they got killed.

The last casualty of the cold war, its loss obscured by the real heroes as well as by the mock-heroics of the actor-politicians going through their repertory, was that unspectacular and much mentioned agency, democracy.

I have already remarked how rarely classical democracy can work when overlaid with superpower policy. But it is worth reminding ourselves, as an antidote against the miasma of sanctimony always rising from the liberal democracies, how illiberal and antidemocratic the cold war encouraged us to be.

Even if we set aside the flagrant corruption and venality in the election of senators and representatives in the United States, or the open contempt shown by government for Parliament in Britain, forty-five years of cold war embedded deep in decision and policy structures habits of lying and deceit, of arrogance on the part of the power elite and toadying by the same elite to the electorate when the time came around. These little ways are precisely those the early republicans[22] and the democratic theorists who succeeded them meant to shackle and disarm.

As has often been pointed out, of course, the U.S. Constitution was devised at a time when a patrician class recently emancipated from monarchism was happily available to run the country. It was never foreseen that the federal system and the electoral process would become so tightly controlled by the ungainsayable obligations of both incumbents and challengers to party and to favored states. Above all, nobody could have intended that corporate, military, and political interests in the domestic society would so decisively be able to tell the president and his administration what to do. They could not have intended that by the twentieth century, after the brilliant exception of Franklin Roosevelt, presidents would be men of such meager gifts.

The record from Truman on is bleak. Each president, because of the nature of the system, was in thrall to his money and his constituency. While it may be true that some did much more harm than others (and that one man, like Bobby Kennedy, as Mailer asserted, might have been a marvelous president and an exception to the rule of mediocrity), this line of kings proved dismally unable to break the cold war's frozen sea of feeling and imagination.

How could they? How could one man, voted for by the minority of the population that bothered to turn out, do very much, lead brilliantly, have the will and vision to change the world for the better? These questions go to the heart of democratic theory and well beyond our present concerns. It will have to do here to point out that at least a president ought to be less of a patron, and less able to appoint all his own court of advisers.

The point at issue, however, is that domestic rhythms locked solidly onto the rhythms of cold war like gear meshes. The drive of each drove the other. The resultant force was ideology.

This book has told many tales of the ways in which the cogs meshed. (Although the tales have of necessity been told from Washington and London; during the four decades so little was really known about Moscow.) A general rule begins to emerge. It is that a superpower's foreign policy cannot be understood without a vivid picture of its domestic and day-to-day politics. A second rule is that it was the competition of ideology that ground the two superpowers together with such a high coefficient of friction for the period of the cold war, ideology being no more and no less than the story about the world that people believed, sometimes fervently, sometimes not. The political history of the epoch is the history of the story of ideology, in all its colorful versions.

Some versions were true, some false; some tall, some dull; some conduced to virtue, some to wickedness. Most ideological tales, we *can* say, were orchestrated, censored, or at least given permission by the chief executive or the first secretary. The position of these leaders in the eye of the ideological storm made them comptrollers of the story of the day, impresarios of the spectacle of command.

For those who keep up a dogged attachment to democracy, the spectacle has been unedifying. The word-and-picture processors gather about the president to be told what to work with. Each day the president or his mouthpieces respond to heavy deference with statement or communiqué. But the "president" himself is mouthpiece for the utterances compiled for him by the story of the day. And yet, what the president says is treated with the utmost respect as decisive and effectual.

The processors go away and repeat the utterances, framing them in the preferred imagery of either power-in-action (pictures of the White House, the Oval Office, Air Force One) or power-in-relaxation (golf course, seaside, Camp David, First Lady). The interpreters separate fact from value and organize the approved response for press and network television, with terse glimpses of urgency (tanks, troops) and the reassuring presence of the good-looking boys and girls in the newsroom.

Such and such are the spectacles of command. News is the most reliable product of the vast and successful communications industry. It is relatively cheap to make, so long as there are not too many hot spots overseas; it lends itself to mass production and networking; it is immediately obsolete—no one will buy yesterday's videotape; and there is insatiable demand for it—it sells and sells.

Making the news has gradually become the president's prime activity and his major contribution to the gross national product, as it has been, in a smaller voice and in a minor key, for British prime ministers when they faithfully repeat their edition of the U.S. president's words for a smaller audience. We, the people, have been expected to receive with due docility and credulity whatever the grandees of power tell us. Television has allowed us to attend the press briefings and interviews with presidents and their secretaries of state, prime ministers, chancellors, and foreign policy luminaries, where we have obligingly listened and mildly believed almost all we have been told. Even when we stopped believing and mistrust of government spread a certain numbness through the collective consciousness, the spectacle continued to reassure us. Even cynicism was assuaged by the sight. The show was going on, and who knew what the Soviet Union might do if it were called off? Think how uneasy we would all have become without those soothing images of calm men in suits getting into aircraft or black limousines, arriving in capitals with Big Ben, the Wall, or the Kremlin behind them, smiling, shaking hands.

Such images begot the same images for four and a half decades. The reassuring ones of serenely confident statesmen alternated with the scary ones of trucks and tanks in blank-windowed streets. Each picture taught us—nearly four generations of cold war conscripts (the founding fathers of the cold war were born before 1900)—that our chosen representatives would stand up for our best values all over the globe, that we needn't worry.

Those same representatives, giddy with power and celebrity as they were sure to become, have thought endlessly about what we think of them. They always sought to organize exactly what *we* are told; presentations for the enemy were always secondary. They tried not to tell us things—about cobalt bombs in Korea, Communists in Whitehall, B-52s over Cambodia—that would upset us. They acted out in public, and for our benefit, those qualities of manly and womanly character on which our great nations so depend: niceness, resolution, efficacity, decisiveness, calm, charm, courage, and a dapper suit.

They took away our intelligence, the information that was ours by right, what little power we had. And we allowed it. We didn't know enough, and we wouldn't learn. A handful of men and women, George Kennan, Neil Sheehan, Joan Didion, Philip Agee, Joan Ruddock, and Edward Thompson among them, have tried to tell us. In the end, quite a large number of people have half-listened to them—enough to dismay our governors, who so badly want to be well thought of if the people really insist on thinking at all.

But none of it says much for our desire to govern ourselves according to the canons of democracy, to be well-informed publics and autonomous electors. In the end it was that queer surge of public opinion over on the eastern side of the iron curtain that recaptured something of the early promise of democracy. But it might never have happened; the cold war still had a lot of running in it.

VIII

The cold war might all never have happened. Its beginning was not inevitable, the dreadful Stalin notwithstanding. He and his killer henchman Beria could indeed have been contained in 1945 when his country lay waste. But fear of communism ran deep in American culture, and the frames of feeling opened up by world war needed another suitable enemy when the war was over. The Soviets fitted the frame perfectly: they asked for it, anyway. Gloomy, suspicious, dreary in appearance, reliably brutal in action, they fitted every caricature that was drawn of them. So prejudice, rigid fearfulness, ignorance, and superstitution continued to drive human affairs for another forty-five years. On the way, the cold war gave rise to a few good yarns, and it gave lots of satisfying opportunity to blow the gaff on lies and duplicity. No doubt there were, as well, fine individuals with noble tales to tell, soldiers of war and peace, diplomats or politicians even. A few noble tales; an ignoble history.

What philosophers call "counterfactuals" is what children call "just suppose." Playing "just suppose" with the cold war can lead to some delightful places, as well as to some terrifying ones. Change the cast a little, or the passions, and the bombs could have gone off over Cuba in 1963 or over Sinai in 1973. Let Bobby Kennedy live through 1968. Let Roosevelt see through an astonishing fourth term to 1948. Deprive the world of two good men: Mikhail Gorbachev and Willy Brandt.

It is easiest to play the game with people, but it can be played as well with events and their consequences. Suppose Germany, like Japan, had refused to let nuclear weapons be housed on its soil; suppose it had been reunited in 1957. Suppose a federated United States of Europe had come into being. Suppose the United States of America had been generously inclined to the highest hopes of the socialist vision, especially when applied to the desperate poverty of small nations. Suppose nuclear weapons had been kept to a minimum.

There is no point in blaming anybody for the way the world turned out.

The force of circumstances is enormous as well as contingent. The plot of the epoch locked together event upon event, and bound it with biographies, obscure and eminent. The plot was the thing, and in it the titanic protagonist and antagonist agonized away until the villain so happily lost. The dismal thing is that the victors then supposed themselves to be completely in the right, the way of life of consumer capitalism to be sweetly vindicated, and the bitter criticism it so deserves to be an act of ingratitude and treason.

At the end of every play the dead characters get up off the stage and take their bow. The casualties of cold war I have listed begin with several million actors dead in the most literal way and quite unable to walk off into the historical wings. The others—political imagination, civil society, progress, socialism, empire, and democracy—can now have another go in a new play with a different plot.

If that play, in its turn, can make play with ideas about the disposal of nuclear weapons and introduce characters capable of envisioning a more cooperative world, it will have made a sound beginning.[23] If the new plot can resolve some of the more repellent aspects of capitalism, if it composes a story in which power is mitigated by minimum justice and a touch of mercy, it will have a happier ending than anyone saw in the twentieth century.

NOTES

Prologue

1. Winston Churchill, *The Second World War*, vol. 6 *Triumph and Tragedy* (London: Cassell, 1954), p. 198.

Chapter 1:   Biography I: The Short Happy Life of Frank Thompson

1. I have reconstructed these remarks. All subsequent Frank Thompson quotations in this chapter are taken from the memorial volume *There Is a Spirit in Europe: A Memoir of Frank Thompson*, ed. E. P. and T. J. Thompson (his brother and mother) (London: Gollancz, 1947).
2. Frank Thompson Papers (E. P. Thompson ©).
3. These details of General Vukmanovic-Tempo's movements are taken from his unpublished reminiscences prepared specifically at the request of Edward Thompson in the 1970s and made available to me out of Edward Thompson's characteristic generosity. Tempo's official memoirs were not published in an English translation until 1990; see S. Vukmanovic, *Struggle for the Balkans*, trans. C. Barnett (London: Merlin, 1990).
4. Thompson, *A Spirit in Europe*.
5. Sue Sumner, "The Lost Loves of Iris Murdoch," *London Mail on Sunday*, 5 June 1988, pp. 17–22.

6. Edward Thompson, *Writing by Candlelight* (London: Merlin, 1978), p. 131.
7. Thompson, *A Spirit in Europe*, p. 59.
8. Ibid., p. 76.
9. Ibid., p. 85.
10. Ibid., p. 102.
11. Ibid., p. 111.
12. Ibid., p. 115.
13. I owe much in what follows to the diligent researches and faithful reconstruction of Thompson's story by R. Stowers Johnson in *Agents Extraordinary* (London: Hale, 1978).
14. Thompson, *A Spirit in Europe*, p. 71.
15. Ibid., p. 169.
16. Ibid., p. 170.
17. Quoted in Edward Thompson's literary papers (see note 2).
18. Thompson Papers.
19. Thompson, *A Spirit in Europe*, p. 43.
20. Thompson Papers.
21. Stowers Johnson, *Agents*, pp. 149–50.
22. Quoted in Thompson, *A Spirit in Europe*, p. 198.
23. Ernest Hemingway, *For Whom the Bell Tolls* (New York: Scribner's, 1940), p. 235.
24. *London News Chronicle*, 8 March 1945, quoted in Thompson, *A Spirit in Europe*, p. 10.
25. Thompson, *A Spirit in Europe*, p. 113.
26. Ibid.

Chapter 2:   Events I: The Casting of the
Iron Curtain, 1945–1947

1. The full text is given in the *Congressional Record*, 79th Cong. 2d sess. A1145–47.
2. Jonathan Lewis and Philip Whitehead, *Stalin: A Time for Judgement* (London: Methuen and Thames TV, 1990), p. 21.
3. Figures quoted in Robert Conquest, *The Harvest of Sorrow* (New York: Oxford University Press, 1986), pp. 299–307.
4. Robert Conquest, *The Great Terror* (London: Macmillan, 1968), pp. 485–86.
5. Lewis and Whitehead, *Stalin*, p. 97.
6. Ibid., p. 168.
7. Described by Hugh Thomas, *Armed Truce: The Beginnings of the Cold War 1945–1946* (London: Hamish Hamilton, 1986), pp. 86–89.
8. The tale of Los Alamos is classically told by Richard Rhodes in *The Making of the Atomic Bomb* (Harmondsworth: Penguin, 1988).
9. Ibid., pp. 703–47.
10. They are reproached for it by Michael Walzer in *Just and Unjust Wars: A Moral Argument with Historical Illustrations* (New York: Basic Books, 1977), pp. 263–68.
11. Rhodes, *Atomic Bomb*, p. 533.

12. Ibid., p. 534.
13. Quoted in ibid., pp. 637–38.
14. Figures from Wilfried Loth, *The Division of the World 1941–1955* (London: Routledge, 1988), pp. 135–39, and Alec Cairncross, *The Price of War* (Oxford: Basil Blackwell, 1986), pp. 189–220.
15. See Alan S. Milward, *The Reconstruction of Western Europe, 1945–1951* (Berkeley: University of California Press, 1984), especially pp. 56–125.
16. See Daniel Yergin, *Shattered Peace: The Origins of the Cold War and the National Security State* (Harmondsworth: Penguin, 1980), p. 314.
17. Ibid., pp. 315–17.
18. Quoted in ibid., p. 327.
19. Quoted in ibid., p. 233.
20. Ibid., pp. 236–37.
21. Here discussed is the version reprinted in George F. Kennan, *American Diplomacy*, expanded ed. (Chicago: University of Chicago Press, 1984).
22. Ibid., pp. 110, 108, 111.
23. Ibid., pp. 117, 125.
24. Ibid., pp. 116–18.
25. Ibid., pp. 126–27.
26. David Horowitz, *From Yalta to Vietnam: American Foreign Policy in the Cold War* (Harmondsworth: Penguin, 1967), p. 65.
27. Quoted in ibid., p. 68.
28. Quoted in *Congressional Record*, 12 March 1947.

Chapter 3: Events II: The Berlin Blockade, 1948–1949

1. Michael J. Hogan, *The Marshall Plan: America and Britain and the Reconstruction of Western Europe 1947–1952* (Cambridge: Cambridge University Press, 1987), pp. 54–57.
2. Ann and John Tusa, *The Berlin Blockade* (London: Hodder and Stoughton, 1988), p. 54.
3. George F. Kennan, *Memoirs 1925–1950*, vol. 1 (Boston: Little, Brown, 1967), p. 292.
4. Ibid., p. 175.
5. Loth, *Division of the World*, p. 203.
6. Dean Acheson, *Present at the Creation: My Years in the State Department* (New York: W. W. Norton, 1969), p. 97.

Chapter 4: Events III: China and the Korean War, 1949–1953

1. See, for instance, the inimitable delirium of Anthony Kubek in his *How the Far East Was Lost: American Policy and the Creation of Communist China 1941–1949* (Washington, D.C.: Intercontex, 1971).
2. I follow pinyin spelling. The classic biography of Mao is, of course, Stuart

Schram's *Mao-Tse-tung* (Harmondsworth: Penguin, 1967). See also Edgar Snow's firsthand and first-rate *Red Star over China* (New York: Random House, 1938).

3. See also the hagiography, Ross Terrill, *Mao: A Biography* (New York: Harper and Row, 1980).

4. For an explanation of this success, see John Dunn, *Modern Revolutions: Analysis of a Political Phenomenon* (Cambridge: Cambridge University Press, 1972), pp. 70–95. See also Edgar Snow, *Red China Today* (Harmondsworth: Penguin, 1970), pp. 25–39.

5. Quoted in Franz Schurmann, *The Logic of World Power: An Inquiry into the Origins, Currents, and Contradictions of World Politics* (New York: Pantheon, 1974), p. 164.

6. See I. F. Stone, *The Hidden History of the Korean War 1950–1951* (Boston: Little, Brown, 1952).

7. Both quoted in Bruce Cumings, *Origins of the Korean War,* 2 vols. (Princeton: Princeton University Press, 1989), pp. 197, 228. See also Fred Halliday and Bruce Cumings, *Korea: The Forgotten War* (New York: Pantheon, 1989).

8. Cumings, *Origins,* vol. 1, p. 49.

9. Ibid., p. 112.

10. Quoted in Daniel Yergin, *Shattered Peace,* p. 259.

11. Max Hastings, *The Korean War* (London: Michael Joseph, 1987), pp. 67–95.

12. Ibid., p. 97.

13. See George Kennan, *Memoirs 1925–1950* (Boston: Little, Brown, 1967), p. 317.

14. Quoted in Halliday and Cumings, *Korea,* p. 128.

15. Quoted in Trumbull Higgins, *Korea and the Fall of MacArthur* (New York: Oxford University Press, 1960), p. 113.

16. Quoted in Halliday and Cumings, *Korea,* p. 152.

17. Quoted in Hastings, *Korean War,* p. 229.

18. See James Cameron's autobiography, *Point of Departure* (London: Panther Books, 1969), p. 171.

19. Halliday and Cumings, *Korea,* p. 197.

Chapter 5:   Biography II: The Diplomat—George Kennan

1. Kennan, *Memoirs,* vol. 1, pp. 291–92.

2. Ibid., p. 294.

3. Unless otherwise noted, all quotations in this chapter are from my interviews with George Kennan during 1989 in Princeton, New Jersey.

4. *Memoirs,* vol. 1, p. 328.

5. Ibid., p. 258.

6. Ibid., pp. 366–67.

7. George F. Kennan, *Memoirs 1950–1963,* vol. 2 (Boston: Little, Brown, 1972), p. 242.

8. Ibid., p. 307.

Chapter 6: Fictions I: Righteousness—*The Magnificent Seven, The Manchurian Candidate, On the Waterfront, Animal Farm, 1984,* and *The Crucible*

1. See Michael Rogin's brilliant *Ronald Reagan: The Movie, and Other Episodes in Political Demonology* (Berkeley: University of California Press, 1987), for its general thesis about cold war movies and the American matriarchy; on *The Manchurian Candidate*, see pp. 252–56.
2. *The Magnificent Seven* prompted one of the best American poets to a long meditation on his country's position; see Edward Dorn, *Gunslinger* (Chicago: Swallow Press, 1971).
3. Arthur Miller, *Timebends: A Life* (London: Methuen, 1987), pp. 328–34.
4. Quoted in Raymond Williams, *Culture and Society 1780–1950* (London: Chatto and Windus, 1959), p. 294.
5. Ibid., p. 293.
6. Richard Rorty, *Contingency, Irony, Solidarity* (Cambridge: Cambridge University Press, 1989), p. 174.
7. Alasdair MacIntyre's formulation in his essay, "A Mistake about Causality in Social Science," in *Philosophy, Politics, and Society*, 2d series, ed. Peter Laslett and W. G. Runciman (Oxford: Basil Blackwell, 1967), p. 68.

Chapter 7: Events IV: The Rosenbergs and the Red Purge, 1950–1953

1. Quoted in David Caute, *The Great Fear: The Anti-Communist Purge under Truman and Eisenhower* (New York: Simon and Schuster, 1978), p. 28.
2. The first of the *grandes peurs* was reported in July 1789 when spontaneous panic seized whole towns, whose inhabitants expected mobs of brigands at any moment: see Simon Schama, *Citizens: A Chronicle of the French Revolution* (New York: Knopf, 1989), chap. 11.
3. Aleksandr Solzhenitsyn, *The Gulag Archipelago 1918–1956* (London: Collins/Fontana, 1971), p. 245.
4. Victor S. Navasky, *Naming Names* (New York: Viking, 1980), pp. 112–20.
5. E. L. Doctorow, *The Book of Daniel* (New York: Random House, 1971), pp. 361–62.
6. See Robert Chadwell Williams, *Klaus Fuchs: Atom Spy* (Cambridge, Mass.: Harvard University Press, 1987).
7. Quoted in Williams, *Fuchs*, p. 33.
8. Quoted in Freeman Dyson, *Disturbing the Universe* (New York: Basic Books, 1979), p. 87.
9. Caute, *Great Fear*, p. 493.
10. Quoted in Navasky, *Naming Names*, p. 130.
11. Arthur Miller, *Timebends: A Life* (London: Methuen, 1987), p. 312.
12. Isaiah Berlin, *Russian Thinkers* (London: Hogarth Press, 1978), pp. 301–2.

13. Quoted in Caute, *Great Fear,* p. 305.
14. R. Rovere, *Senator Joe McCarthy* (New York: Harcourt Brace, 1959), p. 41.
15. Quoted in Caute, *Great Fear,* p. 48.

Chapter 8:   Events V: Budapest, Warsaw, Suez, 1956

1. Strobe Talbott, ed. and trans., *Khrushchev Remembers,* with an introduction and commentary by Edward Crankshaw (Boston: Little, Brown, 1970), p. 339.
2. Crankshaw in Talbott, *Khrushchev Remembers,* pp. x–xii.
3. Talbott, *Khrushchev Remembers,* p. 351.
4. Crankshaw in Talbott, *Khrushchev Remembers,* p. xvii.
5. Quoted in Francois Fejto, *A History of the People's Democracies: Eastern Europe since Stalin* (Harmondsworth: Penguin, 1974), p. 113.
6. Talbott, *Khrushchev Remembers,* p. 426.
7. Czeslaw Milosz, *Collected Poems 1931–1987* (Harmondsworth: Penguin, 1988), p. 116.
8. Leszek Kolakowski, *Marxism and Beyond* (London: Paladin, 1971), p. 164.
9. Quoted in Fejto, *People's Democracies,* p. 106.
10. Anthony Nutting, *No End of a Lesson: The Story of Suez* (London: Constable, 1967), pp. 34–35.
11. Christopher Hitchens, "Mad Dogs and Others: Suez 1956," *Grand Street* 6, no. 1 (1986): 105.
12. Leonard Mosley, *Dulles: A Biography of Eleanor, Allen, and John Foster Dulles and Their Family Network* (London: Hodder and Stoughton, 1978), p. 409.
13. Quoted in S. Roy Fullick and Geoffrey Powell, *Suez: The Double War* (London: Hamish Hamilton, 1979), p. 108.
14. Nutting, *No End,* p. 97.
15. Figure quoted in Fullick and Powell, *Double War,* p. 120.

Chapter 9:   Events VI: The Berlin Wall and the Cuban Missile Crisis, 1961–1962

1. John Le Carré, *Smiley's People* (New York: Knopf, 1978), pp. 326–27.
2. Quoted in Evelyn Lincoln, *My Twelve Years with John F. Kennedy* (New York: David McKay, 1965), p. 274.
3. Quoted in Curtis Cate, *The Ides of August: The Berlin Wall Crisis* (New York: Evans and Company, 1978), pp. 109–10.
4. *Khrushchev Remembers,* p. 460.
5. Hugh Thomas, *Cuba* (London: Eyre and Spottiswoode, 1971), p. 1315.
6. G. Wills, *The Kennedys: A Shattered Illusion* (London: Orbis, 1983), p. 149.
7. Wills, *Kennedys,* p. 144.
8. Robert F. Kennedy, *Thirteen Days: A Memoir of the Cuban Missile Crisis* (New York: W. W. Norton, 1969), p. 85.
9. Elie Abel, *The Missile Crisis* (Philadelphia: J. B. Lippincott, 1966), p. 195.

10. Quoted in Robert A. Divine, ed., *The Cuban Missile Crisis* (New York: Markus Wiener, 1988), p. 36.
11. Kennedy, *Thirteen Days,* p. 67.
12. Ibid., p. 72.
13. John Updike, *Couples* (London: Andre Deutsch, 1968), p. 224.
14. Quoted in Divine, *Missile Crisis,* p. 49.
15. Quoted in ibid., p. 53.
16. Dean Acheson, "Robert Kennedy's Version of the Cuban Missile Affair," *Esquire,* February 1969, p. 77.
17. Roger Hilsman, letter to the editor, *New York Review of Books,* 8 May 1969, p. 37.
18. Arthur Schlesinger, *A Thousand Days: John F. Kennedy in the White House* (London: Andre Deutsch, 1965), p. 716.
19. Schlesinger, *Thousand Days,* p. 714.
20. Quoted in Divine, *Missile Crisis,* p. 39.
21. Quoted in Hugh Sidey, *John F. Kennedy, President* (New York: Atheneum, 1964), p. 127.

Chapter 10:   Biography III: The European—Willy Brandt

1. Unless otherwise noted, all quotations in this chapter are taken from my interview with Willy Brandt, 24 October 1990, University of Warwick, England.
2. Willy Brandt, speech before the SPD annual conference, June 1966, Dortmund, West Germany.

Chapter 11:   Biography IV: The Scientist—Freeman Dyson

1. Unless otherwise noted, all quotations in this chapter are from my interviews with Freeman Dyson, May–June 1989, Princeton, New Jersey.
2. Freeman Dyson, *Disturbing the Universe* (New York: Harper and Row, 1979), p. 38.
3. Freeman Dyson, *Weapons and Hope* (New York: Harper and Row, 1984), p. vi.
4. Lord Solly Zuckerman, "Nuclear Fantasies," *New York Review of Books,* 14 June 1984, p. 5.
5. Dyson, *Disturbing the Universe,* pp. 134–35.
6. Ibid., pp. 135–36.
7. Zuckerman, "Nuclear Fantasies," p. 10.

Chapter 12:   Fictions II: Dead Ends—End-of-the-World Movies

1. Friedrich Nietzsche, *Beyond Good and Evil: Prelude to a Philosophy of the Future* (New York: Penguin, 1973), p. 146.
2. Mrs. W. Orr-Ewing, interview with the editor, in Jack G. Shaheen, ed., *Nuclear War Films* (Carbondale: Southern Illinois University Press, 1978), p. 113.

3. John Keats, *Selected Letters,* ed. F. Page (Oxford: Oxford University Press, 1954), p. 52.

Chapter 13:   Fictions III: Loyalty and Lying—Spy Stories

1. I have been much helped in this chapter by two articles: E. P. Thompson, "A State of Blackmail," in his *Writing by Candlelight* (London: Merlin Press, 1980), and R. W. Johnson, "Subversions," in his *Heroes and Villains: Selected Essays* (England: Harvester Press, 1990), pp. 238–51.
2. Michel Foucault, *Discipline and Punish: The Birth of the Prison* (Harmondsworth: Penguin, 1979), pp. 293–308.
3. Rhodry Jeffreys-Jones, *The CIA and American Democracy* (New Haven: Yale University Press, 1989), p. 71.
4. Ibid., p. 132.
5. Ibid., p. 35.
6. For Angleton's career, see ibid.
7. Arnold Wesker, *Chips with Everything* (London: Jonathan Cape, 1952).
8. The film *Another Country* (1984) was based on the idealized school career of just such a homosexual Communist in the late 1930s.
9. Jim Hunter, *The Flame* (London: Faber & Faber, 1965), p. 16.
10. Chapman Pincher, *Their Trade Is Treachery* (London: Sidgwick and Jackson, 1981), pp. 163, 252–53.
11. Edward Jay Epstein, *Deception: The Invisible War between the KGB and the CIA* (New York: Simon and Schuster, 1989), chaps. 5 and 6.
12. Chapman Pincher, *Inside Story* (London: Sidgwick and Jackson, 1978), p. ix.
13. John Le Carré, *Tinker, Tailor, Soldier, Spy* (New York: Knopf, 1975), p. 345.
14. Fredrick Forsyth, *The Fourth Protocol* (London: Hutchinson, 1984), p. 151.
15. Raymond Williams, *Loyalties* (London: Chatto and Windus, 1985), p. 317.
16. Alan Bennett, *Single Spies* (London: Faber & Faber, 1988), pp. 17–18.
17. Ibid., p. ix.

Chapter 14:   Events VII: Tet, Prague, Chicago, 1968

1. See the deservedly best-selling history by Neil Sheehan, *A Bright Shining Lie: John Paul Vann and America in Vietnam* (New York: Random House, 1988), pp. 102–6.
2. See Frances Fitzgerald, *Fire in the Lake: The Vietnamese and the Americans in Vietnam* (London: Macmillan, 1972), pp. 134, 145.
3. Quoted in Gabriel Kolko, *Anatomy of a War: Vietnam, the United States, and the Modern Historical Experience* (New York: Pantheon, 1985), p. 113.
4. Quoted in ibid., p. 164.
5. A point I take from Franz Schurmann in his classic *The Logic of World Power* (New York: Pantheon, 1974), especially pp. 501–10.
6. William Westmoreland, *Christian Science Monitor,* 27 October 1969, quoted in Noam Chomsky, *At War with Asia* (New York: Random House, 1970), p. 72.

7. Robert McNamara, *The Essence of Security: Reflections in Office* (New York: Harper and Row, 1968), p. 109.

8. Hannah Arendt, *Crises of the Republic* (New York: Harcourt Brace Jovanovich, 1972), p. 34 (Arendt's italics).

9. Samuel Huntington, "The Bases of Accommodation," *Foreign Affairs* 46 (1968): 652.

10. Quoted in Sheehan, *Bright Shining Lie*, p. 699.

11. Philip Jones Griffiths, *Vietnam Inc.* (New York: Collier, 1971), p. 44. The remark was later appropriated by Francis Ford Coppola for the fire-eating colonel of the air cavalry in his film *Apocalypse Now* (see chapter 20).

12. Quoted in Fejto, *People's Democracies*, p. 220.

13. Quoted in David Caute, *Sixty Eight: The Year of the Barricades* (London: Hamish Hamilton, 1988), p. 292.

14. As witness, Daniel Cohn-Bendit, *Obsolete Communism: The Left-Wing Alternative* (Harmondsworth: Penguin, 1969).

15. Norman Mailer, *Miami and the Siege of Chicago* (Harmondsworth: Penguin, 1969), pp. 194–95.

16. Ibid., p. 196.

Chapter 15: Events VIII: The Fall of Allende, SALT 1,
Leaving Vietnam, 1972–1973

1. The best-known biography of Henry A. Kissinger is Seymour B. Hersh's excellent *The Price of Power: Kissinger in the Nixon White House* (New York: Summit/Simon and Schuster, 1983), p. 270. See also the polemical sketches in William Shawcross, *Sideshow: Kissinger, Nixon, and the Destruction of Cambodia* (London: Fontana, 1980), pp. 74–84. Even more biting is Christopher Hitchens, *Prepared for the Worst: Selected Essays and Minority Reports* (New York: Hill and Wang, 1988), pp. 121–31. Kissinger's autobiographical apologetics include *The White House Years* (Boston: Little, Brown, 1979) and *Years of Upheaval* (Boston: Little, Brown, 1982).

2. Hersh, *Price of Power*, p. 263.

3. Cord Meyer, *Facing Reality* (New York: Harper and Row, 1980), p. 185.

4. Quoted in Hersh, *Price of Power*, p. 286.

5. Quoted in Shawcross, *Sideshow*, p. 304.

6. Joan Jara, *Victor* (London: Jonathan Cape, 1983). See also C. D. Perez, *The Murder of Allende* (New York: Harper and Row 1975).

7. Quoted in Joan Jara, "September 11th, 1973,' *Granta* 9 (1983): 121.

8. Jara, "September 11th 1973," p. 121.

9. Ibid., p. 125.

10. Ibid., p. 128.

11. I follow John Newhouse's narrative of SALT in *War and Peace in the Nuclear Age* (New York: Knopf, 1988) but attribute my own interpretation of Nixon's actions.

12. Hersh, *Price of Power,* pp. 188–99.
13. For these figures and an account of the Phoenix Program, see Kolko, *Anatomy of a War,* pp. 397–78, 463–34.
14. H. R. Haldeman, *The Ends of Power* (New York: Dell, 1978), p. 83.
15. Quoted in Kolko, *Anatomy of a War,* p. 443.

Chapter 16:   Events IX: SALT 2 and the Invasion of Afghanistan, 1979

1. See Fred Halliday, *The Making of the Second Cold War* (London: Verso, 1983), p. 108.
2. See Gwyn Prins, ed., *Defended to Death: A Study of the Nuclear Arms Race by the Cambridge University Disarmament Seminar* (Harmondsworth: Penguin, 1983), pp. 114–24 and pp. 300–303.
3. Quoted in R. L. Garthoff, *Détente and Confrontation: American-Soviet Relations from Nixon to Reagan* (Washington, D.C.: Brookings Institution, 1985), p. 602.
4. Comparative figure quoted in Dan Smith, *The Defence of the Realm in the 1980s* (Beckenham: Croom Helm, 1980), pp. 66–105.
5. For the figures, see Barry Rubin's informative *Paved with Good Intentions: The American Experience and Iran* (New York: Penguin, 1980), pp. 90–127.
6. Raja Anwar, *The Tragedy of Afghanistan: A Firsthand Account,* rev. ed. (London: Verso, 1989), p. 232.
7. Ibid., p. 103.
8. Doris Lessing, *The Wind Blows Away Our Words* (New York: Vintage Books, 1987), p. 162.

Chapter 17:   Biography V: The Spy—Philip Agee

1. All quotations in this chapter are taken from my long interview with Philip Agee in New York, in July 1989. Further details of his life are taken from his best-seller *Inside the Company: CIA Diary* (Harmondsworth: Penguin, 1975), and his autobiographical sequel, *On the Run* (Secaucus, N.J.: Lyle Stuart, 1987). I have also referred (on his recommendation) to Joseph Smith, *Portrait of a Cold Warrior* (Secaucus, N.J.: Lyle Stuart, 1976); Victor Marchetti and John Marks, *The CIA and the Cult of Intelligence* (New York: Dell, 1975); Philip Agee and Louis Wolf, *Dirty Work: The CIA in Western Europe* (Secaucus, N.J.: Lyle Stuart, 1978); and John Stockwell, *In Search of Enemies: A CIA Story* (New York: W. W. Norton, 1978).

Chapter 18:   Biography VI: The Journalist—Neil Sheehan

1. See Patrick Cockburn, *Getting Russia Wrong* (London: Verso, 1989), pp. 15–61.
2. See Noam Chomsky and Edward Kellman, *The Political Economy of Mass Media* (Boston: Beacon Press, 1982).
3. Published in book form as Senator Michael Gravel, ed., *The Pentagon Papers,* 4 vols. (Boston: Beacon Press, 1972).
4. All quotations in this chapter are from my interview with Neil Sheehan in Washington in May 1989.

Chapter 19: Fictions IV: Fantasy and Action—Bond, Clancy, Forsyth, Tolkien, and Kundera

1. I take much here from Nicholas Garnham, *Capitalism and Communications: Global Culture and the Economics of Information* (London: Sage Books, 1990), pp. 154–68.
2. Snobbery of caste and nation as the key British export to the United States is analyzed by Christopher Hitchens in *Blood, Class, and Nostalgia: Anglo-American Ironies* (New York: Farrar, Straus, and Giroux, 1990).
3. So sympathetically described by V. S. Naipaul in *A Turn in the South* (New York: Knopf, 1989).
4. J. R. R. Tolkien, *The Return of the King* (London: Allen and Unwin, 1955), p. 103.
5. Milan Kundera, *The Book of Laughter and Forgetting* (Harmondsworth: Penguin, 1983), pp. 62–63.
6. Kundera, *Laughter and Forgetting*, p. 65.
7. Milan Kundera, *The Unbearable Lightness of Being* (New York: Harper and Row, 1984), p. 3.
8. Kundera, *Unbearable Lightness*, pp. 253, 256.
9. Ibid., p. 273.

Chapter 20: Fictions V: Patriotism and Psychosis—The Vietnam War Movies

1. Gilbert Adair, *Hollywood's Vietnam: From* The Green Berets *to* Full Metal Jacket, rev. ed. (London: Heinemann, 1987), p. 36.
2. *The Portable Nietzsche*, ed. Walter Kaufmann (New York: Penguin, 1976), p. 518.
3. Quoted in Lawrence Suid, *Guts and Glory: Great American War Movies* (Reading, Mass.: Addison-Wesley, 1978), pp. 310–11.
4. Adair, *Hollywood's Vietnam*, p. 168.
5. Joseph Conrad, *Heart of Darkness* (1902; reprint, New York: Bantam Books, 1960), p. 84.
6. J. V. Cunningham, *To What Strangers, What Welcome* (Denver: Alan Swallow, 1964), p. 6.
7. Clifford Geertz, "Notes on the Balinese Cockfight," *The Interpretation of Cultures* (New York: Basic Books, 1973), pp. 412–54.
8. The classic review of the term "just war" is Michael Walzer's *Just and Unjust Wars*.
9. Quoted in Adair, *Hollywood's Vietnam*, p. 174.

Chapter 21: Fictions VI: MAD Jokes and the Nuclear Unconscious—*Dr. Strangelove, Woodstock,* and *When the Wind Blows*

1. Milan Kundera, *The Joke* (Harmondsworth: Penguin, 1984).
2. The change is detected and chronicled in Paul Fussell's classic *The Great War and Modern Memory* (Oxford: Oxford University Press, 1975), in its concluding section.

3. Peter Porter, "Your Attention Please," *Collected Poems* (Oxford: Oxford University Press, 1983), p. 38.
4. Its first history was written by Theodore Roszak, *The Making of the Counter Culture* (London: Faber and Faber, 1972).
5. Raymond Briggs, *When the Wind Blows* (London: Hamish Hamilton, 1982).

Chapter 22:   Events X: The Second Cold War, the New Peace Movements, Solidarity, and Ronald Reagan, 1980–1981

1. Prins, *Defended to Death,* pp. 111–14.
2. E. P. Thompson, Public Lectures in Bristol, 11 June 1980.
3. Quoted in Garthoff, *Détente and Confrontation,* p. 972.
4. Quoted in Noam Chomsky, Jonathan Steele, and John Gittings, *Superpowers in Collision: The New Cold War of the 1980s* (Harmondsworth: Penguin, 1984), pp. 75–76.
5. In a memorial edition of *The Spokesman* dedicated to Russell, no. 3 (May 1970): 4.
6. Her Majesty's Stationary Office, *Protect and Survive* (London, 1980), pp. 18–19.
7. Edward Thompson and Dan Smith, eds., *Protest and Survive* (Harmondsworth: Penguin, 1980).
8. Prins, *Defended to Death,* p. 27.
9. The point is made by Michael Walzer in his textbook *Political Action: A Practical Guide to Movement Politics* (New York: Quadrangle, 1971), pp. 28–31.
10. Donald Gress, *Peace and Survival: West Germany, the Peace Movement, and European Security* (Stanford, Calif.: Hoover Institute, 1985), p. 179. It is a pleasure to be able to add that the judicious Dr. Gress includes this paragraph in a chapter called "Eleven Neutral Propositions."
11. *Soviet News,* 12 January 1980.
12. Neal Ascherson, *The Polish August: The Self-Limiting Revolution,* rev. ed. (Harmondsworth: Penguin, 1982).
13. Ibid., p. 124.
14. Ibid., p. 147.
15. Quoted in Halliday, *Second Cold War,* p. 111.
16. Quoted in ibid., p. 15.
17. Quoted in Newhouse, *War and Peace,* p. 337.
18. Quoted in F. H. Knelman, *Reagan, God, and the Bomb* (Buffalo, N.Y.: Prometheus Books, 1985), pp. 182–83.

Chapter 23:   Events XI: The Evil Empire, Star Wars, and Central America, 1983

1. Ronald Reagan's first press conference, 30 January 1981, quoted in Chomsky, Steele, and Gittings, *Superpowers in Collision,* p. 62.
2. Quoted in Garthoff, *Détente and Confrontation,* p. 1010.

3. Geoffrey Barraclough, *From Agadir to Armageddon: Anatomy of a Crisis* (London: Weidenfeld and Nicolson, 1982), p. 170.

4. Ronald Reagan, 1 October 1981, as transcribed by a journalist present as the President spoke in Air Force One. Quoted in Prins, *Defended to Death*, p. 32.

5. Lee Blessing took *A Walk in the Woods* as the title for his 1988 play, which follows the two diplomats to their bench at key moments of the four seasons in Geneva.

6. Newhouse, *War and Peace*, see pp. 352–54 for discussion of 1981 and the SS-20s.

7. Figures in Halliday, *Second Cold War*, chap. 9.

8. Quoted in William Broad, "The Scientists of Star Wars," *Granta* 16 (Science) (1985): 86.

9. Newhouse, *War and Peace*, p. 361.

10. Broad, "Scientists of Star Wars," p. 84.

11. See R. W. Johnson, *Shootdown: The Verdict on KAL 007* (New York: Doubleday, 1985).

12. Seymour M. Hersh, *"The Target Is Destroyed": What Really Happened to Flight 007* (London: Faber and Faber, 1986).

13. Garthoff, *Détente and Confrontation*, p. 1016.

14. Quoted in ibid., p. 1017.

15. James Hinton, *Protests and Visions: Peace Politics in Twentieth-Century Britain* (London: Hutchinson Radius, 1989), p. 183.

16. See Beth Harford and Sally Hopkins, eds., *Greenham Common: Women at the Wire* (London: Women's Press, 1984).

17. Quoted in Peter Davis, *Where Is Nicaragua?* (New York: Simon and Schuster, 1987), p. 74.

18. Ibid., p. 79.

19. From $1.2 billion to $385 million, according to *New York Times*, 23 June 1989.

20. Quoted in Davis, *Where Is Nicaragua?* p. 122.

Chapter 24:   Events XII: Mr. Gorbachev Goes to Reykjavík, 1986

1. *History of Soviet Foreign Policy* (Moscow: Foreign Languages Publishing House, 1981), p. 480. I also take much in this chapter from Fred Halliday's excellent essay, "The Ends of Cold War," *New Left Review* 180 (March-April 1990): 5–24. In the same issue, see also Mary Kaldor, "After the Cold War" and E. P. Thompson's "Comment," pp. 25–40, pp. 139–46, 182.

2. See Roger Poole, *Towards Deep Subjectivity* (London: Allen Lane, 1972), for a discussion of terrorism's signification, pp. 12–43.

3. Quoted in E. P. Thompson and Mary Kaldor, *Mad Dogs: The U.S. Raids on Libya* (London: Pluto Press, 1986), p. 36.

4. This short biography is based on Martin Walker's excellent study *The Waking Giant: The Soviet Union under Gorbachev* (London: Abacus, 1987), pp. 33–52. See also David Lane, *Soviet Economy and Society* (Oxford: Basil Blackwell, 1985).

5. Garry Wills, *Reagan's America* (New York: Viking Penguin, 1988), p. 461.

6. Details in Newhouse, *War and Peace,* pp. 394–404, and see Jonathan Haslam, *The Soviet Union and the Politics of Nuclear Weapons in Europe 1969–87* (London: Macmillan, 1989), pp. 166–70.
7. Quoted in Donald Regan, *For the Record: From Wall Street to Washington* (San Diego: Harcourt Brace Jovanovich, 1988), p. 350.
8. Jane Mayer and Doyle McManus, *Landslide: The Unmaking of the President 1984–1988* (Boston: Houghton Mifflin, 1988), p. 283.
9. Wills, *Reagan's America,* p. 363.

Chapter 25: Biography VII: The Secretary-General—Lord Peter Carrington

1. Lord Peter Carrington, *Reflect on Things Past: The Memoirs of Lord Carrington* (London: Collins, 1988), p. 65.
2. Unless otherwise noted, all quotations in this chapter are from my interview with Lord Peter Carrington, in London during February 1991.
3. Carrington, *Memoirs,* p. 382.
4. Ibid., p. 376.
5. Quoted in ibid., p. 222.
6. Ibid., p. 236.

Chapter 26: Biography VIII: The Dissenter—Edward Thompson

Unless otherwise noted, all quotations in this chapter are from my interviews and conversations with Thompson, held over the years 1987 to 1989 in particular.
1. Edward Thompson, *The Heavy Dancers: Writings on War, Past and Future* (London: Merlin, 1985), p. 188.
2. Ibid., p. 200.
3. Thompson, *Writing by Candlelight,* p. 132.
4. Edward Thompson, "Letter to Leszek Kolakowski," in *The Poverty of Theory* (London: Merlin, 1978), pp. 93–94.
5. Thompson, "Letter," p. 95.
6. Edward Thompson, *William Morris: Romantic to Revolutionary,* rev. ed. (London: Merlin, 1977).
7. Edward Thompson, *Whigs and Hunters: The Origin of the Black Act* (London: Allen Lane, 1975).
8. George Ball, "Summarizing U.S. Policy," *New York Review of Books,* 2 February 1984, p. 35.
9. Reprinted in Thompson and Smith, *Protest and Survive,* p. 35.

Chapter 27: Biography IX: The Novelist—Joan Didion

1. Joan Didion, *Democracy* (New York: Simon and Schuster, 1984), pp. 163–64.
2. Ibid., p. 85.

3. Unless otherwise noted, all quotations in this chapter are from my interview with Joan Didion, conducted in New York, during July 1989.
4. Didion, *Democracy*, p. 188.
5. Ibid., p. 176.
6. Ibid., p. 176.
7. Ibid., p. 83 (Didion's italics).
8. Ibid., p. 53.
9. Joan Didion, *Salvador* (New York: Simon and Schuster, 1983), p. 16.
10. Ibid., p. 34.

Chapter 28: Fictions VII: Mistrust—*Missing, All the President's Men, Gorky Park, Edge of Darkness,* and the Lexicon of Apocalypse

1. John Locke, *Essays on the Law of Nature* (Oxford: Clarendon Press, 1954). See also John Dunn, " 'Trust' in the Politics of John Locke," in *Rethinking Modern Political Theory* (Cambridge: Cambridge University Press, 1985).
2. The quotations are from Alan Bennett, p. 204.
3. Hannah Arendt, "Lying in Politics," in *Crises of the Republic,* p. 41.
4. Vladimir Nabakov, *Lolita* (London: Harmondsworth, 1980), p. 313.
5. Raymond Williams, *Towards 2000* (London: Chatto and Windus, 1983), p. 248.
6. The full text has been published by the scriptwriter, Troy Kennedy Martin, *Edge of Darkness* (London: Faber and Faber, 1990).
7. The relevance of game theory to nuclear weapons is explained by Greville Rumble, *The Politics of Nuclear Defense* (Cambridge, England: Polity Press, 1985), chap. 6.
8. Herbert York, "The Nuclear 'Balance of Terror' in Europe," *Ambio* 4, nos. 5–6 (1975): 203–8.
9. It is described by Desmond Ball in "The Development of the SIOP, 1960–1983," in *Strategic Nuclear Targeting,* ed. Desmond Ball and Jeffrey Richelson (Ithaca: Cornell University Press, 1986).
10. Herman Kahn, *On Thermonuclear War: Thinking about the Unthinkable* (New York: Horizon Press, 1962), pp. 40–96.

Chapter 29: Events XIII: The Final Chorus, 1989

1. I again owe much to the *New Left Review* essays by Fred Halliday, "Ends of Cold War," and Mary Kaldor, "After the Cold War."
2. This summary is taken from Timothy Garton Ash's excellent report, reprinted in *We, the People: The Revolution of '89* (Cambridge: Granta Books, 1990).
3. See G. R. Iliescu, *The Razing of Romania's Past* (London: Architecture and Technology Press, 1990). For a brief, vivid report from the Romanian revolution, see Christopher Hitchens, "On the Road to Timisoara," *Granta* 31 (1990): 129–40.
4. Mercea Dinescu, "Letter," *Granta* 30 (1990): 173.

5. See John Simpson's remarkable report, in *Granta* 28 (1989): 9–26.
6. Ibid., p. 12.
7. Ibid., pp. 21–22.

Chapter 30:   Biography X: The Peace Woman—Joan Ruddock

1. All quotations in this chapter are from my interview with Joan Ruddock, in London during January 1991.

Chapter 31:   The End—An Interpretation

1. See, for example, Piers Brendon, *Ike: The Life and Times of Dwight D. Eisenhower* (London: Secker and Warburg, 1987).
2. For a representative medley of these views, or reports of them, see Charles Krauthammer, "Beyond the Cold War," *New Republic,* 19 December 1988; (unsigned), "The Next War," *New Republic,* 13 February 1989; Hella Pick, "Nato Seeks a New Role," *Guardian* (London), 18 May 1990; John Lewis Gaddis, "Making the Long Peace Last Longer," and "Beyond the Triumph of Liberty," *Guardian* (London), 31 May and 1 June 1990. See also Francis Fukuyama, "The End of History," *National Interest* 15, Summer 1989.
3. Estimates in Cumings and Halliday, *Korea,* pp. 200–201.
4. Estimates in Peter Livingnew and Peter Weiss, eds., *Prevent the Crime of Silence: Reports from the International War Crimes Tribunal* (London: Athen Lane, 1971), pp. 11–56, 383; Kolko, *Anatomy of a War,* p. 200; and Mark Clodfelter, *The Limits of Air Power: The American Bombing of North Vietnam* (New York: Free Press, 1990), pp. 129, 171.
5. Estimates in Francois Ponchaud, *Cambodia Year Zero* (Harmondsworth: Penguin, 1978), pp. 71–92, and William and Shawcross, *Sideshow,* pp. 365–91.
6. Estimates assembled from Peter Dickens, *SAS: The Jungle Frontier: The Borneo Campaign 1963–1966* (London: Army and Armour Press, 1983), p. 217, and S. Tas, *Indonesia: The Underdeveloped Freedom* (New York: Pegasus, 1974), p. 313. See also Anthony Short, *The Communist Insurrection in Malaya 1948–1960* (New York: Crane, Russak, 1975), pp. 295–96, 374–76, 496–503.
7. See Anwar, *Tragedy of Afghanistan,* pp. 254–57 and Lessing, *Words on the Wind,* pp. 157–71.
8. See Basil Davidson, *In the Eye of the Storm: Angola's People* (Harmondsworth: Penguin, 1975), pp. 223–42; Arthur Jay Klinghoffer, *The Angolan War: A Study of Soviet Policy in The Third World* (Boulder: Westview Press, 1980), pp. 1–8; and Joseph Hanlon: *Mozambique: The Revolution Under Fire* (London: Zed Books, 1984), pp. 219–32.
9. See Noam Chomsky, *Turning the Tide: U.S. Intervention in Central America and the Struggle for Peace* (Boston: South End Press, 1985), pp. 3–42, and Holly Sklar, *Reagan, Trilateralism, and the Neo Liberals: Containment and Intervention in the 1980s* (Boston: South End Press, 1986), p. 396.
10. See Conquest, *Harvest of Sorrow,* pp. 299–301.

11. See, for example, Michael Howard, *The Causes of Wars* (London: Allen and Unwin, 1984).

12. All quoted in Gaddis, "Making Long Peace Last."

13. See Anthony Giddens, *The Nation-State and Violence* (Cambridge: Polity Press, 1985), pp. 22–30.

14. See John Dunn, *Western Political Theory in the Face of the Future* (Cambridge: Cambridge University Press, 1979), pp. 55–73.

15. Quoted by Denis Healey, *The Time of My Life* (London: Michael Joseph, 1989), p. 472.

16. See Paul Kennedy, *The Rise and Fall of the Great Powers* (New York: Random House, 1987), pt. 7; Prins, *Defended to Death*, pp. 304–13; and Defense Department, *The Military Balance* (Washington, D.C.: U.S. Government Printing Office, annually).

17. Howard, *Causes of Wars*, pp. 36–64.

18. I borrow here and in the next paragraph from Giddens, *Nation-State*, pp. 325–34.

19. Kennedy, *Great Powers*, pt. 8.

20. Ibid., p. 526.

21. See Fukuyama, "End of History," and his postscript, "History Is Over— Never Mind Iraq," *Guardian* (London), 6 September 1990.

22. See Quentin Skinner's commentary on Lorenzetti's Sienese fresco, "The Business of Good Government," in *Ambrogio Lorenzetti: The Artist as Political Philosopher* (London: British Academy, 1986).

23. As Stephen Cohen did, in a fine essay, "The Next President's Historic Opportunity," *The Nation*, 10 October 1988.

BIBLIOGRAPHY

ABEL, E. *The Missile Crisis*. Philadelphia: University of Pennsylvania Press, 1966.

ACHESON, D. *Present at the Creation: My Years in the State Department*. New York: W. W. Norton, 1969.

ADAIR, G. *Hollywood's Vietnam: From* The Green Berets *to* Full Metal Jacket, rev. ed. London: Heinemann, 1987.

ADAMS, G. *The Iron Triangle: The Politics of Defense Contracting*. New York: Brunswick, Transaction Books, 1981.

AGEE, P. *Inside the Company: CIA Diary*. Harmondsworth: Penguin, 1975.

———. *On the Run*. Secaucus, N.J.: Lyle Stuart, 1987.

———, and L. Wolf. *Dirty Work: The CIA in Western Europe*. Secaucus, N.J.: Lyle Stuart, 1978.

ALI, T. *Street Fighting Years: An Autobiography of the Sixties*. London: Collins, 1987.

ALLEN, V. L. *The Russians Are Coming*. Shipley, Eng.: Moore Press, 1987.

ALLISON, G. T. *Essence of Decision: Explaining the Cuban Missile Crisis*. Boston: Little, Brown, 1971.

ANDERSON, P. "The Figures of Descent." *New Left Review* 161 (January-February 1987): 20–77.

ANWAR, R. *The Tragedy of Afghanistan: A Firsthand Account*, rev. ed. London: Verso, 1989.

ARENDT, H. *On Revolution*. New York: Viking, 1965.

———. *Crises of the Republic*. New York: Harcourt Brace Jovanovich, 1972.

ASCHERSON, N. *The Polish August: The Self-Limiting Revolution,* rev. ed. Harmonds-
worth: Penguin, 1982.
ASH, T. GARTON. *The Uses of Adversity: Essays on the Fate of Central Europe.* Cambridge,
Eng.: Granta Books, 1989.
————. *We, the People: The Revolution of '89.* Cambridge: Granta Books, 1990.
ATTLEE, C. R. *As It Happened.* London: Heinemann, 1954.
BAKER, R. W. *Egypt's Uncertain Revolution under Nasser and Sadat.* Cambridge, Mass.:
Harvard University Press, 1978.
BALL, D., and J. RICHELSON, eds. *Strategic Nuclear Targeting.* Ithaca: Cornell Univer-
sity Press, 1986.
BARKER, E. *British Policy in Southeast Europe in World War Two.* London: Macmillan,
1976.
BARNET, R. J. *Intervention and Revolution: America's Confrontation with Insurgent Move-
ments.* New York: New American Library, 1968.
————. *The Roots of War.* Harmondsworth: Penguin, 1973.
BARNETT, C. *The Audit of War: The Illusion and Reality of Britain as a Great Nation.*
London: Macmillan, 1986.
BARRACLOUGH, G. *From Agadir to Armageddon: Anatomy of a Crisis.* London: Weiden-
feld and Nicolson, 1982.
BELLOW, S. *The Dean's December.* London: Secker and Warburg, 1982.
BENNETT, A. *Single Spies.* London: Faber and Faber, 1988.
BENNETT, T., and J. WOOLLACOTT. *Bond and Beyond: The Political Career of a Popular
Hero.* London: Macmillan, 1987.
BERNAL, M. "Vietnam: A Primer on the War." *Cambridge Quarterly* 3 and 4 (1968):
318–340.
BESCHLOSS, M. R. *May Day: Eisenhower, Khrushchev, and the U2 Affair.* New York:
Harper and Row, 1986.
BIALER, S., ed. *The Domestic Context of Soviet Foreign Policy.* Boulder, Colo.: Westview
Press, 1981.
BISKIND, P. *Seeing Is Believing: How Hollywood Taught Us to Stop Worrying and Love the
Fifties.* New York: Pantheon, 1983.
BLECHMAN, B. M., and S. S. KAPLAN, eds. *Force without War: U.S. Armed Forces as
a Political Instrument.* Washington, D.C.: Brookings Institution, 1978.
BODARD, L. *The Quicksand War: Prelude to Vietnam.* Boston: Little, Brown, 1967.
BRANDT, W. *A Peace Policy for Europe.* London: Weidenfeld and Nicolson, 1969.
BRENDON, P. *Ike: The Life and Times of Dwight D. Eisenhower.* London: Secker and
Warburg, 1987.
BRIGGS, R. *When the Wind Blows.* London: Hamish Hamilton, 1982.
BROAD, W. "The Scientists of Star Wars." *Granta* 16 (Science) (1985): 81–108.
BRZEZINSKI, Z. *Power and Principle: Memoirs of the National Security Adviser 1977–1981.*
New York: Farrar, Straus, and Giroux, 1983.
————. *Game Plan: How to Conduct the U.S.-Soviet Contest.* Boston: Atlantic Monthly
Press, 1986.
CAIRNCROSS, A. *The Price of War.* Oxford: Basil Blackwell, 1986.

CALVINO, I. *Invisible Cities*. London: Picador, 1979.

CAMERON, J. *Point of Departure*. London: Panther, 1969.

CAMPBELL, D. *The Unsinkable Aircraft Carrier*. London: Michael Joseph, 1984.

CANETTI, E. *Crowds and Power*. 1932; reprint, Harmondsworth: Penguin, 1981.

CARE, C. *The Ides of August: The Berlin Wall Crisis 1961*. New York: Evans, 1978.

CARRINGTON, LORD P. *Reflect on Things Past: The Memoirs of Lord Carrington*. London: Collins, 1988.

CARTER, J. *Keeping Faith: Memoirs of a President*. New York: Bantam, 1982.

CAUTE, D. *The Great Fear: The Anti-Communist Purge under Truman and Eisenhower*. New York: Simon and Schuster, 1978.

————. *Sixty Eight: The Year of the Barricades*. London: Hamish Hamilton, 1988.

CAVELL, S. *The World Viewed: Reflections on the Ontology of Film*, rev. ed. Cambridge, Mass.: Harvard University Press, 1979.

CECIL, R. *A Divided Life: A Personal Portrait of the Spy Donald Maclean*. New York: William Morrow, 1989.

CHOMSKY, N. *American Power and the New Mandarins*. New York: Pantheon, 1968.

————. *At War with Asia*. New York: Random House, 1970.

————. *For Reasons of State*. New York: Vintage, 1973.

————. *Turning the Tide: U.S. Intervention in Central America and the Struggle for Peace*. Boston: South End Press, 1985.

————. *The Culture of Terrorism*. London: Pluto Press, 1988.

————, and H. ZINN, eds. *The Pentagon Papers*, vol. 5, *Critical Essays and an Index to Volumes 1–4*. Boston: Beacon Press, 1972.

————, and E. HERMAN. *The Political Economy of Mass Media*. Boston: Beacon Press, 1982.

————, J. STEELE, and J. GITTINGS. *Super Powers in Collision: The New Cold War of the 1980s*. Harmondsworth: Penguin, 1984.

CLARK, I. *Waging War: A Philosophical Introduction*. Oxford: Clarendon Press, 1988.

CLIFFE, L., and B. DAVIDSON. *The Long Struggle of Eritrea*. Nottingham: Spokesman Books, 1988.

CLODFELTER, M. *The Limits of Air Power: The American Bombing of North Vietnam*. New York: Free Press, 1990.

COHEN, S. F. *Sovieticus: American Perceptions and Soviet Realities*. New York: W. W. Norton, 1986.

————. "The Next President's Historic Opportunity." *The Nation*, 10 October 1988.

COHN, N. *The Pursuit of the Millennium*. London: Secker and Warburg, 1957.

COLEMAN, K., and G. HERRING, eds. *The Central American Crisis: Sources of Conflict and the Failure of U.S. Policy*. Wilmington, Del.: Scholarly Resources, 1985.

CONQUEST, R. *The Great Terror*. London: Macmillan, 1968.

————. *The Harvest of Sorrow*. New York: Oxford University Press, 1986.

————, and J.M. WHITE. *What to Do When the Russians Come*. New York: Stein and Day, 1984.

CONRAD, J. *Heart of Darkness* (1902). New York: Bantam, 1960.

CORSON, W. R. *The Armies of Ignorance: The Rise of the American Intelligence Empire.* New York: Dial, 1977.

COSTELLO, J. *Mask of Treachery.* New York: William Morrow, 1989.

CRUZ SMITH, M. *Gorky Park.* London: Collins, 1981.

CUMMINGS, B. *Origins of the Korean War,* 2 vols. Princeton: Princeton University Press, 1989.

———, and F. HALLIDAY. *Korea: The Forgotten War.* New York: Pantheon, 1989.

DAVIDSON, B. *Partisan Picture.* London: Bedford Books, 1946.

———. *In the Eye of the Storm: Angola's People.* Harmondsworth: Penguin, 1975.

DAVIS, P. *Where Is Nicaragua?* New York: Simon and Schuster, 1987.

DEAKIN, F. W. D. *The Embattled Mountain.* Oxford: Oxford University Press, 1971.

DEBORD, G. *Society of the Spectacle,* rev. ed. Detroit: Black and Red, 1970.

DEIBEL, T. L., and J. L. GADDIS, eds. *Containment: Concept and Policy.* Washington, D.C.: Brookings Institution, 1986.

DEUTSCHER, I. *Stalin: A Political Biography.* Harmondsworth: Penguin, 1966.

DICKENS, P. *SAS: The Jungle Frontier: The Borneo Campaign 1963–1966.* London: Army and Armour Press, 1983.

DIDION, J. *Salvador.* New York: Simon and Schuster, 1983.

———. *Democracy.* New York: Simon and Schuster, 1984.

———. *Miami.* New York: Simon and Schuster, 1987.

DIVINE, R. A., ed. *The Cuban Missile Crisis.* New York: Markus Wiener, 1988.

DOCTOROW, E. L. *The Book of Daniel.* New York: Random House, 1971.

DUKES, P. *The Last Great Game: U.S.A. versus U.S.S.R.* London: Pinter, 1989.

DULLES, A. *The Craft of Intelligence.* London: Weidenfeld and Nicolson, 1963.

DUNN, J. *Modern Revolutions: Analysis of a Political Phenomenon.* Cambridge: Cambridge University Press, 1972.

———. *Western Political Theory in the Face of the Future.* Cambridge: Cambridge University Press, 1979.

———. *Rethinking Modern Political Theory.* Cambridge: Cambridge University Press, 1985.

DUPREE, L. *Afghanistan.* Princeton: Princeton University Press, 1980.

DUPUY, T. N. *Elusive Victory: The Arab-Israeli Wars 1941–1974.* Fairfax, Va.: Hero Books, 1984.

DURAS, M. *Hiroshima Mon Amour: Text for the Film by Alain Resnais.* New York: Grove Press, 1961.

DYSON, F. *Disturbing the Universe.* New York: Basic Books, 1979.

———. *Weapons and Hope.* New York: Harper and Row, 1984.

ECO, U. "Narrative Structures in Fleming." In *The Bond Affair,* ed. U. Eco and O. del Buono. London: Macdonald, 1961.

EDEN, A. *Full Circle.* London: Cassell, 1960.

EPSTEIN, E. J. *Deception: The Invisible War between the KGB and the CIA.* New York: Simon and Schuster, 1989.

EUDES, D. *The Partisans and Civil War in Greece 1943–1949.* London: New Left Books, 1971.

FAGEN, R. R. *The Nicaraguan Revolution: A Personal Report*. Washington, D.C.: Institute for Policy Studies, 1981.

FATEMI, F. S. *The U.S.S.R. in Iran*. Cranbury, N.J.: A. S. Barnes, 1980.

FEJTO, F. *A History of the People's Democracies: Eastern Europe since Stalin*. Harmondsworth: Penguin, 1974.

FENTON, J. *Children of War*. Edinburgh: Salamander Press, 1982.

———. *All the Wrong Places: Adrift in the Politics of the Pacific Rim*. New York: Atlantic Monthly Press, 1988.

FESTINGER, L., et al. *When Prophecy Fails: A Social and Psychological Study of a Modern Group That Predicted the Destruction of the World*. New York: Harper and Row, 1964.

FIEDLER, L. A. *An End to Innocence: Essays on Culture and Politics*. New York: Stein and Day, 1972.

FINNIS, J., J. BOYLE, and G. GRISEZ. *Nuclear Deterrence, Morality, and Realism*. Oxford: Clarendon Press, 1988.

FITZGERALD, F. *Fire in the Lake: The Vietnamese and the Americans in Vietnam*. London: Macmillan, 1972.

FOOT, M. R. D. *Resistance: An Analysis of European Resistance to Nazism 1940–1945*. London: Eyre Methuen, 1976.

FORSYTH, F. *The Fourth Protocol*. London: Hutchinson, 1984.

FOUCAULT, M. *Discipline and Prison. The Birth of the Prison*. Hammondsworth: Penguin, 1979.

FRASER, J. *Violence in the Arts*. Cambridge: Cambridge University Press, 1974.

FREEDMAN, L. *Britain and Nuclear Weapons*. New York: Facts on File, 1980.

———. *The Evolution of Nuclear Strategy*. New York Facts on File, 1981.

———. *Atlas of Global Strategy*. New York: Macmillan, 1985.

FUKUYAMA, F. "The End of History?" *National Interest* 15 (Summer 1989): pp. 3–16.

FULLICK, S. R., and G. POWELL. *Suez: The Double War*. London: Hamish Hamilton, 1979.

FUSSELL, P. *The Great War and Modern Memory*. Oxford: Oxford University Press, 1975.

GADDIS, J. L. *Strategies of Containment: A Critical Appraisal of Postwar American National Security Policy*. New York: Oxford University Press, 1982.

———. *The Long Peace: Inquiries into the History of the Cold War*. New York: Oxford University Press, 1987.

GARDNER, L. C. *Architects of Illusion: Men and Ideas in American Foreign Policy, 1941–1949*. Chicago: Quadrangle, 1970.

GARNHAM, N. *Capitalism and Communications: Global Culture and the Economics of Information*. London: Sage Books, 1990.

GARTHOFF, R. L. *Détente and Confrontation: American-Soviet Relations from Nixon to Reagan*. Washington, D.C.: Brookings Institution, 1985.

GEERTZ, C. *Local Knowledge: Further Essays in Interpretive Anthropology*. New York: Basic Books, 1983.

GELLHORN, M. *The Face of War*, rev. ed. New York: Atlantic Monthly Press, 1988.

————. *The View from the Ground*, rev. ed. New York: Atlantic Monthly Press, 1988.

GEORGE, A. L. *Managing U.S.-Soviet Rivalry: Problems of Crisis Prevention*. Boulder, Colo.: Westview Press, 1983.

GERASSI, J., ed. *Revolutionary Priest: The Complete Writings and Messages of Camilo Torres*. New York: Vintage, 1971.

GETTLEMAN, M., et al., eds. *El Salvador: Central America in the New Cold War*. New York: Grove Press, 1981.

GIDDENS, A. *The Nation-State and Violence*. Cambridge, Eng.: Polity Press, 1985.

GIMBEL, J. *The Origins of the Marshall Plan*. Stanford, Calif.: Stanford University Press, 1976.

GLEES, A. *The Secrets of the Service: British Intelligence and Communist Subversion 1939– 1951*. London: Jonathan Cape, 1987.

GOLDMANN, E. F. *The Crucial Decade—and After: America 1943–1960*. New York: Vintage, 1960.

GOLITSYN, A. *New Lines for Old: The Communist Strategy of Deception and Disinformation*. London: Bodley Head, 1984.

GOODWIN, P. *Nuclear War: The Facts*. London: Macmillan, 1981.

GOWING, M. *Independence and Deterrence: Britain and Atomic Energy*, 2 vols. *1945–52*. London: Macmillan, 1974.

Granta 16 (Summer 1985). Special issue—Science.

Granta 28 (Autumn 1989). Special issue—Birthday Special.

Granta 30 (Winter 1990). Special issue—New Europe.

GRAVEL, M. ed. *The Pentagon Papers*, 4 vols. Boston: Beacon Press, 1972.

GRAYSON, B. L., ed. *The American Image of Russia 1917–1977*. New York: Frederick Ungar, 1978.

GREENE, G. *The Quiet American*. Harmondsworth: Penguin, 1960.

GRESS, D. *Peace and Survival: West Germany, the Peace Movement, and European Security*. Stanford, Calif.: Hoover Institute, 1985.

GRIFFITHS, P. J. *Vietnam Inc*. New York: Collier, 1971.

GUTIERREZ, G. *A Theology of Liberation*. New York: Orbis, 1977.

HAIG, A. M., JR. *Caveat: Realism, Reagan, and Foreign Policy*. New York: Macmillan, 1984.

HALBERSTAM, D. *The Best and the Brightest*. Greenwich, Conn.: Fawcett, 1972.

HALLIDAY, F. *The Making of the Second Cold War*. London: Verso, 1983.

————. "The Ends of Cold War." *New Left Review* 180 (March-April 1990): 5–23.

HAMMOND, T. T. *Red Flag over Afghanistan: The Communist Coup, the Soviet Invasion, and the Consequences*. Boulder, Colo.: Westview Press, 1984.

HANLON, J. *Mozambique: The Revolution under Fire*. London: Zed Books, 1990.

HARBUTT, F. J. *The Iron Curtain: Churchill, America, and the Origins of the Cold War*. Oxford: Oxford University Press, 1986.

HARFORD, B., and S. HOPKINS, eds. *Greenham Common: Women at the Wire*. London: Women's Press, 1984.

HARMAN, C. *Bureaucracy and Revolution in Eastern Europe*. London: Pluto Press, 1974.

HARVARD NUCLEAR STUDY GROUP. *Living with Nuclear Weapons*. Cambridge, Mass.: Harvard University Press, 1983.

HASLAM, J. *The Soviet Union and the Politics of Nuclear Weapons in Europe 1969–87*. Basingstoke, Eng.: Macmillan, 1989.

HASTINGS, M. *The Korean War*. London: Michael Joseph, 1987.

HAYDEN, S. *Wanderer*. London: Longmans Green, 1964.

HEALEY, D. *The Time of My Life*. London: Michael Joseph, 1989.

HELLMAN, L. *Scoundrel Time*. Boston: Little, Brown, 1976.

HEMINGWAY, E. *For Whom the Bell Tolls*. New York: Charles Scribner's Sons, 1940.

HENDERSON, N. *The Birth of NATO*. Boulder, Colo.: Westview Press, 1983.

HENRIQUES, R. *One Hundred Hours to Suez: An Account of Israel's Campaign in the Sinai Peninsula*. London: Collins, 1957.

HERBSTEIN, D., and J. EVENSON, *The Devils Are among Us: The War for Namibia*. London: Zed Books, 1989.

HERSEY, J. *Hiroshima*. 1946; rev. ed., Harmondsworth: Penguin, 1981.

HERSH, S. M. *The Price of Power: Kissinger in the Nixon White House*. New York: Summit/Simon and Schuster, 1983.

———. *"The Target Is Destroyed": What Really Happened to Flight 007*. London: Faber and Faber, 1986.

HERZOG, C. *The Arab-Israeli Wars: War and Peace in the Middle East from the War of Independence to Lebanon*. Tel Aviv: Arms and Armor Press, 1982.

HIGGINS, T. *Korea and the Fall of MacArthur*. New York: Oxford University Press, 1960.

HINTON, J. *Protests and Visions: Peace Politics in Twentieth-Century Britain*. London: Hutchinson Radius, 1989.

HITCHENS, C. *Prepared for the Worst: Selected Essays and Minority Reports*. New York: Hill and Wang, 1988.

———. *Blood, Class, and Nostalgia: Anglo-American Ironies*. New York: Farrar, Straus, and Giroux, 1990.

HOFSTADTER, R. *The Paranoid Style in American Politics*. New York: Knopf, 1965.

HOGGART, R. *The Uses of Literacy: Aspects of Working Class Life*. Harmondsworth: Penguin, 1957.

HOROWITZ, D. *From Yalta to Vietnam: American Foreign Policy in the Cold War*. Harmondsworth: Penguin, 1967.

HOWARD, M. *The Causes of Wars*. London: Unwin, 1984.

HOWARTH, P. *Uncercover*. London: Routledge and Kegan Paul, 1980.

HUMPHREY, N., and R. J. LIFTON. *In a Dark Time*. London: Faber and Faber, 1984.

HUNT, M. *Ideology and U.S. Foreign Policy*. New Haven: Yale University Press, 1987.

INGLIS, F. *Popular Culture and Political Power*. Hemel Hempstead: Harvester, 1988.

ISAACSON, W., and E. THOMAS. *The Wise Men: Six Friends and the World They Made*. New York: Simon and Schuster, 1986.

JACKALL, R. *Moral Mazes: Bureaucracy and Managerial Work*. New York: Oxford University Press, 1988.

JAMES, W. *The Varieties of Religious Experience.* 1902; reprint, New York: Collier, 1961.

JAMESON, F. *The Political Unconscious: Narrative as a Socially Symbolic Act.* Ithaca: Cornell University Press, 1981.

JARA, J. *Victor.* London: Jonathan Cape, 1983.

JARVIE, I. C. *Movies in Society.* New York: Basic Books, 1970.

JEFFERSON, L. *John Foster Dulles: The Book of Humor.* New York: St. Martin's Press, 1986.

JEFFREYS-JONES, R. *The CIA and American Democracy.* New Haven: Yale University Press, 1989.

JOHNSON, R. W. *Shootdown: The Verdict on KAL 007.* New York: Doubleday, 1985.

———. *Heroes and Villains.* Hempstead: Harvester, 1990.

KAHN, H. *On Thermonuclear War: Thinking about the Unthinkable.* New York: Horizon Press, 1962.

———. *On Escalation: Metaphors and Scenarios.* New York: Praeger, 1965.

KALDOR, M. *The Baroque Arsenal.* New York: Hill and Wang, 1981.

———, and D. SMITH, eds. *Disarming Europe.* London: Merlin, 1982.

KANER, R. E., ed. *Soviet Foreign Policy in the 1980s.* New York: Praeger, 1982.

KAPUSCINSKI, R. *The Emperor.* London: Pan Books, 1984.

———. *Shah of Shahs.* London: Pan Books, 1986.

———. *Another Day of Life: An Eye Witness Account of Civil War in Angola.* New York: Penguin, 1988.

KEEGAN, J. *The Mask of Command.* Harmondsworth: Penguin, 1988.

KELIHER, J. G. *The Negotiations on Mutual and Balanced Force Reduction: Arms Control in Central Europe.* Headington: Pergamon Press, 1980.

KENNAN, G. F. *Memoirs 1925–1950,* vol. 1. Boston: Little, Brown, 1967.

———. *Memoirs 1950–1963,* vol. 2. Boston: Little, Brown, 1972.

———. *The Nuclear Delusion: Soviet-American Relations in the Atomic Age.* New York: Pantheon, 1983.

———. *American Diplomacy,* exp. ed. Chicago: University of Chicago Press, 1984.

KENNEDY, P. *The Rise and Fall of the Great Powers.* New York: Random House, 1987.

KENNEDY, R. F. *Thirteen Days: A Memoir of the Cuban Missile Crisis.* New York: W. W. Norton, 1969.

KERMODE, F. *The Sense of an Ending: Studies in the Theory of Fiction.* Oxford: Oxford University Press, 1967.

KHRUSHCHEV, N. *Khrushchev Remembers,* trans. and ed. Strobe Talbott, with an introduction and commentary by Edward Crankshaw. Boston: Little, Brown, 1970.

KIDSON, M., and D. SMITH. *The War Atlas: Armed Conflict—Armed Peace.* London: Pan Books, 1983.

KISSINGER, H. *Nuclear Weapons and Foreign Policy.* New York: Anchor, 1957.

———. *The White House Years.* Boston: Little, Brown, 1979.

———. *Years of Upheaval.* Boston: Little, Brown, 1982.

————. *Observations: Selected Speeches and Essays 1982–1984.* Boston: Little, Brown, 1985.

KLINGHOFFER, A. J. *The Angolan War: A Study of Soviet Policy in the Third World.* Boulder, Colo.: Westview Press, 1980.

KNELMAN, F. H. *Reagan, God, and the Bomb.* Buffalo, N.Y.: Prometheus Books, 1985.

KNIGHTLEY, P. *The Second Oldest Profession.* London: Andre Deutsch, 1986.

————. *The Story of Kim Philby.* New York: Knopf, 1989.

KOLKO, G. *Anatomy of a War: Vietnam, the United States, and the Modern Historical Experience.* New York: Pantheon, 1985.

KORN, D. A. *Ethiopia, the United States, and the Soviet Union.* Beckenham: Croom Helm, 1986.

KUBEK, A. *How the Far East Was Lost: American Policy and the Creation of Communist China 1941–1949.* Washington, D.C.: Intercontex, 1971.

KULL, S. *Minds at War: Nuclear Reality and the Inner Conflicts of Defense Policymakers.* New York: Basic Books, 1988.

KUNDERA, M. *The Book of Laughter and Forgetting.* Harmondsworth: Penguin, 1981.

————. *The Joke.* Harmondsworth: Penguin, 1984.

————. *The Unbearable Lightness of Being.* New York: Harper and Row, 1984.

KUNZ, D. "The Economic Diplomacy of the Suez Crisis." In Louis and Owen, *Suez.*

KYLE, K. "Britain and the Crisis, 1955–56." In Louis and Owen, *Suez.*

LAMPHERE, R. J., and T. SCHACHTMAN. *The FBI-KGB War: A Special Agent's Story.* New York: Random House, 1986.

LANE, D. *State and Politics in the U.S.S.R.* Oxford: Basil Blackwell, 1985.

————. *Soviet Economy and Society.* Oxford: Basil Blackwell, 1987.

LAQUEUR, W., and B. RUBIN, eds. *The Israel-Arab Reader: A Documentary History of the Middle East Conflict,* 4th rev. ed. New York: Facts on File, 1985.

LARSON, D. *The Origins of Containment.* Princeton: Princeton University Press, 1982.

LE CARRÉ, J. *Tinker, Tailor, Soldier, Spy.* New York: Knopf, 1975.

————. *Smiley's People.* New York: Knopf, 1978.

————. *A Perfect Spy.* New York: Knopf, 1986.

LEIGH, D. *The Wilson Plot: How the Spycatchers and Their American Allies Tried to Overthrow the British Government.* New York: Pantheon, 1989.

LESSING, D. *The Wind Blows away Our Words.* New York: Vintage, 1987.

LEWIS, J., and P. WHITEHEAD. *Stalin: A Time for Judgement.* London: Methuen and Thames TV, 1990.

LIPPMANN, W. *The Cold War: A Study in U.S. Foreign Policy.* New York: Harper and Row, 1947.

LOONEY, R. E. *Economic Origins of the Iranian Revolution.* Oxford: Pergamon, 1982.

LOTH, W. *The Division of the World 1941–1955.* London: Routledge 1988.

LOUIS, W. R. and R. OWEN, eds. *Suez 1956: The Crisis and Its Consequences.* Oxford: Clarendon Press, 1989.

LUARD, E. *Conflict and Peace in the Modern International System,* rev. ed. London: Macmillan, 1988.

————. *The Globalisation of Politics.* London: Macmillan, 1990.

LUHMANN, N. *Trust and Power.* New York: John Wiley, 1979.

McCOY, A. W., C. READ, and L. ADAM. *The Politics of Heroin in South-east Asia.* New York: Harper and Row, 1972.

MACHIAVELLI, N. *The Discourses,* ed. B. Crick, trans. L. J. Walker. Harmondsworth: Penguin, 1974.

MACINTYRE, A. "A Mistake about Causality in Social Science." In *Philosophy, Politics, and Society,* 2d series, ed. P. Laslett and W. G. Runciman. Oxford: Basil Blackwell, 1967.

MACLEAN, F. *Eastern Approaches.* London: Jonathan Cape, 1949.

MACLEISH, A. *Freedom Is the Right to Choose.* Boston: Beacon Press, 1951.

McMAHON, J. *British Nuclear Weapons: For and Against.* London: Junction Books, 1981.

McNAMARA, R. *Blundering into Disaster: Surviving the First Century of the Nuclear Age.* New York: Pantheon, 1986.

MACSHANE, D. *Solidarity: Poland's Independent Trade Union.* Nottingham: Spokesman Books, 1981.

MAILER, N. *Miami and the Siege of Chicago.* Harmondsworth: Penguin, 1969.

MALE, B. *Revolutionary Afghanistan: A Reappraisal.* London: Croom Helm, 1982.

MANDELSTAM, N. *Hope against Hope.* Glasgow: Collins Harvill, 1971.

————. *Hope Abandoned.* Glasgow: Collins Harvill, 1974.

MARCHETTI, V., and J. MARKS. *The CIA and the Cult of Intelligence.* New York: Dell, 1975.

MARQUEZ, G. G. "Operation Carlota." *New Left Review* 101–2 (1977): 123–37.

MARSHALL, B. *Willy Brandt.* London: Sphere Books, 1990.

MARTIN, T. K. *Edge of Darkness.* London: Faber and Faber, 1990.

MATHESON, N. *The "Rules of the Game" of Superpower Military Intervention in the Third World 1975–1980.* Washington, D.C.: University Press of America, 1982.

MAYER, J., and D. McMANUS. *Landslide: The Unmaking of the President 1984–1988.* Boston: Houghton Mifflin, 1988.

MEDVEDEV, Z. "Soviet Power Today." *New Left Review* 179 (January-February 1990): 65–80.

————, and R. MEDVEDEV. *Khrushchev: The Years in Power.* New York: Columbia University Press, 1976.

MELMAN, S. *Pentagon Capitalism.* New York: McGraw-Hill, 1970.

MILLER, A. *The Crucible.* New York: Bantam, 1959.

————. *Timebends: A Life.* London: Methuen, 1987.

MILWARD, A. S. *The Reconstruction of Western Europe, 1945–1951.* Berkeley: University of California Press, 1984.

MINNION, J., and P. BOLSOVER. *The CND Story: The First Twenty-five Years.* London: Allison and Busby, 1983.

MOSLEY, L. *Dulles: A Biography of Eleanor, Allen, and John Foster Dulles, and Their Family Network.* London: Hodder and Stoughton, 1978.

MUIR, E. *An Autobiography.* London: Methuen, 1964.

Mus, P. *Ho Chi Minh: le Vietnam, l'Asie.* Paris: Seuil, 1971.

Muslow, B. *Mozambique: The Revolution and Its Origins.* Boulder, Colo.: Westview Press, 1985.

Myrdal, A. *The Game of Disarmament: How the United States and Russia Run the Arms Race.* New York: Pantheon, 1976.

Naipaul, V. S. *A Turn in the South.* New York: Knopf, 1989.

Navasky, V. *Naming Names.* New York: Viking, 1980.

Newhouse, J. *War and Peace in the Nuclear Age.* New York: Knopf, 1988.

Nietzsche, F. *The Portable Nietzsche,* ed. Walter Kaufmann. New York: Penguin, 1976.

Nutting, A. *No End of a Lesson: The Story of Suez.* London: Constable, 1967.

Nye, J. S., Jr., ed. *The Making of America's Soviet Policy.* New Haven: Yale University Press, 1984.

Orwell, G. *1984.* Harmondsworth: Penguin, 1950.

Ottaway, M., and D. Ottaway. *Ethiopia: Empire in Revolution.* New York: Africana Publishing, 1978.

Oye, K., R. J. Lieber, and D. Rothchild, eds. *Eagle Defiant: United States Foreign Policy in the 1980s.* Boston: Little, Brown, 1983.

Park, W. *Defending the West: A History of NATO.* Brighton: Wheatsheaf, 1986.

Partisan Review 24 (1957). Symposium—The Crisis in Communism.

Partisan Review 29 (1962). Symposium—The Cold War and the West.

Pincher, C. *Their Trade Is Treachery.* London: Sidgwick and Jackson, 1981.

Pipes, R. *Survival Is Not Enough: Soviet Relations and America's Future.* New York: Simon and Schuster, 1984.

Pollack, J. D. *The Lessons of Coalition Politics: Sino-American Security Relations.* Santa Monica, Calif.: Rand, 1984.

Ponchaud, F. *Cambodia Year Zero.* Harmondsworth Penguin, 1978.

Porter, B. D. *The U.S.S.R. in Third World Conflict: Soviet Arms and Diplomacy in Local Wars 1945–1980.* Cambridge: Cambridge University Press, 1984.

Posner, C., ed. *Reflections on the Revolution in France: 1968.* Harmondsworth: Penguin, 1970.

Pressen, R. W. *John Foster Dulles: The Road to Power.* New York: Free Press, 1982.

Prins, G., ed. *Defended to Death: A Study of the Nuclear Arms Race by the Cambridge University Disarmament Seminar.* Harmondsworth: Penguin, 1983.

————, ed. *The Nuclear Crisis Reader.* New York: Vintage, 1984.

Race, J. *War Comes to Long An: Revolutionary Conflict in a Vietnamese Province.* Berkeley: University of California Press, 1972.

Radosh, R., and J. Milton. *The Rosenberg File: A Search for the Truth.* New York: Holt, Rinehart and Winston, 1983.

Raina, P. *Political Opposition in Poland 1954–1977.* London: Poets and Painters Press, 1978.

Ranelagh, J. *The Agency: The Rise and Decline of the CIA from Wild Bill Donovan to William Casey.* New York: Simon and Schuster, 1986.

REARDEN, S. L. *The Evolution of American Strategic Doctrine*. Boulder, Colo.: Westview Press, 1984.

REGAN, D. *For the Record: From Wall Street to Washington*. San Diego: Harcourt Brace Jovanovich, 1988.

RHODES, R. *The Making of the Atomic Bomb*. Harmondsworth: Penguin, 1988.

ROGIN, M. *The Intellectuals and McCarthy: The Radical Specter*. Cambridge, Mass.: MIT Press, 1967.

————. *Ronald Reagan: The Movie, and Other Episodes in Political Demonology*. Berkeley: University of California Press, 1987.

RORTY, R. *Contingency, Irony, Solidarity*. Cambridge: Cambridge University Press, 1989.

ROSZAK, T. *The Making of the Counter Culture*. London: Faber and Faber, 1972.

ROVERE, R. *Senator Joe McCarthy*. New York: Harcourt Brace, 1959.

RUBIN, B. *Paved with Good Intentions: The American Experience and Iran*. New York: Penguin, 1981.

RUMBLE, G. *The Politics of Nuclear Defence*. Cambridge, Eng.: Polity Press, 1985.

RUSTON, R. *A Say in the End of the World: Morals and British Nuclear Weapons Policy 1941–1987*. Oxford: Clarendon, 1989.

RYAN, A. *Bertrand Russell: A Political Biography*. Harmondsworth: Penguin, 1990.

SADAT, A. *In Search of Identity: An Autobiography*. New York: Harper and Row, 1978.

ST. JOHN, J. *Day of the Cobra: The True Story of KAL 007*. Nashville: Thomas Nelson, 1984.

SALISBURY, H. *American in Russia*. New York: Harper and Row, 1955.

SAVIGEAR, P. *Cold War or Détente in the 1980s: The International Politics of American-Soviet Relations*. Brighton: Wheatsheaf, 1987.

SCHEER, R. *With Enough Shovels: Reagan, Bush, and Nuclear War*. New York: Random House, 1982.

SCHELL, J. *The Real War*. New York: Pantheon, 1987.

SCHILLING, W. R., P. Y. HAMMOND, and G. H. SNYDER. *Strategy, Politics, and Defense Budgets*. New York: Columbia University Press, 1962.

SCHLESINGER, A. *A Thousand Days: John F. Kennedy in the White House*. London: Andre Deutsch, 1965.

————. "Origins of the Cold War." *Foreign Affairs* 46 (October 1967): 22–52.

SCHRAM, S. *Mao-Tse-tung*. Harmondsworth: Penguin, 1967.

SCHURMANN, F. *The Logic of World Power: An Inquiry into the Origins, Currents, and Contradictions of World Politics*. New York: Pantheon, 1974.

SEN, A. *Poverty and Famines: An Essay on Entitlement and Deprivation*, rev. ed. Oxford: Clarendon Press, 1982.

SHAHEEN, J. G., ed. *Nuclear War Films*. Carbondale: Southern Illinois University Press, 1978.

SHAWCROSS, W. *Sideshow: Kissinger, Nixon, and the Destruction of Cambodia*. London: Fontana, 1980.

SHEEHAN, N. *A Bright Shining Lie: John Paul Vann and America in Vietnam*. New York: Random House, 1988.

SHORT, A. *The Communist Insurrection in Malaya 1948–1960.* New York: Crane, Russak, 1975.

SHUTE, N. *On the Beach.* New York: Signet, 1958.

SKINNER, Q. *Ambrogio Lorenzetti: The Artist as Political Philosopher.* London: British Academy, 1986.

SKLAR, H. *Reagan, Trilateralism, and the Neoliberals: Containment and Intervention in the 1980s.* Boston: South End Press, 1986.

———. *Washington's War on Nicaragua.* Boston: South End Press, 1988.

SMITH, S. *Doubletalk: The Story of the First Strategic Arms Limitation Talks.* New York: Doubleday, 1980.

SNOW, E. *Red Star over China.* New York: Random House, 1938.

———. *Red China Today.* Harmondsworth: Penguin, 1970.

SOLZHENITSYN, A. *The Gulag Archipelago 1918–1956.* London: Collins/Fontana, 1971.

———. *The Mortal Danger: How Misconceptions about Russia Imperil America (with Replies),* 2d ed. New York: Harper and Row, 1981.

SOMERVILLE, K. *Angola: Politics, Economics, and Society.* London: Frances Pinter, 1986.

STAFFORD, D. *Britain and European Resistance 1940–45.* London: Macmillan, 1980.

STEEL, R. *Walter Lippmann and the American Century.* Boston: Little, Brown, 1981.

STEPHANSON, A. *George Kennan and the Art of Foreign Policy.* Cambridge, Mass.: Harvard University Press, 1989.

STOCKWELL, J. *In Search of Enemies: A CIA Story.* New York: W. W. Norton, 1978.

STONE, I. F. *The Hidden History of the Korean War 1950–1951.* New York: Monthly Review Press, 1952.

STOWERS JOHNSON, R. *Agents Extraordinary.* London: Hale, 1978.

SUID, L. H. *Guts and Glory: Great American War Movies.* Reading, Mass.: Addison-Wesley, 1978.

SWEET-ESCOTT, B. *Baker Street Irregular.* London: Methuen, 1965.

SZULC, T. *The Illusion of Peace: Foreign Policy in the Nixon Years.* New York: Viking, 1978.

TALBOTT, S. *Deadly Gambits: The Reagan Administration and the Stalemate in Nuclear Arms Control.* New York: Knopf, 1984.

———. *The Master of the Game: Paul Nitze and the Nuclear Peace.* New York: Knopf, 1988.

TANHAM, G. K., and D. J. DUNCANSON. "Some Dilemmas of Counterinsurgency." *Foreign Affairs* 48, no. 1 (1969): 113–122.

TAS, S. *Indonesia: The Underdeveloped Freedom.* New York: Pegasus, 1974.

TERRILL, R. *Mao: A Biography.* New York: Harper and Row, 1980.

THOMAS, H. *Cuba.* New York: Harper and Row, 1967.

———. *Armed Truce: The Beginnings of the Cold War, 1945–1946.* London: Hamish Hamilton, 1986.

THOMPSON, E. P. *The Making of the English Working Class.* London: Gollancz, 1963.

————. *William Morris: Romantic to Revolutionary,* rev. ed. London: Merlin Press, 1977.

————. *The Poverty of Theory.* London: Merlin, 1978.

————. *Writing by Candlelight.* London: Merlin, 1980.

————. *Zero Option.* London: Merlin, 1982.

————. *Double Exposure.* London: Merlin, 1985.

————. *The Heavy Dancers: Writings on War, Past and Future.* London: Merlin, 1985.

————, and T. J. THOMPSON, eds. *There Is a Spirit in Europe: A Memoir of Frank Thompson.* London: Victor Gollancz, 1947.

————, and D. SMITH, eds. *Protest and Survive.* Harmondsworth: Penguin, 1980.

————, F. HALLIDAY, and R. BAHRO. *Exterminism and Cold War.* London: Verso, 1982.

————, and M. KALDOR. *Mad Dogs: The U.S. Raids on Libya.* London: Pluto Press, 1986.

THUBRON, C. *Behind the Wall: A Journey through China.* New York: Atlantic Monthly Press, 1988.

TOLKIEN, J. R. R. *The Return of the King.* London: Allen and Unwin, 1955.

TOULOUSE, M. G. *The Transformation of John Foster Dulles: From Prophet of Realism to Priest of Nationalism.* Macon, Ga.: Mercer University Press, 1985.

TRUNSKI, S. *Grateful Bulgaria.* trans. J. Penchera and A. Todurov. Sofia: Sofia Press, 1979.

TUCKER, R. C., ed. *Stalinism: Essays in Historical Interpretation.* New York: W. W. Norton, 1977.

TUSA, A., and J. TUSA. *The Berlin Blockade.* London: Hodder and Stoughton, 1988.

TUVESON, E. L. *Redeemer Nation: The Idea of America's Millenial Role.* Berkeley: University of California Press, 1968.

U.S. GOVERNMENT PRINTING OFFICE. *The Military Balance,* annually.

VANCE, C. *Hard Choices: Critical Years in America's Foreign Policy.* New York: Simon and Schuster, 1983.

VAN NESS, P. *Revolution and Chinese Foreign Policy.* Berkeley: University of California Press, 1970.

VUKMANOVIC-TEMPO, S. *Struggle for the Balkans.* London: Merlin, 1990.

WALZER, M. *Political Action: A Practical Guide to Movement Politics.* Chicago: Quadrangle, 1971.

————. *Just and Unjust Wars: A Moral Argument with Historical Illustrations.* New York: Basic Books, 1977.

WATERBURY, J. *The Egypt of Nasser and Sadat: The Political Economy of Two Regimes.* Princeton: Princeton University Press, 1983.

WEST, N. *Molehunt: Searching for Soviet Spies in MI5.* New York: William Morrow, 1989.

WHEELER-BENNETT, J. W., and A. NICHOLLS. *The Semblance of Peace: The Political Settlement after the Second World War.* London: Macmillan, 1972.

WILLIAMS, R. *Culture and Society 1780–1950.* London: Chatto and Windus, 1959.

————. *Towards 2000.* London: Chatto and Windus, 1983.

WILLIAMS, R. C. *Klaus Fuchs: Atom Spy.* Cambridge, Mass.: Harvard University Press, 1987.

WILLIAMS, W. A. *Empire as a Way of Life.* New York: Oxford University Press, 1980.

WILLS, G. *The Kennedys: A Shattered Illusion.* London: Orbis, 1983.

————. *Reagan's America.* New York: Viking Penguin, 1988.

WILSON, J. C. *Vietnam in Prose and Film.* Jefferson, N.C.: McFarland and Co., n.d.

WINKS, R. *Cloak and Gown: Scholars in the Secret War 1939–1961.* New York: William Morrow, 1987.

WOLFE, A. *The Rise and Fall of the "Soviet Threat": Domestic Sources of the Cold War Consensus.* Washington, D.C.: Institute for Policy Studies, 1979.

YERGIN, D. *Shattered Peace: The Origins of the Cold War and the National Security State.* Harmondsworth: Penguin, 1980.

ZUCKERMAN, LORD S. *Nuclear Illusion and Reality.* New York: Vintage, 1983.

————. *Star Wars in a Nuclear World.* London: W. Kimber, 1986.

A BRIEF GUIDE TO THE READING

PART ONE

The first year and a half of the cold war is retold with great vividness and judgment by Daniel Yergin (1980). Lewis Gaddis (1982) tells the story but from a more strategic perspective. Gaddis also, in my view, vitiates his exemplary first book by apologetics in his later collection of essays (1987). Hugh Thomas (1986) contributes a characteristically stout and gripping read from a position of vehement defense of the cold war and its necessity, while Wilfried Loth lays a sharp and convincing argument from the "revisionist" side placing more blame than is tasteful at the door of the United States.

The history of the Bulgarian partisans is partially told by the sometime commander of all Balkan partisan activity, Vukmanovic-Tempo (1990). His story is given a British twist (from the officer class) by Fitzroy Maclean (1949). The most exciting autobiography from SOE adventures is Basil Davidson's (1946). Frank Thompson's personal papers (1947) are, alas, no longer in print.

The best-known biographer of the horrors of Stalinism is probably Robert Conquest (1968, 1986). The best biographer of Stalin is undoubtedly Deutscher (1966). I, in addition, drew extensively upon Lewis and Whitehead

(1990), the companion volume to Whitehead's remarkable TV documentary with its entirely new footage from the Terror and afterward.

For the history of the atomic bomb I relied almost entirely on Rhodes's classic (1988) with some help from Marjorie Gowing's account (1974). For a history of the Berlin Blockade I relied on the Tusas's perhaps too resolutely anti-Russian tract (1988). For the Korean war I relied primarily upon I. F. Stone's two astonishing volumes of criticism of U.S. policy, culled entirely from the public records of the day (1952). I also relied more upon Bruce Cumings' magisterial volumes (1989) than on Hastings' readable, but decidedly favoritist, narrative (1987). The literature on China is, of course, enormous, and I am a tyro among it. For Mao I have used Schram's well-known textbook (1967), more generally, I have counted on Edgar Snow (1970), and Jonathan Spence (1990).

Reminiscences by the leading political actors of the opening years of cold war are rightly dominated by the two volumes of George Kennan's *Memoirs* (1967, 1972). They are also well supplemented by Acheson's more waspish and self-satisfied contribution (1969). Nobody, however, could write adequately about the last years of international politics in which Britain took a genuinely efficacious part without constant recourse to Bullock's classic biography of Bevin (1983).

Lastly, Rogin (1987) here, as later, is powerfully (if a bit schematically) suggestive about cold war movies in general.

PART TWO

The many histories of cold war purges in the United States yield first place for immediacy and comprehensiveness to Caute (1978) and Navasky (1980). For forensic judiciousness I have used Radosh (1983), and for the imaginative accusation flung at the head of the Statue of Liberty, I have used E. L. Doctorow (1971). 1956 sowed a harvest of dragons' teeth in the library, among which Fejto (1974) is the steadiest, Hitchens (1990) the most cutting, and Healey (1989) the most passionate. Nutting's marvelous essay (1967) uncovered many truths on the Suez Crisis. The consequences of the crisis were well documented by Correlli Barnett (1986) from the British Right. Khrushchev's own garrulous reminiscences (1970) cover all the action after Stalin's death and, racy and candid as they are, are also an ingenuous and indispensable sourcebook of Soviet pathology.

The Cuban crisis has become the most studied battle honor of the epoch, and has also turned into the historical barrier across which revisionists (or those who blame the United States) and pietists (or those who put everything at the door of the Kremlin) shout loudest and most unheedingly. Divine's anthology of papers (1988) is excellent, Bobby Kennedy's memoir (1969) famously unbuttoned, Keegan's fulsome endorsement (1988) of John Kennedy's every move embarrassing, Kahn's tough consequentialism (1962, 1965) terrifying.

Sometime over the years chronicled in part two, nuclear strategic thinking (if the word will pass) became fully institutionalized. I have found Newhouse's textbook (1988) always useful and pretty reliably cold-war mongering. Freedman (1981) is also an honestly pro-nuclear history of policy. Schilling and company (1962) were also very helpful. Henry Kissinger (1957). Other contributions to the theory of exterminism will follow in their turn.

Willy Brandt still awaits the great biography he surely deserves. Freeman Dyson (1979, 1984) has proved his own, attractive biographer. Chiliasm and eschatology have, understandably, long preoccupied critics of culture, and none (to my taste) has been more interesting about the formal aspects of apocalypse than Frank Kermode (1967). Cohn's (1957) history of millennarians is a deserved best-seller.

I could not have hoped to offer very much more than a few perfunctory tips on the subject of espionage. John Le Carré is its fictional doyen, as everyone agrees; his oeuvre already extends far beyond the limits of this book. Dulles is the most self-betraying, Jeffreys-Jones the most piously respectable, Knightley the most readable, and R. W. Johnson the most pungent. Other reading in this connection crops up more directly with Philip Agee's contributions below.

PART THREE

As we approach the present, and the scale of the American commitment to the world's agenda grows more enormous, the bibliography lengthens at a great rate.

In the case of Vietnam, it quickly became interminable (though never indifferent), and at first divided largely into defenders and opponents of the war—the local version of revisionists and pietists among cold war historians. Of the books written out of the heat of battle, Frances Fitzgerald's prizewinner (1972) remains a magnificent introduction to the history and a convincing polemic against the war. To have brought off such a triumph so close to the events is indeed rare. Noam Chomsky (1968, 1970, 1973) still seems to me to be as accurate and stinging now when I re-read him as then when he first called the intelligentsia to account by his single-handed condemnation of American policy and policy-makers. David Halberstam (1972) turned

Chomsky's insights into the U.S. mandarinate into a full-scale history of the new class. Melman (1970) writes a critical primer for its political economy, which Prins (1983) later brings up to date.

Kolko's (1985) magisterial and compelling study is surely the major work of structural history to come out of the war and its generation in the United States, and Sheehan's (1988) provides the equivalent in a narrative idiom. For the *annus mirabilis et horribilis* itself, 1968, the chronicle of the global year is briskly told by Caute (1988), commemorated by Tariq Ali (1987), and locally rendered with egotistical genius by Norman Mailer (1969). Hannah Arendt's essays (1972) on what the year and its aftermath meant to the Constitution are monuments in the genre of what one might dub the archaeology of ignorance, while Schell (1987) is marvelously vivid about the immediate matters of life and death. Griffiths's anthology of photographs (1971) from the Magnum collective is unerasable from the memory.

Insofar as the years after 1968 until Carter are dominated by the fearsome duo, then books by and about Henry Kissinger (1979, 1982) and Richard Nixon are indispensable to the historian, in spite of and perhaps because of Kissinger's monstrous ego, or Nixon's mendaciously self-protective flair and originality. Hersh's biography of Kissinger (1983) is unyieldingly harsh and might best be qualified by Healey's (1989) and Carrington's (1988) unignorable admiration for the man. But Hersh's account of the Chilean coup is corroborated by Joan Jara (1983) on the spot, as well as by the Brookings Institute's official historian, Garthoff (1985), whose enormous history of U.S. foreign policy in the seventies and eighties is witty, unflagging, and judicious.

The terrible history of Cambodia in these years is bleakly told by Ponchaud (1978) and passionately by Shawcross (1980). James Fenton (1988) was our man-on-the-spot in Saigon and Phnom Penh. Davidson (1975) was in bloody old Angola, and Hanlon (1990) describes Mozambique at the same moment. Anwar (1989) is the most thorough history of the Afghan invasion, and an excellent piece of narrative history into the bargain, and Kapuscinski (1984, 1986, 1988) turned up punctually at each and every desert revolution.

In this section, of course, Agee adds substance to the prodigal spy literature, not only with his own books (1976, 1978, 1987) but also in prompting much timely muckraking. The most thorough raking of the muck, however, was done in the full glare of the lights by the Senator Gravel edition of *The Pentagon Papers* (1972).

Fraser (1974) is an excellent guide to the artistic violence which exploded in parallel to the inartistic violence of the period, whether in Hanoi or Kent State College. Adair (1989) has some good anecdotes about the relevant movies. Susan Sontag famously spotted that the times had brought camp to birth, and Bennett's anthology (1987) is a choice example of the genre. Roszak

(1972) is only one of many bards singing of *Woodstock*, but apart from Freud's classic, only Rogin (1987) is good on the jokes.

PART FOUR

By the time we reach the 1980s, the market for political memoirs has grown at a tropical rate. The historian must turn up the assertively ill-written pages of Scheer (1982), Haig (1984), Regan (1988), and their like in a repellent genre for fear of missing an unmissably juicy anecdote. But the serious reporters of the day, all of them still cold warriors. Include Newhouse (1988), as already noticed, and throngs of nuclear war theorists, whose bookmaking lives became busiest in the decade in which their fantasies became plainly incredible. But the honest chronicler will find Freedman (1982), Haslam (1989), and, of course, Richard Pipes (1984), the most chilling expositors of the facts as well as of the rational lunacies of their dark trade. On the other side, though no less chilling, is Alva Myrdal (1976) and her well-merited Nobel Prize, together with the extraordinary casebook of Steven Kull (1988); easily the best book on the somber history of nuclear alchemy has already been praised. It is the fine work of the Cambridge Disarmament seminar under Gwyn Prins's editorship (1983).

Neal Ascherson made the history of Solidarnosc (1982) his own, in Anglophone countries at least, but Central American political history is, being so much more various, also more crowded and contradictory. As the text says, Joan Didion (1983, 1987) is an unexampled guide to its contemporary horribleness, Chomsky's (1985, 1988) a vehement indictment of American incompetence and spendthrift cruelty, Sklar (1986) his powerful pupil, and Coleman and company (1985) telling judges of each disaster. Marquez is, however, the supreme fictionalist of this little moment of cold war.

As to the texts of mistrust, first (naturally) Locke and then Luhmann (1979) tell us of the necessity of trust; Troy Kennedy Martin (1990) has printed in full his gripping script, and the leading apocalyptics have been cited.

PART FIVE

Finally, Timothy Garton Ash shot to stardom as our not always attractive guide to the revolutions of 1989, and the quarterly *Granta* proved a remarkable storehouse for impressions of the day.

Necessarily, there are books which have accompanied my chronicle at its every turn, whether as prompts to thought or critics of what I have written.

Schurmann's (1974) remarkable work is one such and Milward's (1984) is another, for although it ostensibly stops at 1957, the weight of the book's argument rolls it forward to the millennium. I have also turned back time and again to Emma Rothschild and the truly astonishing punctuality with which she assessed U.S. defense budgets year by year in the well-loved pages of the *New York Review*. Fredric Jameson (1981) was my constant if difficult guide to form in the narratives of the day, and Tony Giddens's (1985) too-little regarded insights remain to warn us of a dangerous future in a world where the rich are still so inexcusably congratulating themselves on their duty-less privileges.

Richard Hoggart's (1957) great work was both master and friend, and taught me to solve my difficulties with value and experience; Hitchens (1988, 1990) kept my blood up; Edward Thompson (1980, 1982, 1985, etc., etc.), historian of some of the trickiest moments of the epoch, as well as orator, poet, and comrade, walked always and inquiringly ahead of me; of us all.

INDEX